AF544953

ND | Springer Series in **Nonlinear Dynamics**

ND | Springer Series in **Nonlinear Dynamics**

Solitons – Introduction and Applications
Editor: M. Lakshmanan

What Is Integrability?
Editor: V.E. Zakharov

Rossby Vortices, Spiral Structures, Solitons
Astrophysics and Plasma Physics in Shallow Water Experiments
By M.V. Nezlin and E.N. Snezhkin

Algebro-Geometrical Approach to Nonlinear Evolution Equations
By E. D. Belokolos, A. I. Bobenko, V. Z. Enolskii, A. R. Its and V. B. Matveev

Darboux Transformations and Solitons
By V. B. Matveev and M. A. Salle

Optical Solitons
By F. Abdullaev, S. Darmanyan and P. Khabibullaev

Wave Turbulence Under Parametric Excitation
Applications to Magnetics
By V. S. L'vov

Kolmogorov Spectra of Turbulence I Wave Turbulence
By V. E. Zakharov, V. S. L'vov and G. Falkovich

Nonlinear Processes in Physics
Editors: A. S. Fokas, D. J. Kaup, A. C. Newell and V. E. Zakharov

Important Developments in Soliton Theory
Editors: A. S. Fokas and V. E. Zakharov

Bifurcation and Chaos Theory and Applications
Editor: J. Awrejcewicz

J. Awrejcewicz (Ed.)

Bifurcation and Chaos

Theory and Applications

With 135 Figures

Professor Jan Awrejcewicz

Technical University of Łódź, Division of Dynamics and Control (K-13)
Stefanowskiego 1/15, 90-924 Łódź, Poland

Volume Editor

Professor Miki Wadati

University of Tokyo, Faculty of Science, Department of Physics
7-3-1 Hongo, Bunkyo-ku, Tokyo 113, Japan

ISBN 3-540-58531-1 Springer-Verlag Berlin Heidelberg New York

CIP data applied for

Typesetting: Camera ready copy from the author/editor using a Springer T_EX macro package
SPIN: 10424400 55/3140 - 5 4 3 2 1 0 - Printed on acid-free paper

Preface

This volume of the Springer Series in Nonlinear Dynamics addresses the wide spectrum of researchers working on nonlinear dynamical systems. It is the result of an idea to put together some important results from the different branches of science and engineering encompassed by bifurcation and chaos phenomena. This approach of tracking an essential nonlinear behaviour in science and engineering is certainly needed to display the impact of both these fundamental areas of human endeavor. It is expected that this will result in new ideas taken from one branch being seeded into another. It should also bring a fresh breeze to the essential foundations laid by Poincaré, Newton, Laplace, Lyapunov, Duffing, van der Pol and others. The name of Poincaré does not appear accidentally as the first one. His interest in nonlinear dynamical systems was embedded in a philosophy encompassing a group of sciences that are today independent. The approach presented here is in fact intuitively very similar. We believe and expect that great progress can result from such an impact on engineering and science and from collecting relevant overviews in one volume. We encourage the reader to think and interpret the result in an interdisciplinary manner. This volume can perhaps prove that not isolation but integration of many behaviors helps us to better understand the individual phenomena analysed. If one goes abroad for a long time and comes back to one's home country, one sees and thinks about one's home country in a different, more objective way. This observation also describes the philosophy behind this book and a major motivation for compiling it.

Developing this idea a little bit more we can say this book is aimed at everyone who is interested in nonlinear science and engineering. But a fundamental mathematical knowledge is needed to understand perhaps 80% of its contents. However, the book is written to be understood by students, both graduate and undergraduate, researchers and teachers from applied mathematics, theoretical and applied physics, and also engineering sciences such as mechanics, electrical and electronic engineering, civil engineering and computer science.

The intention of this volume on bifurcation and chaos, which consists of invited contributions, is to illustrate that bifurcation and chaos phenomena represent a challenging area in applied mathematics, theoretical and applied physics, and engineering science. We hope to show that analogous bifurcation and chaotic phenomena are often observed in quite different fields of investigations and that they are universal features of nonlinear dynamical systems. The book is intended to guide the reader through the most up-to-date research on bifurcation and chaos in one unique volume. It should benefit the reader not only in the particular field of special interest, but it will also help in transplanting ideas taken from another field.

I am very grateful to all contributors to this book for their willingness to prepare their individual chapters as well as their patience in waiting for a final version of the project. Prof. M. Wadati, Series Editor, and Dr. A.M. Lahee, Springer Editor, are thanked for their time and consideration paid to the preparation of this volume. Finally, I would like to extend my thanks to Dr. J. Mrozowski and Dr. A. Ligier for assistance in preparing the camera-ready copy.

Jan Awrejcewicz, Lódź

Table of Contents

List of Contributors

J. Awrejcewicz
Division of Dynamics and Control,
Technical University of Lódź,
Stefanowskiego
90–924 Lódź, Poland

A.K. Bajaj
School of Mechanical Engineering
Purdue University,
West Lafayette,
IN 47907–1288, U.S.A.

S.I. Chang
School of Mechanical Engineering
Purdue University,
West Lafayette,
IN 47907–1288, U.S.A.

B.Y. Chirikov
Budker Institute of Nuclear Physics
630090 Novosibirsk, USSR

P. Davies
School of Mechanical Engineering
Purdue University,
West Lafayette,
IN 47907–1288, U.S.A.

A. Kanasugi
Department of Electrical and
Electronic Engineering,
Saitama University
Shimo-Ohkubo
Urawa, 338 Japan

H. Kawakami
Department of Electrical and
Electronic Engineering,
Tokushima University,
Tokushima-shi, 770 Japan

M. Morisue
Department of Electrical and
Electronic Engineering,
Saitama University
Shimo-Ohkubo
Urawa, 338 Japan

J. Mrozowski
Division of Dynamics and Control,
Technical University of Lódź,
Stefanowskiego
90–924 Lódź, Poland

M. Potier-Ferry
Laboratoire de Physique et
Mécanique des Matériaux
Université de Metz
Ile du Sauley
57045 Metz, France

G. Rega
Dipartimento di Ingegneria
delle Strutture,
delle Acque e del Terreno
Università dell'Aquila,
L'Aquila, Italy

R. Stoop
Swiss Federal Institute of Technology (ETH),
CH–8092 Zurich, Switzerland

J. Tani
Institute of Fluid Science,
Tohoku University,
2-1-1 Katahira, Aoba-ku,
Sendai, 980 Japan

G.S. Whiston
National Power PLC,
Bilton Centre,
Cleeve Road, Leatherhead, Surrey,
KT22 7SE United Kingdom

J. Wu
Department of Mechanics,
Peking University,
Beijing 100871, PRC

T. Yoshinaga
School of Medical Sciences,
Tokushima University,
Tokushima-shi, 770 Japan

K. Zhou
Department of Mechanics,
Peking University,
Beijing 100871, PRC

Introduction

This book presents and discusses a collection of the research results from physical and engineering science written by leading scientists in the special branches of nonlinear dynamical systems. The book has a fractal-like construction. Almost all of the independent chapters review in depth the present status of research activities from a field under consideration, first by providing the reader with an elementary introduction and then through the presentation of the latest concepts of the considered topics and finally introducing the original theoretical, computational or experimental results. The general construction of the book is prepared in a similar manner. In the first chapters, new theoretical achievements in the theory of chaos and its associated bifurcations are outlined. Then the numerical algorithms for the effective study of high dimensional Hopf bifurcation problems and codimension two bifurcations supported by theoretical considerations are provided. The last chapters of the book are focused more on engineering bifurcation and engineering chaos, covering the various problems relevant to the field of electrical and electronics engineering, mechanical engineering and biomechanical engineering. It should be emphasized that each of the above mentioned chapters brings more light to the individually analyzed, special deterministic, dynamical engineering systems, and is strongly supported by the theoretical and numerical studies. Again we appeal to the fractal-like construction of the book. All of the new aspects and results of dynamical behaviour found in the special systems taken from electrical and mechanical engineering enrich and increase the general level of understanding the dynamical behaviour of the physical systems showing many universal properties.

The structure of the book lends itself to a wide variety of potential readers and it will be of interest to graduate students, teachers and researchers of theoretical and applied physics, electrical and electronic engineering and mechanical engineering. The book is recommended for advanced engineers, involved particularly in investigation of electrical circuits, in the performance of superconducting electronic devices such as the Josephson junctions, in vibro-impact dynamics, in analysis of electro-mechanical systems with magnetic force and with magnetic levitated masses, in dynamical analysis of curved beams, shallow arches and suspended cables, in flutter and chaotic oscillations of rectangular plates. Readers will certainly be provided with a digest of fundamentals and status of the subject up to the latest and most advanced topics from the point of view of nonlinear dynamics, bifurcation and chaos.

It should be mentioned that all of the considered engineering problems can be useful in application and, in all topics considered, special new results are outlined. Readers oriented either in the theory or application of chaotic oscillations and associated bifurcation will certainly be satisfied with the material covered by the book. Almost all of the chapters include a brief introduction to the investigated problems and, moreover, in all chapters an independent theory is established and further illustrated either by numerical or both numerical and experimental analysis. The chapters are equally accessible to those oriented either towards theory or applications. All of the chapters can be read independently and the reader is not forced to work through all of them in sequence.

The book is aimed at an audience with a fundamental knowledge of mathematics and nonlinear dynamics including stability theory. It can serve as well as an education for more theoretically advanced engineers.

The first chapter written by B.V. Chirikov is motivated by a number of unsolved problems of nonintegrable quantum dynamics and is mainly addressed to mathematicians and theoretical physicists. A new definition of quantum chaos is given, which is supported by a new modified concept of mixing. The analogy between dynamical localization in momentum space and the Anderson localization in disordered solids (static theory) is discussed. Additionally, it is shown that the definition of quantum chaos can be successfully applied to any linear classical waves. The concise and logic style of writing should attract a theoretically oriented reader interested in open questions of chaotic dynamics.

In the second chapter prepared by R. Stoop the thermodynamic formalism has been applied to the characterization of large classes of chaotic dynamical systems. The behaviour of chaotic systems can be described by the scaling properties of a finite number of elements. Usually the observation is directed towards the phase space. The observation object is defined by the probability measure and only the location of points in the considered space is analyzed. Observation of amplitude changes along the time axis defines the second most popular method of tracing chaotic phenomena.

The process of deriving macroscopic quantities from microscopic behaviour is called the thermodynamic formalism because it is in common use in thermodynamics. In order to obtain a macroscopic characterization and to obtain averaged information on the scaling behaviour, procedures similar to those of statistical mechanics are used.

Based on the generalized thermodynamic formalism, a more refined description of the properties of dynamical systems has been given. With the help of appropriate models, theoretical tools have been outlined which permit one to predict and understand the problems arising during the numerical characterization of the scaling behaviour in dissipative dynamical systems. With the same tools, different combinations of first-order phase transition, which appear for generic dynamical systems, were elucidated.

The next chapter, written by Wu Jike and Zhou Kun, deals with new numerical methods for high-dimensional Hopf bifurcation problems. So far almost all numerical methods now available for Hopf bifurcation problems are only effective for low-dimensional problems. With increasing dimensionality of the system analyzed, computation time increases strongly. For this reason, a new numerical method with low computational complexity is presented for the computation of static bifurcation, Hopf bifurcation and for tracing a closed orbit on a large scale.

An equivalent relationship between invariant manifolds of the vector fields and the solution manifolds of systems of nonlinear equations is given. The study of the bifurcated solution is turned into finding a pair of complex conjugate eigenvalues with maximum norm of matrix. The problem of following the closed orbits on a large scale is equivalent to tracking the solution manifold of a nonlinear equation system in higher dimensional space. Examples provided show the efficiency of the proposed method.

There are researchers who think that computer simulation should never come before a rigorous mathematical analysis. G.S. Whiston addresses his research to these people, who are involved in many different fields of nonlinear dynamical systems. Based on a good mathematical background, clearly written, and containing interesting results of a hard pure way of thinking, it carefully provides the reader with more complete information pertinent to whole classes of the vibro-impact systems. It is also expected that the results obtained could be valid for the high-dimensional complicated systems. Such vibro-impact discontinuities arise from the lack of "structural stability" of zero velocity or "grazing impacts" and lead to a "loss of predictability", where arbitrarily close initial conditions diverge in discrete steps across discontinuity. The proposed theory concentrates on description of the singularity structure of linear autonomous systems and is applicable to systems of arbitrary dimension. First construction of a vibro-impact system with a single amplitude constraint is shown. The possible "trapping" of grazing impact leads to the construction of a high dimensional singularity surface. This construction is clarified by an example with three-dimensional "Poincaré section" and a nondifferentiable two-dimensional singularity surface. Then a geometry of this surface is discussed using the methods of catastrophe theory. Finally, unfolding theory is applied to analyze the nature of the discontinuities of the Poincaré map.

The compact theory introduced by the author shows that nondifferentiability in vibro-impact dynamics can lead to breakdown of the global stable manifolds and to the generation of homoclinic tangles by the translation of segments of stable and unstable manifolds across each other.

The vibro-impact dynamics plays an important role engineering systems. Many types of nuclear reactors are prone to heat exchanger tube fretting wear through vibro-impact/slide induced by fluid-structural interaction. Also, in many machines, for many types of industrial plant, a vibro-impact response can occur which is very harmful. Thus, a detailed analysis is needed to solve the engineering problem and certainly this chapter helps in this matter.

It is trivial to say that periodic orbits play a fundamental role in nonlinear dynamical systems. The bifurcations of periodic solutions are known as codimension one bifurcations. If a periodic solution satisfies two bifurcation conditions, then the bifurcation is referred to as a codimension two bifurcation. Codimension one and two bifurcation of periodic orbits is very important. These types of problems appear in physics and engineering, where the investigated dynamical systems are governed by ordinary nonlinear differential equations. The chapter prepared by H. Kawakami and T. Yoshinaga, is motivated by an attempt to introduce a sufficient and relatively simple numerical technique for finding bifurcation points, which separate a qualitatively different dynamical behaviour. In almost all cases research attention is focused on establishing the critical parameters, which separate the different phase flow described by the investigated physical system. Here the computational algorithm is very efficient because it is based on shooting and Newton's methods (it is easy to find strong nonlinear periodic orbits).

Another aspect is the investigation of nonlinear phenomena for parameters close to codimension two bifurcations including jump and hysteresis phenomena, frequency entrainment, chaotic synchronization, chaotic states with both stable and unstable invariant sets. All the mentioned nonlinear behaviours, or even more, are found in two appropriate electrical engineering examples.

The application of standard chaos concepts to the real world for the example of the superconducting devices of Josephson junctions used in an ultra fast computer and ultra weak magnetic detector certainly explains the motivation of the contribution by M. Morisue and A. Kanasugi. The chapter is specially accessible to researchers and advanced students of electrical and electronic engineering. The clear style of writing and ease in formulation of the solution to the stated problems should certainly find a resonance in the potential readers. Periodic, quasi-periodic and chaotic orbits are analyzed mainly numerically for the Josephson autonomous, nonautonomous and distributed parameter circuit. The occurrence of chaotic noise exceeds the thermal noise in Josephson junction by several orders of magnitude, and therefore the analysis is not only academic but the results obtained allow one to avoid chaotic behaviour, which is fundamental for practice. An additional new result should be pointed out: a kind of periodic oscillation followed by relaxation oscillation instead of a non-latching periodic oscillation (reported earlier) has been found. This chapter can be treated as a bridge between the theoretical results of chaos and highly advanced technology.

The chapter by J. Tani includes an intersection of two new interdisciplinary fields of research: applied electromagnetic and chaotic deterministic behaviour. The research and methodology results presented are motivated by expanding evolutionary technologies of the magnetoelastic, magnetic, levitation as well as magnetoelastic postbuckling systems. The contribution addresses engineers and advanced students working and interested in electromagnetic field—materials interactions on the one hand, and those involved the application of chaos concepts on the other hand. Both newcomers and advanced researchers will certainly find useful the analytical methods supported by the center manifold theory, the mathematical formulation of the governing equations as well as numerical tools used to prove the existence of chaos and bifurcation in the systems with magnetic force. Without any doubt newcomers will benefit from clearness of presentation, while advanced researchers could be positively surprised by the obtained results.

The numerical evidence for chaos in wire spring system shows that for excitation by alternating current the chaotic behaviour is always transient, whereas for the transverse support excitation nontransient chaos occurs for some range of the excitation frequency.

A system with two parallel wires carrying currents, a multi-well potential magnetoelastic system and a system with magnetic levitation are considered. In the case of a multi-well potential magnetoelastic system, numerical simulation is confirmed by experimental results. The results show that the magnitude of the chaotic vibrations is larger than the resonance vibrations. The reason is that the resonance corresponds to the regular vibrations around one equilibrium point and chaos corresponds to the random-like vibration around two equilibrium points with jump phenomena.

Three kinds of simple system with magnetically levitated mass are analyzed on the basis of the center manifold theory. They include a magnetically levitated mass with the mass excitation, the same mass with horizontal elastic beam excitation and the levitated mass with support excitation. In all cases considered, numerical evidence of chaos has been given in the three different systems mentioned above. The range of chaos for the first two systems is very narrow, and the structure of the strange attractor is very different. Additionally, in this chapter, some special bifurcation diagrams exhibiting a variety of nonlinear phenomena are presented.

Curved beams, shallow arches, and suspended cables can be reduced to the consideration of asymmetric (Helmholtz) and symmetric (Duffing) type oscillators. Bifurcation and chaos in the Helmholtz-Duffing oscillator is analyzed in depth by G. Rega. He demonstrates that the presence of both even and odd nonlinearities causes the occurrence of a rich and complex set of responses. This simple model exhibits multiple coexisting periodic attractors, strange chaotic attractors occurring in a various ranges of control parameter values, different routes to chaos from periodic solutions, and regular and fractal basin boundaries. This oscillator serves as an example for discussing three fundamental questions: 1. Diagnosis problems which are concerned with the implementation and correct use of dynamic measures; 2. Scenario problems which refer to construction of behaviour charts in control parameter space and of basins of attraction in initial parameter space; 3. Prediction problems which are concerned with determination of critical bounds for the occurrence of bifurcations and chaos in regions of regular nonlinear response. This chapter includes photographs which certainly help to give better insight into the geometrical analysis of the nonlinear oscillator considered.

The chapter written by S.I. Chang, A.K. Bajaj and P. Davies, aims to focus the attention of the reader to the validity of application of the approximate averaging and multiple-scale technique to the nonlinear analysis of continuous systems which are governed by partial differential equations. The chapter is specially addressed to researchers and engineers from the field of mechanical engineering, but it could be also of interest for theoretical and applied physicists and all nonlinear dynamics researchers. The reason is mainly that the presentation focuses on many interdisciplinary interesting nonlinear phenomena, such as multiple solutions, jumps, subharmonic and superharmonic resonances, and amplitude modulated motions including period doubling and chaos. The most important results, however, are related to the interaction analysis among the modes through the internal and external resonances. It has been clearly shown and illustrated that most interesting resonances arise due to the exchange of energy between the modes in internal resonance, whereas through the external resonance energy can be supplied to the modes in internal resonance. Averaging procedures are supported by sufficient numerical techniques used to solve the amplitude differential equations.

In particular, however, the chapter should interest a large number of mechanical engineering researchers. First, the nonlinear behaviours of structures, including strings, beams, arches, plates and shells are well reviewed, and then some open problems are outlined, discussed and illustrated. Second, a general way of including important nonlinearities, such as geometric, inertial, those due to material properties, or arising because of damping mechanism or boundary conditions, is clearly presented.

A careful bifurcation analysis of the averaged equations is carried out as a function of the excitation amplitudes and frequency, and as a function of the damping present in the plate. Various saddle-node, pitchfork, and Hopf bifurcation sets are constructed, and it is shown that the response of the plate depends very significantly on the mode which is directly excited. The limit cycle, as well as chaotic solutions, of the averaged equations are shown to predict qualitatively similar-modulated motions for the original two-mode model for neighbouring values of parameters.

Finally, the chapter is clearly written and well embedded into the cited references, which certainly helps the reader in understanding and getting new ideas for his own research.

In the chapter prepared by J. Awrejcewicz, J. Mrozowski and M. Potier-Ferry, attention is focused on the nonlinear dynamics of a rectangular plate driven by the aerodynamic force action. First, the validity of the technical problems caused by dynamic instability of the rectangular plates is outlined and then the governing equations are derived. The geometrical nonlinearities due to the rotation of the structure elements are taken into account and three partial differential equations with the appropriate boundary conditions are obtained. A particular solution including the expected torsional effects together with the usual bending effects is sought. Then the assumed solution is put to the governing equation set and, after applying the Galerkin procedure, the set of two second order ordinary nonlinear differential equations is obtained.

In the second part of the conribution, analytical and numerical analysis of the obtained equations is given. It has been found for the Hopf bifurcation curve that, with the increase of one of the two control parameters, the periodic orbit found earlier becomes unstable and then (with a further change of control parameter) a period doubling scenario leading to chaos is observed. Within this scenario the development of the strange chaotic attractor accompanied by the inverse bifurcation cascade is discussed in some detail. Also intermittency chaotic phenomena are illustrated and analyzed. Some conclusions related to real plate behaviour are formulated.

Quantum Chaos and Ergodic Theory

B.Y. Chirikov

Budker Institute of Nuclear Physics
630090 Novosibirsk, USSR

Abstract

The conception of quantum chaos is described in some detail. The most striking feature of this novel phenomenon is in that all the properties of classical dynamical chaos are retained but, typically, on finite and different time scales only. The necessary reformulation of the ergodic and algorithmic theories, as parts of the general theory of dynamical systems, is discussed. A number of specific unsolved problems is listed.

1. Introduction

This paper is primarily addressed to mathematicians with the main purpose of explaining new physical ideas in the so-called *quantum chaos* which has recently been attracting ever growing interest of many researchers [1–5, 10].

The breakthrough in understanding of this phenomenon has been achieved, particularly, due to a new philosophy accepted, explicitly or more often implicitly, in most studies of quantum chaos. Namely, the whole physical problem of quantum dynamics was separated into two different parts: (i) the proper quantum motion described by a specific dynamical variable $\phi(t)$ which obeys, e.g., the Schrödinger equation, and (ii) the quantum measurement including ψ collapse which, as yet, has no dynamical description. In this way one can single out the vague problem of the fundamental randomness in quantum mechanics which is related to the second part only, and which in a sense is foreign to the proper quantum system. The remaining first part then fits perfectly the general theory of dynamical systems.

The importance of quantum chaos is not only in that it represents a new unexplored field of nonintegrable quantum dynamics with many applications, but also, and this is most interesting for the fundamental science, in reconciling the two seemingly different dynamical mechanisms for the statistical laws in physics.

Historically, the first mechanism is related to the *thermodynamic limit* $N \to \infty$ in which the completely integrable system becomes chaotic for typical (random) initial conditions (see, e.g.,[6]). A natural question—what happens for large but finite number of freedoms N—has still no rigorous answer but the new phenomenon of quantum chaos, at least, presents an insight into this problem too. We call this mechanism, which is equally applicable in both classical and quantum mechanics, the *traditional statistical mechanics* (TSM).

The second (new) mechanism is based upon the strong (exponential) local instability of motion characterized by positive Lyapunov's exponent $\Lambda > 0$ [6, 7]. It is not at all restricted to large N, and is possible, e.g., for $N > 1$ in a Hamiltonian system. However, this mechanism has been considered, until recently, in the classical mechanics only. We term this the *dynamical chaos* as it does not require any random parameters or any noise in the equations of motion.

The quantum system bounded in phase space has a discrete energy (frequency) spectrum and is similar, in this respect, to the finite-N TSM. Moreover, such quantum systems are even completely integrable in the Hilbert space (see, e.g. [3]). Yet, the fundamental correspondence principle requires the transition to classical mechanics, including dynamical chaos, in the *classical limit* $q \to \infty$, where q is some quasi-classical parameter, e.g., the quantum number n (the action variable, $\hbar = 1$). Again, a natural physical conjecture is that for finite but large q there must be some chaos similar to finite-N TSM. Yet, in a chaotic quantum system the number of degrees of freedom N does not need to be large similarly to the classical chaos. The quantum counterpart of N is q, both quantities determining the number of frequencies which control the motion. Thus, mathematically, the problem of quantum chaos is the same as that for the finite-N TSM.

The main difficulty here (especially for mathematicians) is that the both problems suggest some chaos in the discrete spectrum which is completely contrary to the existing theory of dynamical systems and to the ergodic theory where such a spectrum corresponds to the opposite limit of regular motion.

The ultimate origin of the quantum integrability is discreteness of the phase space (but not, as yet, of the space-time!) or, in the modern mathematical language, the noncommutative geometry of the former.

As an illustration I will make use of the simple model described classically by the *standard map* (SM) [7, 8]:

$$\bar{n} = n + k\sin\theta \ ; \qquad \bar{\theta} = \theta + T\bar{n} \tag{1}$$

with action-angle variables n, θ, and perturbation parameters k, T. The quantized standard map (QSM) is given by [9, 10]

$$\bar{\psi} = \exp(-\mathrm{i}k\cos\bar{\theta})\exp\left(-\mathrm{i}\frac{T}{2}\hat{n}^2\right)\psi \ , \tag{2}$$

where the momentum operator $\hat{n} = -\mathrm{i}\partial/\partial\theta$. To provide the complete boundedness of the motion we consider SM on a torus of circumference (in n)

$$L = \frac{2\pi m}{T} \tag{3}$$

with integer m to avoid discontinuities. The quasi-classical transition corresponds to quantum parameters $k \to \infty$, $T \to 0$, $L \to \infty$ while classical parameters $K = kT = \text{const}$, and $m = LT/2\pi = \text{const}$ remain unchanged.

QSM models the *energy shell* of a conserved system which is the quantum counterpart of the classical energy surface.

In the studies of dynamical systems, both classical and quantal, most problems unreachable for rigorous mathematical analysis are treated "numerically" using the computer as a universal model. With all obvious drawbacks and limitations such "numerical experiments" have very important advantage as compared to the laboratory experiments, namely, they provide the complete information about the system under study. In quantum mechanics this advantage becomes crucial as in the laboratory one cannot observe (measure) the quantum system without a radical change of its dynamics.

2. Definition of Quantum Chaos

The common definition of classical chaos in physical literature is the *strongly unstable motion*, that is one with positive Lyapunov's exponents $\Lambda > 0$. The Alekseev-Brudno theorem then implies that almost all trajectories of such a motion are unpredictable, or random (see [11]). A similar definition of quantum chaos, which still has adherents among both mathematicians as well as a few physicists, fails because, for the bounded systems, the set of such motions is empty due to the discreteness of the phase space and, hence, of the spectrum.

The common definition of quantum chaos is *quantum dynamics of classically chaotic systems* whatever this might happen to be. Logically, this is a simple and clear definition. Yet, in my opinion, it is completely inadequate from the physical viewpoint just because such a chaos may turn out to be a perfectly regular motion as, for example, in case of the *perturbative localization* [12]. In QSM this corresponds to $k \leq 1$ when all quantum transitions are suppressed independent of classical parameter K which controls the chaos.

I would like to define quantum chaos in such a way as to include some essential part of classical chaos. The best definition I have managed to invent so far reads: the *quantum chaos is statistical relaxation in a discrete spectrum.* This definition is certainly in contradiction to the existing ergodic theory as the relaxation (particularly, correlation decay) requires the mixing, hence, a continuous spectrum. In what follows I will try to explain a new, modified, concept of mixing which is necessary to describe the peculiar phenomena of quantum chaos.

3. The Time Scales of Quantum Dynamics

The first numerical experiments with QSM already revealed the quantum diffusion in n close to the classical one under conditions $K \geq 1$ (classical stability border) and $k \geq 1$ (quantum stability border) [9]. Further studies confirmed this conclusion and showed that the former followed the latter in all details but on a *finite time interval* only [10, 13]. The latter fact was the clue to understanding the dynamical mechanism of the diffusion, which is apparently an aperiodic process, in a discrete spectrum. Indeed, the fundamental uncertainty principle implies that the discreteness of the spectrum is not resolved for sufficiently short time intervals. Whence, the estimate for the *diffusion (relaxation) time scale*:

$$t_R \sim \rho_0 \leq \rho\,. \tag{4}$$

Here ρ is the density of (quasi)energy levels, and ρ_0 is the same for the *operative eigenstates* which are actually present in the initial quantum state $\psi(0)$. In QSM the quasi-energies are determined mod $2\pi/T$ and, surprisingly, $\rho = LT/2\pi = m$ is a classical parameter (3). As to ρ_0, it depends on the dynamics and is given by the estimate [10, 13]:

$$\frac{\rho_0}{T} \sim \frac{t_R}{T} \equiv \tau_R \sim D \equiv \frac{\langle(\Delta n)^2\rangle}{\tau} \leq \frac{m}{T}\,. \tag{5}$$

Here τ is discrete map's time (the number of iterations), and D is the classical diffusion rate. This remarkable expression relates an essentially quantum characteristic (τ_R) to the classical one (D). The latter inequality in Eq. (5) follows from that in Eq. (4), and is explained by the boundedness of QSM on a torus.

In the quasi-classical region $\tau_R \sim k^2 \to \infty$ (see Eq.(1)) in accordance with the correspondence principle. Yet, the transition to the classical limit is (conceptually) difficult to understand (and still more to accept) as it involves two limits ($k \to \infty$ and $t \to \infty$) which do not commute. The second limit is related to the existing ergodic theory which is asymptotic in t. Meanwhile the new phenomenon of quantum chaos requires the modification of the theory to a finite time which is a difficult mathematical problem still to be solved. The main difficulty is in that even the distinction between the two opposite limits in the ergodic theory—discrete and continuous spectra—is asymptotic only.

In a relatively new *algorithmic theory* of dynamical systems the finite-time trajectories are also considered but, as yet, with the strongest statistical property—the randomness—only, which is generally unnecessary for a meaningful statistical description.

Besides the relatively long time scale (5) there is another one given by the estimate [14, 10]

$$t_r \sim \frac{\ln q}{\Lambda} \to \frac{T|\ln T|}{\ln(K/2)} \tag{6}$$

where q is some (large) quasi-classical parameter, and where the latter expression holds for QSM. It may be termed the *random time scale* since here the quantum motion of a narrow wave packet is as random as classical trajectories according to the Ehrenfest theorem. This was well confirmed in a number of numerical experiments [15]. The physical meaning of t_r is in the fast spreading of a wave packet due to the strong local instability of classical motion.

Even though the random time scale t_r is very short it grows indefinitely in the quasi-classical region ($q \to \infty$, $T \to 0$), again in agreement with the correspondence principle.

The big ratio t_R/t_r implies another peculiarity of quantum diffusion: it is dynamically stable as was demonstrated in striking numerical experiments [16].

4. The Quantum Steady State

As a result of quantum diffusion and relaxation some steady state is formed whose nature depends on the *ergodicity parameter*

$$\lambda = \frac{l_s}{L} \simeq \frac{D}{L}\,, \tag{7}$$

where l_s is the so-called localization length (see Eq.(10) below). If $\lambda \gg 1$ the quantum steady state is close (on average) to the classical statistical equilibrium which is described by ergodic phase density $g_{\mathrm{cl}}(n) = \mathrm{const}$ (for SM on a torus) where n is continuous variable. In quantum mechanics n is integer, and the quantum phase density $g_q(n,\tau)$ in the steady state fluctuates [17, 5], the ergodicity description can be given by relation

$$g_q(n) = \overline{|\psi_s(n,\tau)|^2} = \frac{1}{L}\,, \tag{8}$$

where the bar denotes time averaging.

According to numerical experiments the ergodicity does not depend on the initial state which implies that all eigenfunctions $\phi_m(n)$ are also ergodic, on average, with Gaussian fluctuations [17, 5]:

$$\langle|\phi_m(n)|^2\rangle = \frac{1}{L}\,. \tag{9}$$

This is always the case sufficiently far in the quasi-classical region as $\lambda \sim k^2/L \sim Kk/m \to \infty$ with $k \to \infty$ ($K = kT$ and $m = LT/2\pi$ remain constant) in accordance with Shnirelman's theorem [18].

An interesting unsolved problem is the microstructure of ergodic eigenfunctions, particularly, the so-called 'scars' [29] which reveal the set of classical periodic trajectories (see [30] for the theory of scars).

Finite fluctuations (9) show that a single chaotic quantum system, described by $\psi_s(n,\tau)$, represents, in a sense, finite statistical ensemble of $M \sim L$ "particles". The fluctuations can result in partial recurrences toward the initial state but the recurrence time is much longer as compared to the relaxation time scale τ_R and sharply depends on the recurrence domain.

If $\lambda \ll 1$ the quantum steady state is qualitatively different from the classical one. Namely, it is localized in n within the region of size l, around the initial state if the size of the latter $l_0 \ll l_s$. Numerical experiments show that the phase space density, or the *quantum statistical measure*, is approximately exponential [10, 13]

$$g_s(n) \simeq \frac{1}{l_s} \exp\left(-\frac{2|n|}{l_s}\right) ; \qquad l_s \simeq D \tag{10}$$

for initial $g(n,0) = \delta(n)$. The quantum ensemble is now characterized by $M \sim l_s \sim k^2$ "particles".

The relaxation to this steady state is called *diffusion localization*, and it is described approximately by the diffusion equation [19, 28]

$$\frac{\partial g}{\partial \tau'} = \frac{1}{2} \frac{\partial}{\partial n} D \frac{\partial g}{\partial n} \pm \frac{\partial g}{\partial n} \tag{11}$$

for initial $g(n,0) = \delta(n)$, where the signs "$\pm$" correspond to $n \neq 1$, and where new time

$$\tau' = \tau_R \ln\left(1 + \frac{\tau}{\tau_R}\right) \tag{12}$$

accounts for the discrete motion spectrum [20]. The last term in Eq. (11) describes "backscattering" of ψ wave propagating in n which eventually results in the diffusion localization. The fitting parameter $\tau_R \sim 2D$ was derived from the best numerical data available (see Ref. [21], where a different theory of diffusion localization was also developed).

5. Concluding Remarks

In conclusion I would like to briefly mention a few important results for unbounded quantum motion. In SM this corresponds to $L \to \infty$. First, there is an interesting analogy between dynamical localization in momentum space and the celebrated Anderson localization in disordered solids which is a statistical theory. It was discovered in [22] and essentially developed in [23]. The analogy is based upon (and restricted by) the equations for eigenfunctions. The most striking (and less known) difference between the two problems is in the absence of a diffusion regime in $1D$ solids [24]. This is because the energy level density of the operative eigenfunctions in solids

$$\rho_0 \sim \frac{ldp}{dE} \sim \frac{l}{u} \sim t_R \tag{13}$$

which is the localization (relaxation) time scale, is always of the order of the time interval for a free spreading of the initial wave packet at characteristic velocity u.

Another similarity between the two problems is in that the Bloch extended states in a periodic potential correspond to a peculiar quantum resonance in QSM for rational $T/4\pi$ [9, 10].

An interesting open question is the dynamics for irrational Liouville's (transcendental) $T/4\pi$.

As was proved in [25] the motion can be unbounded in this case unlike a typical irrational value. The latter is the result of numerical experiments, no rigorous proof of localization for $k \gg 1$ has been found as yet.

In [28] the conjecture is put forward, supported by some semiqualitative considerations, that depending on a particular Liouville's number the broad range of motions is possible, from a purely resonant one ($|n| \sim \tau$) down to complete localization ($|n| \leq l$).

If quantum motion is not only unbounded but its rate in unbounded variables is exponential, then "true" chaos (not restricted to a finite time scale) can occur. A few exotic examples together with considerations from different viewpoints can be found in [10, 26]. However, such chaos does not seem to be a typical quantum dynamics.

The final remark is that the quantum chaos, as defined in Sect. 2, comprises not only quantum systems but also any linear, particularly classical, waves [27]. So, it is essentially the *linear wave chaos*. Moreover, a similar mechanism also works in completely integrable nonlinear systems like the Toda lattice, for example [31]. From a mathematical point of view all these new ideas require reconsideration of the existing ergodic theory. Perhaps it is better to say that a new ergodic theory is wanted which, instead of benefiting from the asymptotic approximation ($|t| \to \infty$ or $N \to \infty$), could analyze the finite-time statistical properties of dynamical systems. In my opinion, this is the most important conclusion emerging from first attempts to comprehend quantum chaos.

References

1. Proc. Les Houches Summer School on Chaos and Quantum Physics, Elsevier 1991
2. F. Haake: Quantum signatures of chaos. Springer 1991
3. B. Eckhardt: Phys. Reports **163**, 205 (1988)
4. G. Casati and L. Molinari: Suppl. Prog. Theor. Phys. **98**, 287 (1989)

5. F.M. Izrailev: Phys. Reports **196**, 299 (1990)
6. I. Kornfeld, S. Fomin and Ya. Sinai: Ergodic theory. Springer 1982
7. A. Lichtenberg, M. Lieberman: Regular and stochastic motion. Springer 1983; G.M. Zaslavsky: Chaos in dynamic systems. Harwood 1985
8. B.V. Chirikov: Phys. Reports **52**, 263 (1979)
9. G. Casati et al: Lecture Notes in Physics **93**, 334 (1979)
10. B.Y. Chirikov, F.M. Izrailev and D.L. Shepelyansky: Sov. Sci. Rev. C**2** (1981) 209; Physica D **33**, 77 (1988)
11. V.M. Alekseev and M.V. Yacobson: Phys. Reports **75**, 287 (1981)
12. E.V. Shuryak: Zh. Eksp. Teor. Fiz. **71**, 2039 (1976)
13. B.V. Chirikov, D.L. Shepelyansky: Radiofizika **29**, 1041 (1986)
14. G.P. Berman and G.M. Zaslavsky, Physica A **91**, 450 (1978)
15. M. Toda and K. Ikeda: Phys. Lett. A **124**, 165 (1987); A. Bishop et al: Phys. Rev. B **39**, 12423 (1989)
16. D.L. Shepelyansky, Physica D **8**, 208 (1983); G. Casati et al: Phys. Rev. Lett. **56**, 2437 (1986)
17. F.M. Izrailev: Phys. Lett. A **125**, 250 (1987)
18. A.I. Shnirelman: Usp. mat. nauk **29**, No 6, 181 (1974); On the asymptotic properties of eigenfunctions in the regions of chaotic motion, addendum in: V.F. Luzutkin: The KAM theory and asymptotics of spectrum of elliptic operators, Springer 1991
19. B.V. Chirikov, CHAOS **1**, 95 (1991)
20. B.V. Chirikov: Usp. fiz. nauk **139**, 360 (1983); G.P. Berman, F.M. Izrailev: Operator theory: advances and applications, **46**, 301 (1990)
21. D. Cohen: Quantum chaos, dynamical correlations and the effect of noise on localization, 1991 (unpublished)
22. S. Fishman et al: Phys. Rev. A **29**, 1639 (1984)
23. D.L. Shepelyansky, Physica D **28**, 103 (1987)
24. E.P. Nakhmedov et al: Zh. Eksp. Teor. Fiz. **92**, 2133 (1987)
25. G. Casati and I. Guarneri: Comm. Math. Phys. **95**, 121 (1984)
26. S. Weigert: Z. Phys. B **80**, 3 (1990); M. Berry: True quantum chaos? An instructive example. Proc. Yukawa Symposium, 1990; F. Benattz et al: Lett. Math. Phys. **21**, 157 (1991)
27. B.V Chirikov: Linear chaos. Preprint INP 90–116, Novosibirsk 1990
28. B.V Chirikov: Chaotic quantum systems. Preprint INP 91–83, Novosibirsk 1991
29. E. Heller: Phys. Rev. Lett. **53**, 1515 (1984)
30. E.B. Bogomolny: Physica D **31**, 169 (1988); M. Berry: Proc. Roy. Soc., London, A **423**, 219 (1989)
31. J. Ford et al: Prog. Theor. Phys. **50**, 1547 (1973)

On the Complete Characterization of Chaotic Attractors

R. Stoop

Swiss Federal Institute of Technology (ETH),
CH–8092 Zürich, Switzerland

Abstract

The paper proves that a unified so-called thermodynamic formalism (both phase space and time space) can describe large classes of chaotic dynamical systems. It includes problems of scaling behaviour (scale invariance and nonunified thermodynamic formalism), the unified approach (the generalized entropy function and hyperbolic models with complete grammars) and proposed extensions (convergence properties, influence of grammar, nonhyperbolicity and phase transitions).

1 Introduction

The aim of the 'thermodynamic formalism' by Ruelle [1] is to deduce from microscopic knowledge of axiom-A systems the macroscopic behavior. As the most prominent example, occurrence or absence of phase transitions should be predicted. The task involves an averaging process over a certain (canonical, microcanonical, grandcanonical,..) ensemble and an explicit probability measure. A unique role among the possible probability measures is played by the Gibbs measure: The Gibbs measure ρ_G is the measure which describes the thermal equilibrium (for large systems). It can be formulated in the form of a variational principle. For the specific systems considered by Ruelle in his investigations the result followed that phase transitions were not possible.

In the following, the thermodynamic formalism has been applied to the characterisation of large classes of chaotic dynamical systems. It was discovered that for the probabilistic approach in the phase-space as well as for the dynamical approach in time-space the formalism could successfully be applied with little modification. After investigating a variety of systems (of either mathematical or experimental nature), it was surprisingly found that phase transitions seemed to be a common rather than a rare phenomenon. This fact indicated that for nature a larger class of systems could be relevant than those for which Ruelle's beautiful results apply.

So far, no attempts have been made to unify the two approaches. For a more refined and complete description of chaotic dynamical systems, this unified approach proved to be a tool of high value, especially for the investigation of phase transitions. Also, from the unified approach, the two 'traditional' approaches can be recovered. The purpose of the present contribution is to present first a short introduction to this elegant means and then to continue with the discussion of some more advanced topics, such as phase-transitions and questions related to nontrivial grammars. Last, but not least, the approach is able to give a deeper insight into a number of other problems associated with the scaling behavior of dynamical systems.

2. Scaling Behavior

2.1 Scale Invariance

At the birth of the 'chaotic business' was the observation that even a low-dimensional, simple system was able to produce very complex behavior because of the possibility to switch between different unstable states or elements. Apparently random, the motion is, however, of purely deterministic nature. The finiteness of the 'generating elements' leads, without many further conditions, directly to the scaling property of such systems. The geometric expression of the scaling property are the self-similar structures which appear in phase- and in time-space. In simple terms, an object can be called self-similar, if it does not change its appearance under magnification. Consider the simple self-similar object of a Koch-curve K. K can be interpreted as the graph of a relation: $x = t$, $y = F_{\mathrm{Koch}}(t)$, $t \in [0,1]$. To obtain a simple nontrivial example, starting from the unit interval $[0,1]$, the middle third is removed and replaced by two sides of an equilateral triangle, pointing outwards. In the next step this procedure is repeated, for each of the new intervals of length $\frac{1}{3}$ in comparison to the previous generation, and so on. The final self similar curve evolves as the asymptotic object. Although a general Koch-curve is not a single-valued function, it is easy to see that with $\mu_0 = 3^{-n}$, $n = 0, 1, 2, \ldots$, $F_{\mathrm{Koch}}(t)$ satisfies the homogeneity condition

$$F_{\mathrm{Koch}}(\mu_0 t) = \mu_0^a F_{\mathrm{Koch}}(t), \tag{1}$$

with the scaling exponent $a = 1$, at each point of the graph. More generally, a function F is said to be scaling or scale invariant [2], if $F(\mu t) = \mu^a F(t)$, $\forall \mu > 0$. Functions with that property are called homogeneous functions (the power-law functions, for example) [3]. Necessary conditions to be satisfied by a scaling function F are the positiveness of $F(\epsilon)$ and the existence of the limit

$$a = \lim_{\epsilon \to 0} \frac{\log(F(\epsilon))}{\log(\epsilon)}. \tag{2}$$

Equivalently the last property is written as $F(\epsilon) \sim \epsilon^a$.

As pointed out before, for a chaotic system the behavior can be described from the scaling properties of a finite number of elements, observation of coarse-scale resolution provided. In the most popular case the observation is directed towards the phase-space, which yields Cantor sets for the simplest chaotic systems. In this case, the probability measure is the object of observation, and only the location of points in space is considered (the temporal order of the generated points is ignored). A second popular object of observation is the speed of the change of the amplitude of adjacent points along the time axis. This object of observation is directly related to the instability properties of the system. The basic quantity to be considered is then the slope of the dynamical map or, more conveniently, the logarithm of the slope. To provide a simple example for illustration we consider the supercritical tent map (Fig. 1a). As the result of the first observation process, the Cantor set shown in Fig. 1b evolves (this drawing corresponds to very coarse resolution!). As the result of an observation from the second point of view, the result shown in Fig 1c is obtained. Now, each element of the asymptotic set can be labelled by the sequence of the unstable elements which led to its appearance—examples are indicated in Fig. 1b. In this way, each finite observation process is related to a finite sequence of symbols on which the dynamical map acts as the shift map (symbolic dynamics). If the observational process is divided into a number of observations of shorter length, a set of finite sequences is obtained. In order to capture the asymptotic behavior of the system, the length of observation should be as long as possible. In practical applications, this requirement is contradicted by the finite resolution and data storage capacity of computers. Generically, not all combinations of symbols appear for a given system. On the contrary, the structure of the allowed combinations is another characteristic of the system which is referred to as the grammar of the system relative to a given symbolic representation. This topic, the measurement of the complexity of a dynamical system, is presently one subject of research efforts.

Notwithstanding the question of whether in the process of infinite resolution at each location in phase-space or at any starting point in time-space the same result would be obtained (this is essentially the question of ergodicity), a set of finite-resolution observations yields information on the system. This information can be used to predict the asymptotic properties of the system. But, furthermore, also the fluctuation behavior around the asymptotic properties is a characteristic of a particular system. As a consequence, from the observation of finite resolution in phase- or in time-space, using an averaging process, characteristic asymptotic properties should be worked out. For the first point of view, the self similar structures in the phase-space can be described with the help of a logarithmic scaling exponent of the measure. The *local crowding index* $\alpha(x_0, \epsilon)$ is then introduced [4] as

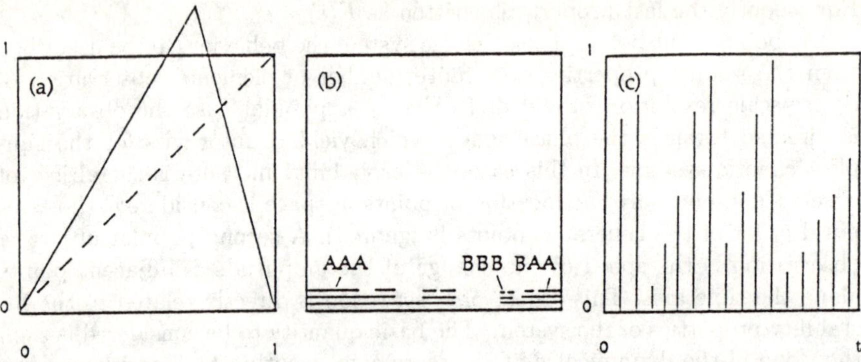

Fig. 1. Supercritical tent map (a), generated Cantor set (b), temporal sequence of points (c)

$$\alpha(x_0, \epsilon) = \frac{\log P(B(x_0, \epsilon))}{\log \epsilon} , \tag{3}$$

where $P(B(x_0, \epsilon))$ denotes the probability of the ball $B(x_0, \epsilon)$ of radius ϵ and center x_0 with respect to the natural measure $P(B(x_0, \epsilon)) = \int_{B(x_0,\epsilon)} d\rho(x)$. In time-space the *effective Lyapunov exponents* of k steps [5] are defined as the logarithmic stretching rates

$$\lambda_{ik} = \frac{\log(DFa_i^k(x_0))}{k} \tag{4}$$

(the index i refers to the ith direction in space). They describe the local dynamical scaling properties of the support.

In order to obtain a macroscopic characterization, an averaging process must now take place. Due to the close relationship with the method used in thermodynamics to derive macroscopic quantities from microscopic behavior, this process is called the thermodynamic formalism.

2.2 Non-unified Approach

To obtain averaged information on the scaling behavior, we proceed in analogy to the procedures of statistical mechanics and in accordance with the thermodynamic approach. The scaling due to the 'temporal' evolution of the support is described by the probability [6, 7]

$$P_{\mathrm{T}}(x, n) \sim T^{1/n \log(DFa^{n,+}(x))} , \tag{5}$$

where $T := e^{-n}$. Then the average with respect to time

$$\lim_{n \to \infty} \frac{\log\langle (DFa^{n,+})^{-(\beta-1)} \rangle}{T} := \Lambda(\beta) \tag{6}$$

is evaluated. An entropy-like quantity

$$\phi(\lambda) = \beta\lambda - \Lambda(\beta) \tag{7}$$

can be associated through a Legendre transformation, where

$$\lambda(\beta) = \frac{d\Lambda(\beta)}{d\beta} \, . \tag{8}$$

In complete analogy we have

$$P\big(B(x,\epsilon)\big) \sim \epsilon^{\alpha(x)} \, , \tag{9}$$

and the corresponding averaged quantity is then obtained as

$$\lim_{\epsilon\to 0} \frac{\log\langle P(B(x,\epsilon))^{q-1}\rangle}{\log \epsilon} = \tau(q) \, , \tag{10}$$

which describes the scaling of the measure.

$$f(\alpha) = \alpha q - \tau(q) \, , \tag{11}$$

where

$$\alpha(q) = \frac{d\tau(q)}{dq} \tag{12}$$

is the corresponding entropy-like quantity.

In the following, $\phi(\lambda)$ and $f(\alpha)$ are referred to as the scaling functions of the scaling of the support and of the measure, respectively. Other common expressions are spectrum of Lyapunov exponents and spectrum of generalized dimensions. In order to give some more insight and to point out exemplarily the independence of the two approaches, let us consider a number of simple examples (Fig. 2 and Fig. 3).

We recall that the SRB-measure [1,8] of a system is the Gibbs state μ_ϕ for $\phi = -\log|DFa|$ (the Gibbs state is the unique invariant probability measure μ_ϕ, which satisfies $c_1 \le \frac{\mu_\phi(J_{i=(1,\ldots,M)^n})}{\exp(-nP+S_n\phi(x_i))} \le c_2$). Here, c_1 and c_2 are positive constants and $S_n\phi(x) := \phi(x) + \phi(Fa(x)) + \ldots + \phi(Fa^{n-1}(x))$. $P = \lim_{n\to\infty} \frac{1}{n} \log \sum_{i=(1,\ldots,M)^n} \exp(S_n\phi(x_i))$ is called the (topological) pressure. x_i is a point of the interval which leads to the maximum of $S_n\phi$ on J_i. As is also well known, for $n \to \infty$ the Lyapunov exponent approaches one single value and does not depend on the starting point. Likewise, for $\epsilon \to 0$ the same local dimension is obtained, irrespective of the central point. Hence, the measurement of values different from the information dimension or Lyapunov exponent, respectively, must be interpreted in terms of large deviation theory [9–11]. The entropy-like quantities $f(\alpha)$ and $\phi(\lambda)$ induce in this sense a deviation function.

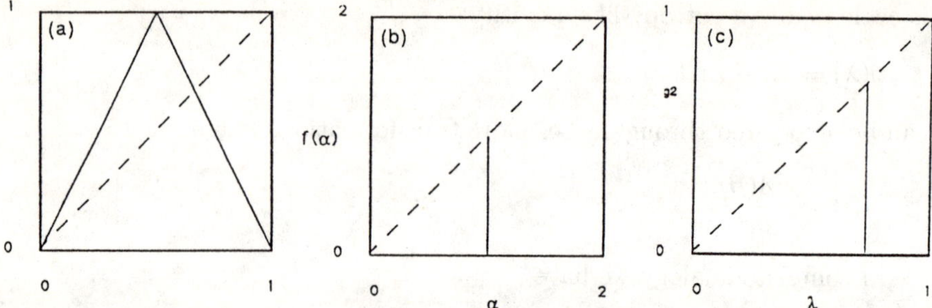

Fig. 2. Fully developed, symmetric tent map (a), fractal characterization (b), temporal characterization (c)

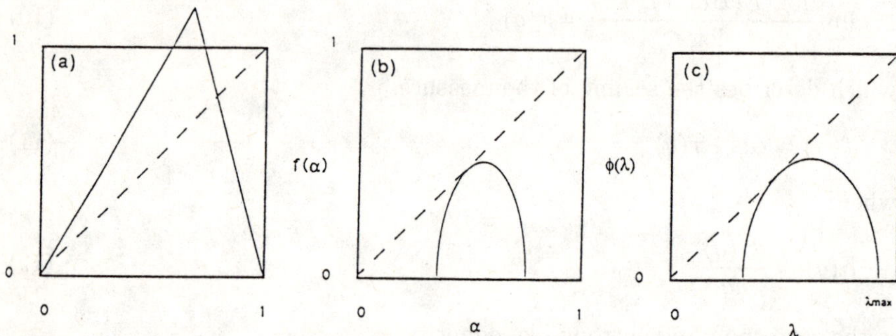

Fig. 3. Asymmetric supercritical tent map (a), fractal characterization (b), temporal characterization (c)

3. Unified Approach

3.1 The Generalized Entropy Function

In this section, the two points of view will be combined. The procedure results in an elegant, unified approach. The connection between the two approaches is indicated by the fact that both points of view use the same symbolic description. As a consequence, the new formalism is not represented by a product of the old approaches. In contrast to the non-unified approach, the scaling behavior will no longer be described with the help of one single variable. Instead, the scaling of the support and the scaling of the measure are taken into account at the same time. In the sense of large deviation theory, our starting point amounts to a level-3 description of the dynamical system. The partition which is used is related to the isotherm-isobar ensemble [12] in statistical mechanics.

A *generalized partition function* for the attractor A is then defined in the following way [13]:

$$Z_{\mathrm{G}}(q,\beta,n) = \sum_{j\in(1,\ldots,M)^n} \ell_j^{\beta} p_j^{q} \,. \tag{13}$$

Here p_j is the probability of falling in the jth region of the partition (depending on the iteration n), ℓ_j describes the size of the region and β and q can be called "filtering exponents". For higher-dimensional cases see [14]. For simplicity it is assumed above that a generating partition of M elements has been found. The evaluation of the average of the scaling quantities then becomes simple, since only the scaling properties of a finite number of elements in the generating partition have to be evaluated. Although the starting point is the same as in [13], a more general formalism can be developed and a different interpretation can be given. In the following I owe much to arguments given in Ref. [15] where, however, neither the connection of the present formalism with the scaling function of Lyapunov exponents nor with the information-theoretic entropies was established. Let us introduce for the elements of Z_{G} a "local scaling assumption" with respect to iteration. It is assumed that ℓ and p scale with n in the following way:

$$\ell_j = \mathrm{e}^{-n\epsilon_j} \,, \tag{14}$$

$$p_j = \ell_j^{\alpha_j} \,. \tag{15}$$

The generalized partition function is then obtained as

$$Z_{\mathrm{G}}(q,\beta,n) = \sum_{j\in(1,\ldots,M)^n} \mathrm{e}^{-n\epsilon_j(\alpha_j q+\beta)} \,, \tag{16}$$

and the *generalized free energy*

$$F_{\mathrm{G}}(q,\beta) = \lim_{n\to\infty} \frac{1}{n} \log Z_{\mathrm{G}}(q,\beta,n) \tag{17}$$

can be derived.

Alternatively, the generation process of the system can be described for hyperbolic problems as a fixed-point or eigenvalue equation, in our case by means of the iterative version of the generalized or bivariate Frobenius-Perron equation [7, 16]

$$\lambda(q,\beta)Q_{n+1}(x,y) = \sum_{\epsilon=0,1,\ldots} \frac{Q_n(f_\epsilon^{-1}(x), g_\epsilon^{-1}(y))}{|f'(f_\epsilon^{-1}(x))|^{\beta}|g'(g_\epsilon^{-1}(y))|^{q}} \,. \tag{18}$$

Here, ϵ labels the choice being made between the M inverses of the maps f, g. The sum is performed for f and g over the same symbolic substrings. Starting from any smooth density $Q_0(x,y)$ in $(0,1)\times(0,1)$, a unique eigenvalue $\lambda(q,\beta)$ assures convergence towards a finite $Q(x,y)$. The dependence on

x, y disappears for large n, for x, y in the invariant set. The free energy arises from the largest eigenvalue of the associated Frobenius-Perron operator

$$[LQ](x,y) = \lambda(q,\beta)Q(x,y)\ . \tag{19}$$

Therefore, in order to derive the generalized free energy or Gibbs potential, the relation

$$\lambda(q,\beta) = \exp(F_{\rm G}(q,\beta)) \tag{20}$$

can be used. A *generalized entropy* $S_{\rm G}(\alpha,\varepsilon)$ is related to a "global scaling assumption": it is assumed that the number of regions whose scaling exponents are between (α,ε) and $(\alpha+d\alpha,\varepsilon+d\varepsilon)$ can be written as

$$N(\alpha,\varepsilon)d\alpha d\varepsilon \sim {\rm e}^{nS_{\rm G}(\alpha,\varepsilon)}d\alpha d\varepsilon\ , \tag{21}$$

in the limit $n\to\infty$. With the help of the generalized entropy, the generalized partition function can be written as an integral in the following way:

$$Z_{\rm G}(q,\beta,n) \sim \int d\varepsilon \int d\alpha {\rm e}^{n[S_{\rm G}(\alpha,\varepsilon)-(\alpha q+\beta)\varepsilon]}\ . \tag{22}$$

Using a saddle point approach in the limit $n\to\infty$, the relation

$$F_{\rm G}(q,\beta) = S_{\rm G}(\langle\alpha\rangle,\langle\varepsilon\rangle) - (\langle\alpha\rangle q + \beta\langle\varepsilon\rangle \tag{23}$$

is obtained, where $\langle\varepsilon\rangle$, $\langle\alpha\rangle$ lead to the maximum of the bracket [ ] above. This shows that $F_{\rm G}(q,\beta)$ and $S_{\rm G}(\alpha,\varepsilon)$ are connected via an (unusual) generalization of a two-dimensional *Legendre transformation.* $\langle\varepsilon\rangle$ and $\langle\alpha\rangle$ can be calculated from the free energy $F_{\rm G}(q,\beta)$ as

$$\langle\varepsilon\rangle = -\frac{\partial}{\partial\beta}F_{\rm G}(q,\beta)\ , \tag{24}$$

and

$$\langle\alpha\rangle = -\frac{\partial}{\partial q}\frac{F_{\rm G}(q,\beta)}{\langle\varepsilon\rangle}\ . \tag{25}$$

The generalized entropy $S_{\rm G}$ can be derived in a similar way as

$$S_{\rm G}(\langle\alpha\rangle,\langle\varepsilon\rangle) = F_{\rm G}(q,\beta) - q\frac{\partial F_{\rm G}(q,\beta)}{\partial q} - \beta\frac{\partial F_{\rm G}(q,\beta)}{\partial\beta}\ . \tag{26}$$

To focus on the dependence of the entropy on the scaling exponent ε alone, an additional entropy $S_{\rm G}(\varepsilon)$ can be introduced, according to

$${\rm e}^{nS_{\rm G}(\varepsilon)} = \int d\alpha {\rm e}^{n[S_{\rm G}(\alpha,\varepsilon)]}\ . \tag{27}$$

$S_{\rm G}(\varepsilon)$ can be obtained from $S_{\rm G}(\alpha,\varepsilon)$ with the help of the condition $q=0$. We note that this entropy function coincides with the entropy of the thermodynamical formalism defined by Oono and Takahashi [17].

3.2 Hyperbolic Models with Complete Grammars

In this section the new concepts are applied to simple attractors and repellers. A few model cases are discussed to show the characteristic behavior of the different entropy functions.

An attractor is called *self-similar* [18] if a partition can be found such that in any step i of the n steps, the scaling region can be divided into the M subregions of the same scaling behavior as the old region. More explicitly, it is required that the partition function Z_{G} can be factorized into an n-fold product of the form

$$Z_{\mathrm{G}}(q,\beta,n) = \sum_{j\in(1,\ldots,M)} \ell_j^\beta p_j^\alpha \cdot \sum_{j\in(1,\ldots,M)} \ell_j^\beta p_j^\alpha \cdot, \ldots, \cdot \sum_{j\in(1,\ldots,M)} \ell_j^\beta p_j^\alpha \qquad (28)$$

(n times), such that $Z_{\mathrm{G}}(q,\beta,n)$ can be written in a simpler way as

$$Z_{\mathrm{G}}(q,\beta,n) = Z_{\mathrm{G}}(q,\beta,1)^n \; . \qquad (29)$$

Thus, the requirement of self-similarity is equivalent to the existence of a generating partition. However, the formalism can be generalized to cover systems and partitions for which the partition function converges towards a self similar system

$$Z_{\mathrm{G}}(q,\beta,n) \to \widetilde{Z_{\mathrm{G}}}(q,\beta,n) \; . \qquad (30)$$

In this case, the attractor is called asymptotically self-similar. If the scaling exponents of a system are not identical with respect to the different directions in the phase-space, the attractor is called *self-affine* [18]. For completeness, let us point out that with the identification: $-n\varepsilon$ = *energy* = E; $n\varepsilon\alpha$ = *volume/temperature* = $V/k_{\mathrm{B}}T$; $-\beta$ = *inverse temperature* = $1/k_{\mathrm{B}}T$; q = *pressure* = p, the analogy with statistical mechanics can be made explicit [7, 19] (k_{B} denotes Boltzmann's constant). In this isotherm-isobar partition, n can be identified with the number of spins with as many possible states as given by the associated symbolic dynamics. In this way, a mapping to a spin-chain is established. It can be shown that all thermodynamic relations follow. Note that E and V are not independent and that the sum extends over the symbolic sequences S_n. In this way, a covering of the attractor with finite diameters is chosen, irrespective of the probability of S_n itself.

As the simplest characteristic application we discuss the "Cantor construction", and references are given for the two-dimensional Baker map. In the following section, we generalize these models by focusing on incomplete grammars and nonhyperbolicity.

The uniform Cantor set

$$C := \left\{ x \in \mathbb{R} \mid x = 2\sum_{i=1}^{\infty} s_i 3^{-i} \; , \; s_i \in \{0,1\} \right\} \qquad (31)$$

can be taken as a simple model of a dynamical system (Fig. 1). For $n \to \infty$ a closed, totally disconnected, perfect set is obtained (totally disconnected: containing no intervals; perfect: every point in the set is an accumulation point or limit point of other points in the set). These properties serve as a general definition for a Cantor set [20]. In the multi-dimensional, non-isotropic case this definition generalizes to the notion of a Cantor direction [21].

Let us assume that the measure of the middle third is distributed to the remaining two thirds. Then it is easily found that the partition function Z_{G} has the form

$$Z_{\mathrm{G}}(q,\beta,n) = \sum_{k=1}^{2^n} \left(3^{-n\beta}(1/2)^{nq}\right) , \tag{32}$$

and the free energy F_{G} can be expressed as

$$F_{\mathrm{G}}(q,\beta) = \log(3^{-\beta}2^{-q+1}) . \tag{33}$$

A particularly interesting case is the case of a *zero of* $F_{\mathrm{G}}(q,\beta)$. If q is interpreted as the independent and β as the dependent variable, respectively, then let us denote this zero by $\beta_0(q)$. For $n \to \infty$, $\beta_0(q)$ leads to a maximum $\neq \infty$ of the partition function. Hence it can be taken as a generalization for $q \neq 0$ of the Hausdorff dimension [22]. As a different generalization of the Hausdorff dimension, the quantity $-\beta_0/(q-1)$ has also been considered [23]. Figure 4 shows a plot of both quantities. Notice that for the present case $-\beta_0/(q-1)$ is constant. Fractals with that property are called uniform. We note that $-\beta_0(q)$ is often denoted by $\tau(q)$ [13]. The scaling function $f(\alpha)$ can be obtained with the help of a Legendre transformation (c.f. eq. (22)).

The derivative of the free energy with respect to β,

$$-\frac{\partial F_{\mathrm{G}}(q,\beta)}{\partial \beta} , \tag{34}$$

evaluated at $q = 1$ and $\beta = 0$, gives the time average of the logarithmic stretching rates, the Lyapunov exponent of the system, where some ambiguity with respect to the sign of the Lyapunov exponent seems to be implicit. The sign can, however, be understood by observing that for an ordinary dynamical system the contracting direction corresponds to attraction under backward iteration in time.

In our model, the escape rate κ is zero. If the measure of the middle third is not redistributed it is seen that the escape rate has the value

$$\kappa = \log(3) - \log(2) . \tag{35}$$

The generalized entropy is different from zero only for $\varepsilon = \log(3)$ and $\alpha = \log(2)/\log(3)$. There, the value $S_{\mathrm{G}}(\alpha,\varepsilon) = \log(2)$ is assumed.

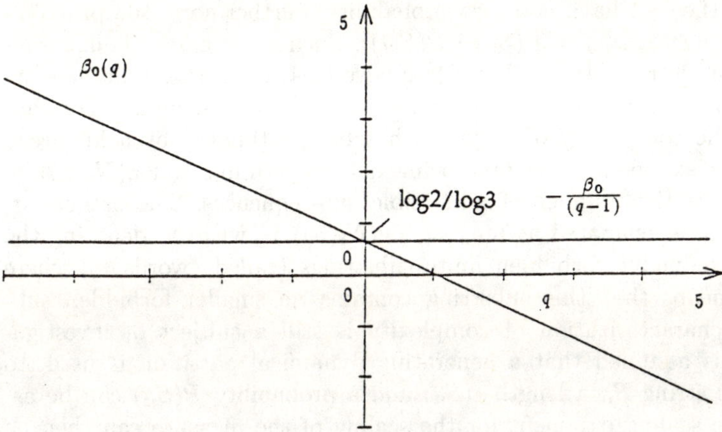

Fig. 4. Functions β_0 and $-\beta_0/(q-1)$ for the uniform Cantor set (see text)

A more realistic picture of the Cantor structures that are observed in nature can be given by the generalization to non-uniform Cantor sets, i.e., if at least two different stretching factors are present, $\ell_1 = 1/4$, $\ell_2 = 2/5$, e.g. In this case a non-uniform Cantor set [20] is obtained. If the probabilities are normalized (they need, however, not be proportional to the length of the pieces, e.g. $\ell_1 = 1/4$, $\ell_2 = 2/5$; $p_1 = 2/5$, $p_2 = 3/5$), then the limiting set is called a *non-uniform Cantor set with measure.* It is seen that for the partition function and the free energy the expressions

$$Z_{\mathrm{G}}(q,\beta,n) = \sum_{k=1}^{n}(b_{n,k})((\ell_1^\beta p_1^q)^{n-k}(\ell_2^\beta p_a^q)^k \tag{36}$$

and

$$G_{\mathrm{G}}(q,\beta) = \log(\ell_1^\beta p_1^q + \ell_2^\beta p_2^q) \tag{37}$$

are obtained. The zero of $F_{\mathrm{G}}(q,\beta)$ is evaluated from the solution of the equation

$$\ell_1^{\beta_0} p_1^q + \ell_2^{\beta_0} p_2^q = 1 \tag{38}$$

for $\beta_0(q)$. Again, $\beta_0(q)$ is a nonlinear function of q. As pointed out before, the scaling behavior of a system is captured in the entropy function $S_{\mathrm{G}}(\alpha,\varepsilon)$. For the non-uniform Cantor set with measure, this function is displayed in Fig. 5. Note that the support of $S_{\mathrm{G}}(\alpha,\varepsilon)$ is not a linear manifold. To explain this, and in order to obtain some other insight, let us interpret the two-scale Cantor set in the framework of symbolic dynamics. The system has a *complete* symbolic tree, i.e. all sequences of the symbols are allowed, they all have non-zero probability. Furthermore, the probabilities factorize ($P(S_{n+n'}) = P(S_n) * P(S_{n'})$). Such a system, although having maximum metric entropy (log(2)), is nevertheless a rather uninteresting one, as it is not able to produce surprises, it is not very complex. For that reason [24] the complexity of a system has been estimated by a hierarchy of complexity exponents, the first being defined as $\lim_{n\to\infty} \log(N_{\mathrm{a}}(n))/n$, where $N_{\mathrm{a}}(n)$ is the number of admissible sub-sequences. The second order complexity is evaluated as $\lim_{n\to\infty} \log(N_{\mathrm{f}}(n))/n$, with N_{f} denoting the number of irreducible forbidden finite substrings (called "words"), where irreducible means that this substring contains no smaller forbidden substring. The characterization of complexity is still a subject of investigations. As it is assumed that a generating dynamical partition is used, to any symbolic string S_n a length $\epsilon(S_n)$ and a probability $P(S_n)$ can be associated. The scaling exponent for the scaling of the measure can then be defined as $\alpha(S_n) = \log(P(S_n))/\log(\epsilon(S_n))$, $n \to \infty$. Of course, this exponent has the same value as the corresponding exponent in the phase-space. Observe that for composed strings of length $n + n'$, it is generally found that $P(S_{n+n'}) \neq P(S_n) * P(S_{n'})$. The scaling exponent α, however, is still found through the expression $\alpha(S_{n+n'}) = \log(P(S_{n+n'}))/\log(\epsilon(S_n) * \epsilon(S_{n'}))$. Even if the probabilities factorize, the former relation is generally nonlinear. This fact, of course, is responsible for the curvilinear nature of the support of S_{G} in Fig. 5. Ascending functions are convex over the positive axis, descending are concave.

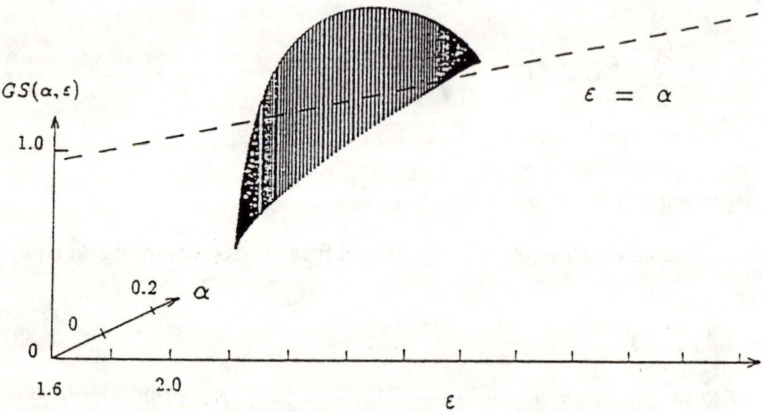

Fig. 5. $S_{\mathrm{G}}(\alpha, \varepsilon)$ for the non-uniform Cantor set with measure. $S_{\mathrm{G}}(\alpha, \varepsilon)$ is shown on the (ε, α)-plane, viewed from an angle of 45°. The support of $S_{\mathrm{G}}(\alpha, \varepsilon)$ is still one-dimensional, but no longer a linear manifold

An important and yet simple example of a one-dimensional Cantor set is furnished by the escape from a *strange repeller*, a topic which is discussed in the Refs. [25–27].

So far, the scaling exponent α for the scaling of the measure was either constant or there was a bijective relationship between α and ε. This leads to a kind of degeneracy which can be removed if we consider the three-scale Cantor set with measure. (The "ternary" Cantor set is of relevance, e.g., for the scaling behavior of the postcritical regime of the circle map, where the tree is not complete [28].)

In Figs. 6a-c we display the scaling properties of this system. As can be seen, the support of S_G is no longer a one-dimensional manifold. The borders of S_G and the supports of the different functions which describe the various aspects of the scaling behavior of the system are curved.

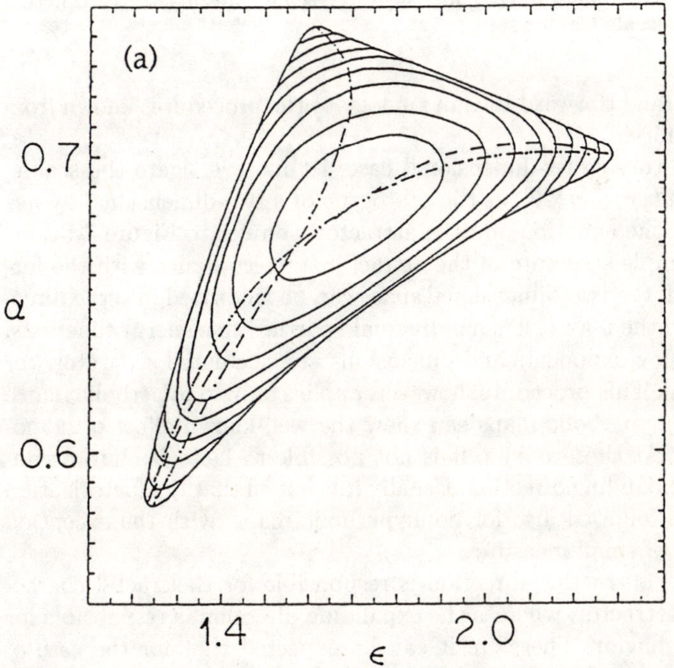

Fig. 6. (a) Support of $S_G(\alpha, \varepsilon)$ in the $\varepsilon-\alpha$ plane for the three-scale Cantor set with $p_1 = 0.2$, $p_2 = 0.3$, $p_3 = 0.5$, $\ell_1 = 0.1$, $\ell_2 = 0.2$, $\ell_3 = 0.3$. Contour lines are shown, increasing in steps of 0.1. Indicated lines: supports of the functions $S_G(\varepsilon)$ (double-dotted—dashed), $f(\alpha)$ (dashed-dotted) and $g(\Lambda)$ (dashed). $g(\Lambda)$ is the Legendre transform of the dynamical Renyi entropies $K(q)$ [26, 34]

In this way, starting from a generalized entropy function built on both distributions of α and ε, the calculation of one-dimensional models can be

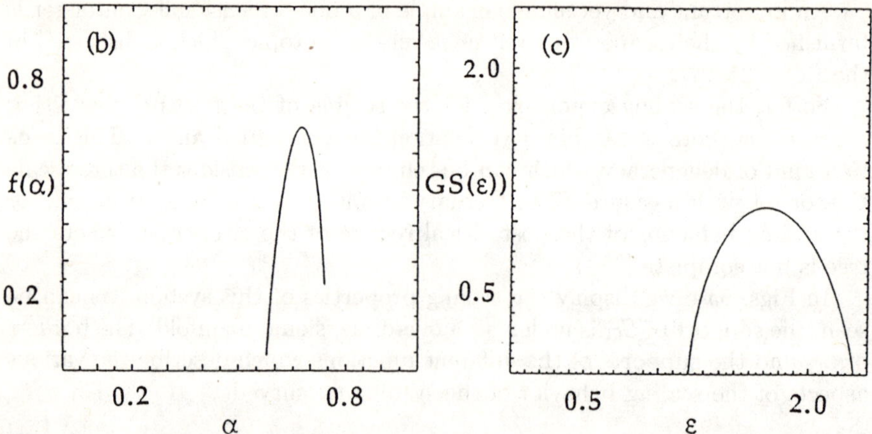

Fig. 6 Continued: (b) function $f(\alpha)$ for the three-scale Cantor set; (c) function $S_{\mathrm{G}}(\varepsilon)$ for the three-scale Cantor set

performed in a straightforward way, in analogy to the procedures known from statistical mechanics.

To cover also the higher-dimensional case, let us investigate the scaling behavior of the *baker map* [20], as the prototype of a two-dimensional hyperbolic system. For the equation and the attractor we refer to Figure 7 below.

Due to the simple structure of the attractor, in accordance with the formalism developed, the two-dimensional space can be factorized into the directions of the x- and the y-axes. Then individual "partial" free energy functions, entropies, Lyapunov exponents and dimensions are calculated separately for each direction [14]. This procedure, however, applies only to hyperbolic maps. For example, nonhyperbolic maps can show the well-known effect of homoclinic tangencies. At those points, it is not possible to factorize into a contracting and an expanding direction. Usually it is hoped that the factorization procedure can be followed also for nonhyperbolic maps, with the exception of a set of points of small measure.

Basically, the contracting direction is responsible for the fractal characterization of the attractor, whereas the expanding direction is responsible for the dynamical behavior. Therefore it can be expected that for the case of constant Jacobian maps, a close relationship between the scaling functions of fractal dimensions and of Lyapunov exponents exists. In a more general context, this is not the case.

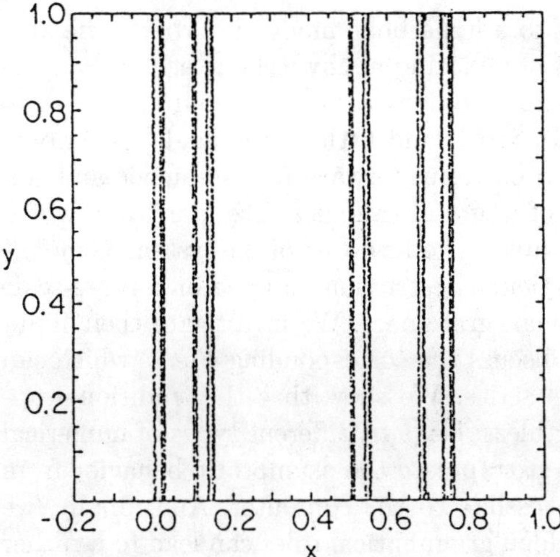

Fig. 7. Strange attractor of the baker map: $(x_{n+1}, y_{n+1}) = (\xi_a x_n, y_n/a)$, $y_n \le a$, $(x_{n+1}, y_{n+1}) = (1/2 + \xi_b x_n, (y_n - a)/b)$; $a = 2/5$, $b = 3/5$, $\xi_a = 1/5$, $\xi_b = 7/20$. Note that $\xi_a + \xi_b < 1$

4. Extensions

4.1 The Need for Extensions

So far, the present outline has not covered the nonhyperbolic maps, whose characteristic effects are often easily detected in experimental settings. Furthermore, the folding of the maps is generically incomplete. This means that models of real-world processes are equipped with nontrivial grammars, which constitute a major problem for the evaluation of the asymptotic scaling functions from experimental data. In addition, we will make clear that for numerical reasons instead of the canonical ensemble, the grand canonical ensemble can be advantageous.

4.2 Convergence Properties

As it is common use to define a chaotic attractor as the closure of its non-forbidden periodic orbits [20], it should not be too surprising that the grammatical rules, which allow only a part of the periodic orbits to exist, have a non-negligible influence on the associated scaling functions. Whereas the appearance of nonhyperbolic contributions can be traced in the scaling functions derived from time series without knowledge of the underlying grammar, the former influence is far less easy to investigate. In the following we intend to make clear that such an influence cannot be neglected: it has a large impact on the form of the associated scaling functions. Although in the generic case of experimental time series a large number of nonhyperbolic points is to be expected, we restrict ourselves to a hyperbolic model, in order to be able to work out this influence against dominating nonhyperbolic effects.

Traditionally, the 'thermodynamic averages' are calculated from the 'canonical' partition function (eqs. (13), (16)), and little use is made of periodic orbits. However, a more recent approach put forward by Cvitanović and collaborators [29] suggested the use of a 'grand-canonical-like' partition, which explicitly takes account of the unstable periodic orbits of the system. Generalizing our model, we consider a particular restriction imposed on a three-scale Cantor set by simple, but nontrivial, grammars. We investigate their influence on the scaling behavior and discuss the corresponding effects which can be observed in experimental time series. We show that the traditional approach, although not entirely worthless, leads to different types of numerical problems, so that it is difficult to extrapolate the asymptotic behavior from finite substrings of the length accessible to the computer. And, finally, we point out that the discovery of hidden grammatical rules can lead to a rather drastic change of the scaling functions.

4.2.1 The Model

In order to demonstrate the generic convergence properties we consider the following situation: As an appropriate model for a dissipative dynamical system, we consider the three-scale Cantor set and use for the symbolic description the symbols A, B, and C. We impose a grammar by the requirement that a threefold repetition of one symbol, C, cannot occur. For the symbols A, B, and C, the length scales and the associated probabilities are specified. At length $n = 1$, the following length scales and probabilities are given: $l_A = 0.1$, $l_B = 0.2$, $l_C = 0.3$; $p_A = 0.2$, $p_B = 0.3$, $p_C = 0.5$. We then presume that a hierarchical discovery is being made in the following sense: While at level 2 the branching probabilities remain, at level 3 the particularity of the sequence CCC, which is not allowed, is detected. For simplicity we assume that the effect of this discovery is that the probability of the forbidden substring is proportionally distributed to the neighboring branches CCA and CCB.

In contrast to what could be expected, it is not possible to exclude at a given level n once and for all the generation of the forbidden substring: The generation of the fractal is 'not closed' with respect to the iteration number n. The hope is then that an increase of the length n could be a remedy for the situation. Taking into account the structure of the grammar, it is easily seen that this expectation can only partially be justified. In order to investigate the convergence of the approach, we now start from the ensemble of all possible substrings of length n and calculate the generalized entropy function $S_{\mathrm{G}}(\alpha, \varepsilon)_n$. Then we derive the entropy-like functions, $S_{\mathrm{G}}(\varepsilon)_n$, e.g., for different lengths n.

4.2.2 Results

In Sect. 3.2, the form of the generalized entropy function of the unrestricted Cantor set has already been discussed. While the restriction is imposed, upon increasing the length of the substrings considered, the support of the entropy function converges towards the modified, asymptotic form (Fig. 8) which has been calculated using the zeta function approach.

We obtain a rather oscillatory convergence (Fig. 9) which is due to the fact that different symbolic substrings could not yet be excluded at the given level n [30,31] as a function of n.

As can be seen, this convergence suffers from limitations imposed by the computational tools. For example, the level which can be achieved numerically is limited by the computation time used for the summation of the partition function. The use of high values of the weighting exponents q and β causes severe numerical problems (in our computations we use a range from -30 to 30 for both parameters). However, this method is nevertheless capable of unveiling some of the important, invariant properties of the system. For instance the maximum of $S_{\mathrm{G}}(\varepsilon)$ is rather well-approximated by $S_{\mathrm{G}}(\varepsilon)_n$ (c.f. Fig. 10), whereas the r.h.s. behavior of the scaling function is only captured with limited accuracy.

4.2.3 Influence of the Grammar

So far we pointed out the intrinsic superiority of the 'grand-canonical-like' approach for the present problem. What happens if instead of element C the same exclusion rule is applied to the elements A and B, respectively? We obtain asymptotically the supports of the generalized entropy functions as shown in Figs. 11(a) and 11(b). From these figures, the effects caused in the specific scaling functions can already be estimated. In the Fig. 11(c) and Fig. 11(d) we display the associated scaling functions $S_{\mathrm{G}}(\varepsilon)$.

Let us point out that these effects are very similar to those observed during the evaluation of the 'dynamical' or 'geometrical' scaling functions $\phi(\lambda)$ from experimental time series. Upon increasing the length of the substrings considered, different shapes of the associated scaling function $\phi(\lambda)$ are observed, seemingly contradicting the law of independent averages. As has been shown, this behavior is due to the nontrivial role of the grammar of the system. At least for experimental systems with a finite grammar, the approach via zeta functions leads to an accuracy of the scaling function not known to be attained by the traditional approach. For infinite grammars, however, the situation is far less favorable — the cycle expansions converge typically but poorly.

Our results indicate that the use of the zeta function approach for the evaluation of the thermodynamic averages can be superior to the more traditional ones. Much of the success of this approach, however, depends on the fact that the alphabet associated with our model can be reformulated to

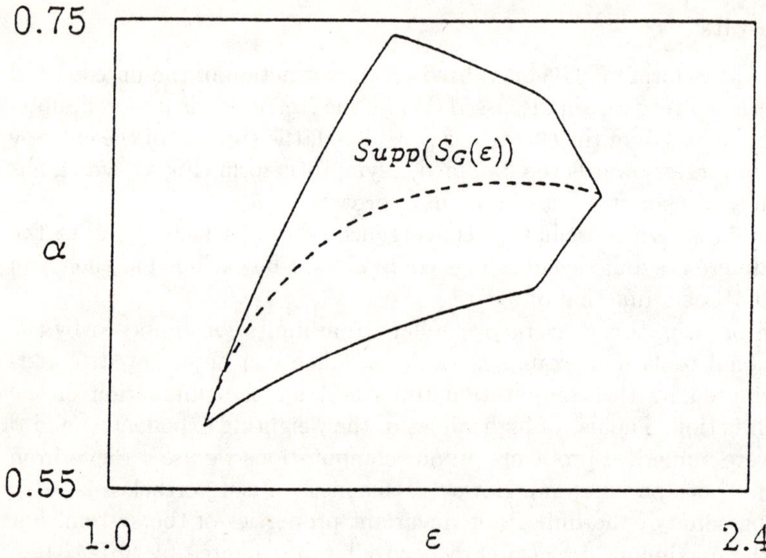

Fig. 8. Asymptotic form of the generalized entropy function, calculated from the zeta-function approach. The support of $S_{\mathrm{G}}(\alpha, \varepsilon)$ is shown, i.e., the region in the (α, ε)-plane for which $S_{\mathrm{G}}(\alpha, \varepsilon)$ is a nonzero. The dashed line indicates the support of $S_{\mathrm{G}}(\varepsilon) = S_{\mathrm{G}}(\alpha, \varepsilon)|_{q=0}$

constitute a complete one. As a consequence, for an experimental setting or a model with a complex grammar, the application of this powerful tool may not be straightforward. Therefore, the evaluation via the more traditional approach, with all its disadvantages, will often be a reasonable alternative if complete as possible information about the scaling behavior of a system is to be extracted.

4.3 Nonhyperbolicity and Phase Transitions

4.3.1 Non-unified Approach

It is well-known that phase transitions can be observed if the nonunified approach is used. As the origin for such effects, phenomena of nonhyperbolicity, as the most important the homoclinic tangency points, and the coexistence of attractors (crisis) have been worked out [32, 33]. Therefore, a discussion of the nonanalyticies of the entropy-like functions is essential. To this end two remarks are necessary. Firstly, assume that $F_s(V, T)$ is continuous, but not differentiable at a point P (the same then holds for the associated Gibbs free energy at point P^*). In this case $\Lambda(\beta)$ changes discontinuously. In view of eqs. (6) and (7) (where now the infinimum over β must be taken) this leads to a linear part in $\phi(\lambda)$. An analogous situation holds for $\lambda(q)$ and $f(\alpha)$.

Secondly, since the entropy functions of individual "simple systems" are strictly convex, the thermodynamic formalism developed will assign a convex

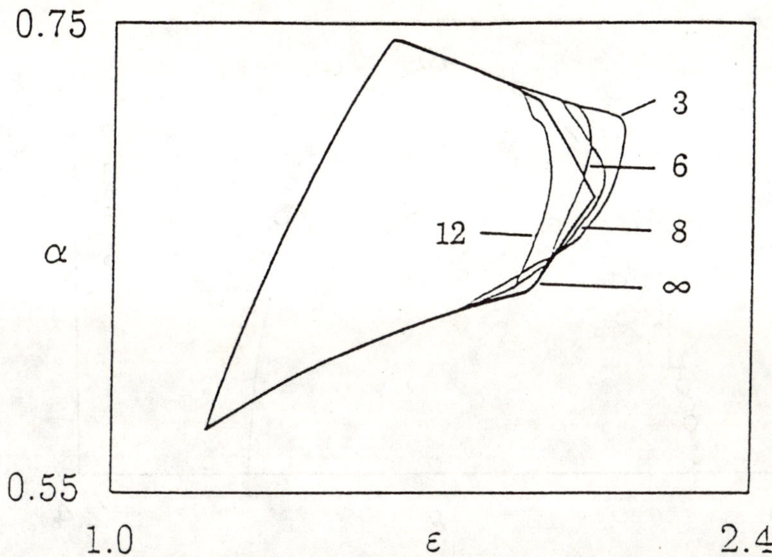

Fig. 9. Convergence towards the asymptotic generalized entropy function. Shown are the supports of the entropy function as a function of the level n. The oscillation is due to the fact that at a given level not all sequences which are forbidden for a higher levels can be excluded without suppressing other allowed sequences

(although no longer strictly convex) function to a "mixture" consisting of a linear combination of two or more simple systems. For uniformly scaling systems this then leads to first-order phase transitions. Higher order phase transitions may be obtained using non-uniformly scaling systems and imposing appropriate coexistence conditions. A characteristic picture obtained for the mixture of two systems is shown in Fig. 12.

The two-humped curve is obtained from a calculation based on a histogram (eq. (4)), while the non-convex part of this curve is substituted by the segment AB when using the thermodynamic formalism. The endpoints (A, B) of the solid line correspond to "pure phases". The change from A to B is made at fixed "temperature" $\beta(\lambda)(q(\alpha))$. The points with $\lambda(\alpha)$ below the solid line correspond to mixed states. States with different entropies can coexist at the same temperature.

As pointed out before, the case of constant Jacobian can induce a relationship between the dynamical and the static entropy-like spectra which cannot be expected for the general case. The scaling function of the support and the scaling function of the measure are in principle independent. They do not necessarily share the analyticity properties. In situations where a phase-transition-like effect occurs, the whole range of scaling behavior can be separated into regions of qualitatively different behavior, due to distinct dominant dynamical processes or topological phenomena. The different regions can, for example, be characterized by qualitatively different eigenfunctions of

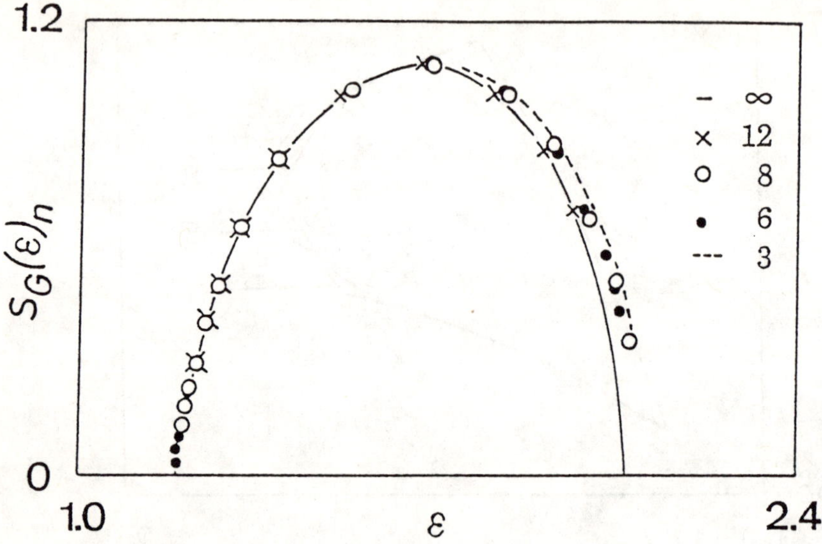

Fig. 10. Convergence of the entropy functions $S_{\mathrm{G}}(\varepsilon)_n$ towards the asymptotic entropy function $S_{\mathrm{G}}(\varepsilon)_{n=\infty}$. The convergence is limited by poor resolution for large n (see text)

the generalized Frobenius-Perron equation. In this way, a better qualitative picture of the different competing scaling processes can already be obtained. For the example of the logistic map it has been shown that the associated eigenfunctions have different symmetry properties [26]: For $\beta < \beta_c$ the eigenfunctions are symmetric ("disordered phase"), while for $\beta > \beta_c$ the symmetry is broken ("ordered phase"), a fact which is very reminiscent of what happens in most real phase transitions.

In one- and two-dimensional *hyperbolic* maps, phase-transition-like effects are not possible, as can be shown in analogy to one-dimensional spin systems, provided that the generating partition has not infinitely many elements and if there are no long-time correlations between the symbols (a situation which corresponds to infinitely many spin states or long-range interactions). For higher-dimensional systems, however, this is no longer true. In the next paragraph, these items are discussed by use of the non-hyperbolic model from a unified point of view.

4.3.2 Unified Approach

The Model

As the specific model for our investigation we consider a system for which the support is generated by a hyperbolic map, whereas the measure attributed to the support is given by a nonhyperbolic map (described in an explicit way by eq. (18)). As a very simple example of a hyperbolic support map we take the tent map

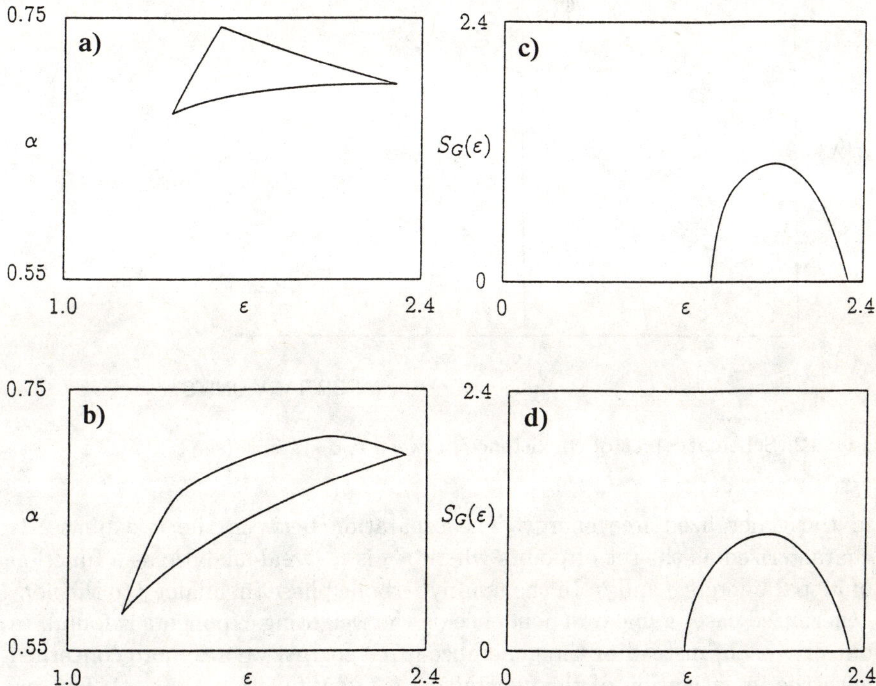

Fig. 11. Support of the generalized entropy function for the case of forbidden substring *AAA* (a); Support of the generalized entropy function for the case of forbidden substring *BBB* (b); $S_{\mathrm{G}}(\varepsilon)$ obtained for the models with forbidden substrings *AAA* (c); $S_{\mathrm{G}}(\varepsilon)$ obtained for the models with forbidden substrings *BBB* (d)

$$\begin{aligned} f : x &\mapsto \frac{x}{\ell_1}\,, \quad \text{for } x \in [0, \ell_1/(\ell_1+\ell_2)]\,, \\ x &\mapsto \frac{(1-x)}{\ell_2}\,, \quad \text{for } x \in [\ell_2/(\ell_1+\ell_2), 1]\,, \end{aligned} \tag{39}$$

where $\ell_1 = \frac{2}{9}$ and $\ell_2 = \frac{2}{7}$. The map of the measure is, instead, given by

$$g : x \mapsto 4(1-x)x\,, \tag{40}$$

the fully developed logistic map. Note that in this example, no restriction is imposed on the associated binary grammar.

As pointed out in Sect. 2.1, in order to derive the generalized free energy or Gibbs potential, the relation

$$\lambda(q,\beta) = \exp\left(F_{\mathrm{G}}(q,\beta)\right) \tag{41}$$

is used. The properties of the eigenvalues and associated eigenfunctions can be investigated, using the method of Ref. [35]. It was shown in Ref. [16] that the hyperbolic and the nonhyperbolic contributions can clearly be distinguished

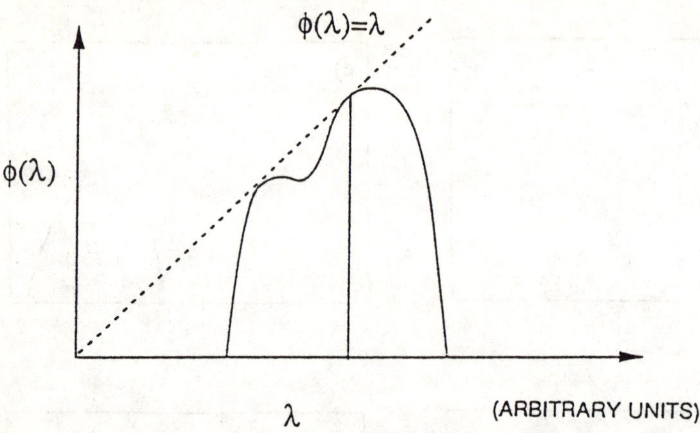

Fig. 12. Schematic plot of coexistence between two systems (see text)

in the generalized free energy. The separation between the two phases is characterized as the set of points where F_{G} is not real-analytic as a function of its parameters q and β. In the non-hyperbolic phase (in analogy to the non-generalized case) a linear dependence on the weighting exponents is found. In our approach, instead of the generalized free energy, we are more concerned with the investigation of the generalized entropy function $S_{\mathrm{G}}(\alpha, \varepsilon)$. For our model, S_{G} will show characteristic, generic properties.

Manifestation of Phase Transitions in F_{G}

For hyperbolic systems, the application of the generalized Frobenius-Perron equation is straightforward. To get rid of finite-n corrections, the difference between two neighboring levels can be considered: $\log \lambda(q, \beta) \simeq \log Z_n(q, \beta) - \log Z_{n+1}(q, \beta)$. Typically, exponential convergence is encountered. For nonhyperbolic systems, the equation still can be used, but its derivation needs some more care. The nonhyperbolicity of the measure or of the support (in our example of the measure map) usually comes from the existence of a local maximum of these maps or from singularities in their invariant sets. This property leads to a phase-transition-like behavior of $F_{\mathrm{G}}(q, \beta)$; around the points of nonanalytic behavior of $F_{\mathrm{G}}(q, \beta)$ the convergence is then found to be much worse.

The appearance of points of nonanalytical behavior can generically be explained as follows. In addition to the eigenvalue $\lambda_{\mathrm{hyp}}(q, \beta)$ obtained from smooth, singularity-free eigenfunctions $Q(x, y)$, the use of singular initial functions leads to new eigenvalues $\lambda_{\mathrm{nonhyp}}(q, \beta)$. The whole system chooses then its free energy from the requirement $F_{\mathrm{G}} = \max(F_{\mathrm{G\,hyp}}, F_{\mathrm{G\,nonhyp}})$, where both contributions to the generalized free energy are smooth functions of their arguments. As the two individual functions intersect, a first-order phase tran-

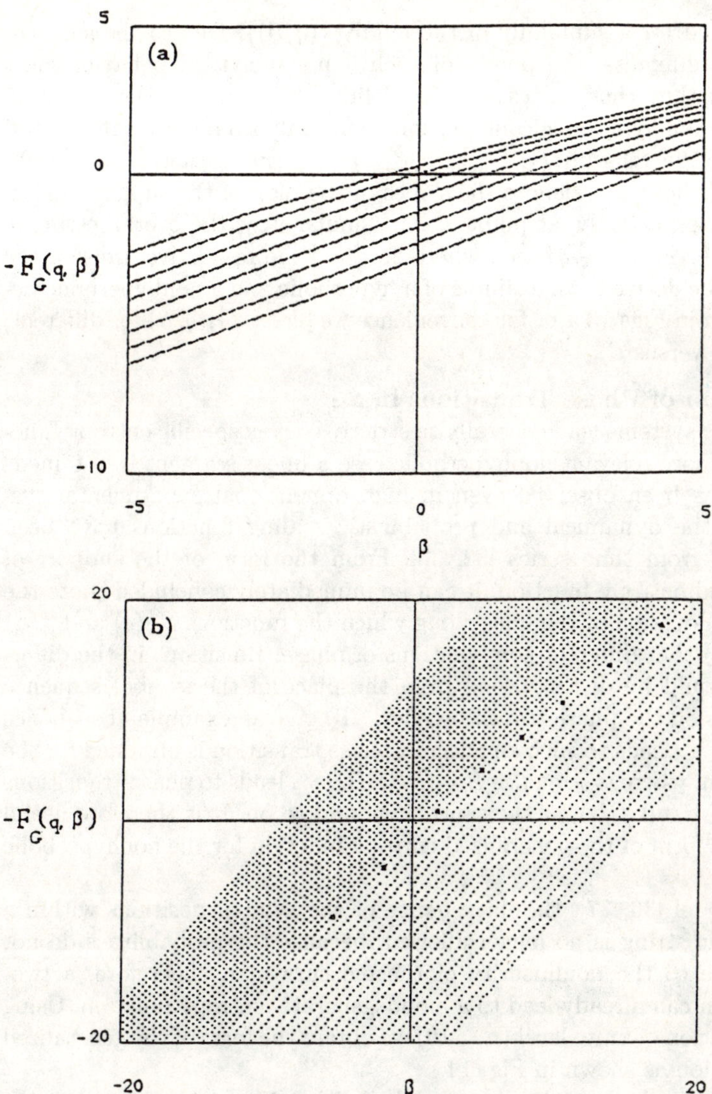

Fig. 13. a) Generalized free energy for a hyperbolic system (tent maps for the support and the measure). The figure shows the lines obtained from fixed equidistant q-values, starting from $q = -5$ (bottom curve) and ending at $q = 2$ (top curve). No phase-transition-like effect is observed; b) generalized free energy of the nonhyperbolic system described in the text. The existence of a critical line is indicated by heavy dots. The q-values, held fixed for each line, range from -20 to 20. No account has been taken of logarithmic corrections. An approximation of level $n = 10$ has been used

sition is created for a subfamily of the family $\{(q, \beta)\}$. Due to the smoothness of the arguments, the points of such a nonanalytical behavior must be connected; they thus form the 'critical line' [16]. Since in the case relevant to us the nonhyperbolic contribution stems only from the neighborhood of the origin, the relation $F_{\mathrm{G\ nonhyp}}(q, \beta) = -\beta \log c - q \log c'$ can be obtained easily, where c, c' denote the slopes of the map of the support and of the measure, respectively, at point 0. In a similar way, the overall estimate $F_{\mathrm{G}}(q, \beta) \geq -\frac{1}{z}\beta \log c - \frac{1}{z'} q \log c'$, where z, z' are the respective order of the maxima, can be derived. An example of a hyperbolic and a nonhyperbolic behavior is given in Figs 13 a,b (for convenience we plot $-F_{\mathrm{G}}(q, \beta)$ for different, fixed q-values versus β).

Manifestation of Phase Transitions in S_{G}

Hyperbolic systems lead generally to strictly convex specific entropy functions. In the more relevant nonhyperbolic case a linear segment is obtained. This effect has been observed even in high dimensional experimental systems, where the dynamical and probabilistic scaling functions have been reconstructed from time series [32, 33]. From the form of the support of the generalized entropy function, it can be immediately concluded (note the characteristic location of the lines along which the functions $S_{\mathrm{G}}(\varepsilon)$ and $f(\alpha)$ are evaluated) that different combinations of phase transitions in the different spectra could be distinguished from the place of the symbol sequence which triggers off the phase transition effect. To give an example, if sequence $\bar{A} = AAAAA\ldots$ leads to such an effect, a phase transition is obtained for the $f(\alpha)$ spectrum, whereas sequence $\bar{B} = BBBBB\ldots$ leads to phase-transitions for $S_{\mathrm{G}}(\varepsilon)$, $\phi(\lambda)$ and $f(\alpha)$, at the same time, and so on. It is therefore useful to discuss the form of the generalized entropy function for the nonhyperbolic model.

In our model [36, 37], the interchange of the symbol positions within a given symbolic string is no longer allowed, because the probabilities do not commute, due to the nonlinearity of the measure map. Therefore, a two-symbol system can already lead to a 'nondegenerate' entropy function. Using an approximation of finite level ($n = 10$) we obtain the form of the generalized entropy function as shown in Fig. 14.

Emanating from the top segment labelled by N, a 'sheet' touches the hyperbolic part along the critical line L. In the same figure, the support of the hyperbolic aspect of the system is shown by the dotted line. Whenever a specific entropy function crosses the critical line, a phase transition effect is observed.

For our example, from the form of the generalized entropy function, it follows that the more specific scaling functions $S_{\mathrm{G}}(\varepsilon)$, $g(\Lambda)$ and $f(\alpha)$ inherit the phase transition effect. For convenience, the spectra $S_{\mathrm{G}}(\varepsilon)$ and $f(\alpha)$ are shown in Fig. 15.

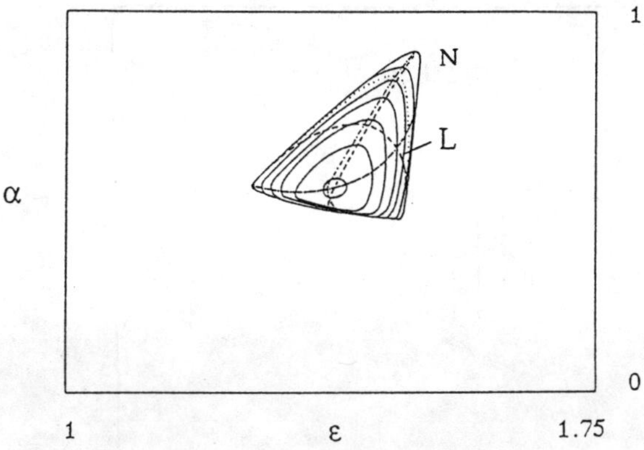

Fig. 14. Generalized entropy function $S_G(\alpha, \varepsilon)$ for the nonhyperbolic model, using an approximation of level $n = 10$. The critical line (dashed curve L), the support of the hyperbolic contribution (dotted line), the support of $S_G(\varepsilon)$, $f(\alpha)$ and $g(\Lambda)$ (double-dotted-dashed, dashed-dotted and dashed, respectively) are indicated

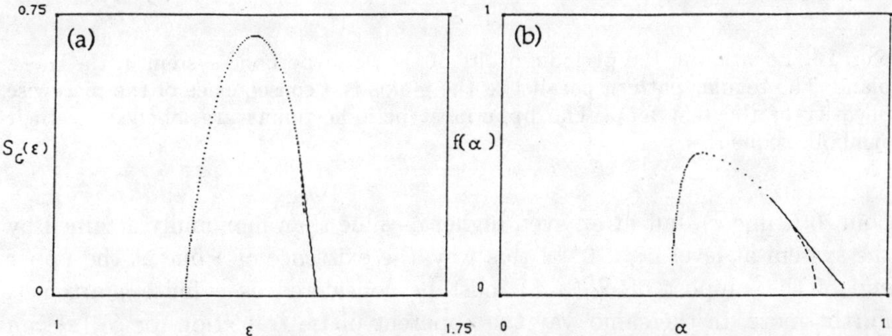

Fig. 15. (a) 'Dynamical' or 'geometrical' scaling function $S_G(\varepsilon)$ for the nonhyperbolic model and (b) probabilistic scaling function $f(\alpha)$ for the nonhyperbolic model. The dashed branches indicate the hyperbolic contribution

Note the linear parts for both scaling functions $S_G(\varepsilon)$ and $f(\alpha)$. In accordance with the thermodynamic formalism, this effect is interpreted as first-order phase transitions.

Periodic Orbits

For an improved understanding of the generalized entropy function, let us make a comparison with the support of the periodic orbits of the system (Fig. 16).

As is easily found, the endpoints of the line N are determined by the periodic orbits with the symbol sequences $1\bar{0} = 1000000\ldots$ (left end) and $\bar{0} = 000000\ldots$ (right end point). These symbol sequences correspond to the endpoints 1 and 0 of the unit interval. For the system considered, in the limit of $n \to \infty$, these points can be expected to converge to a point located above

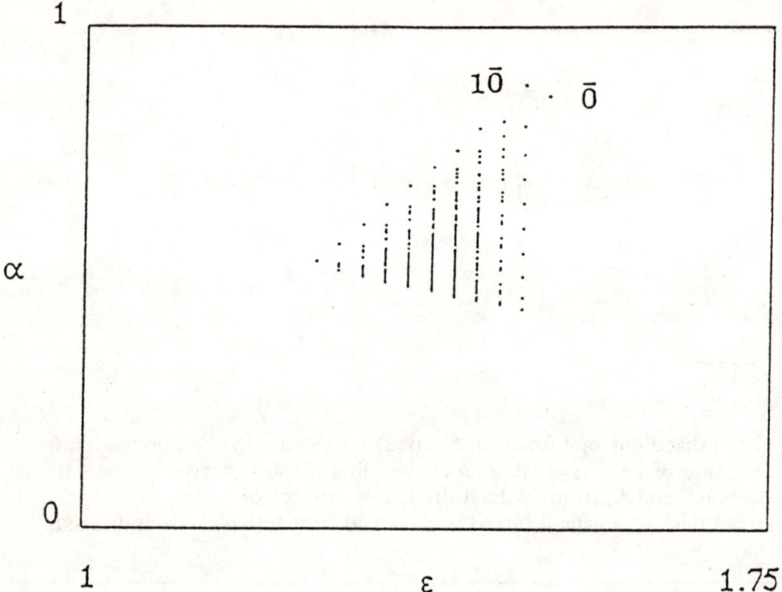

Fig. 16. Location of the periodic orbits of the nonhyperbolic system in the $\alpha - \varepsilon$ plane. The regular pattern parallel to the y-axis is a consequence of the piecewise linearity of the tent map. The uppermost periodic points are labelled by their symbolic sequences

point $\bar{0}$ (same ε), but at an even higher α-value than maximally attained by the system at level $n = 10$. In this way, the existence of a line at the upper end of the support of $S_{\mathrm{G}}(\alpha, \varepsilon)$ must be considered as a finite-n artefact. Furthermore, in the same way the apparent phase transition for $S_{\mathrm{G}}(\varepsilon)$ can be shown to be an artificial effect due to the approximation of the finite level n. From the approximation of a finite level alone, a phase transition is suggested (Fig. 15a)).

Asymptotic Situation

With the help of periodic orbits and numerical calculations, the asymptotic behavior can be worked out. A careful asymptotic investigation comes up with the insight illustrated in Fig. 17 a, b: The critical line is located exactly above the bottom border of the support of the generalized entropy. Therefore, $S_{\mathrm{G}}(\varepsilon)$, being confined to this line itself in the asymptotic limit, shows no phase transition.

Different Combinations of Phase Transitions

Now, let us investigate how this nonhyperbolic system changes if a new hyperbolic element is added [38]. In this way, a nonhyperbolic three-scale Cantor model is created. With the same technique if compared with the two-scale model, the associated generalized entropy function is calculated. However, for this example the scaling properties are chosen in such a way that

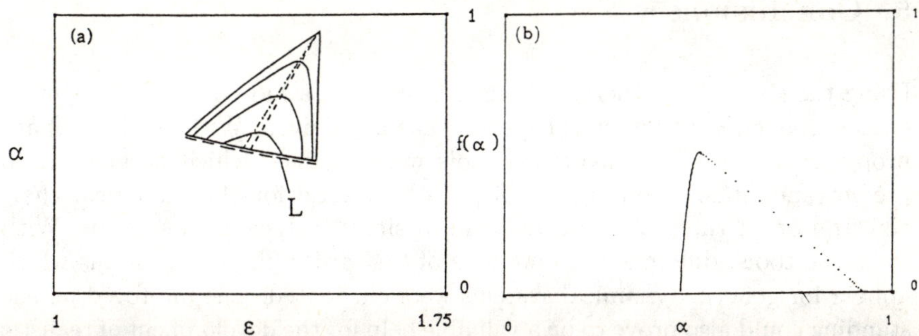

Fig. 17. Asymptotic behavior of the nonhyperbolic two-scale model: a) support of the generalized entropy $S_{\mathrm{G}}(\alpha,\varepsilon)$ for the nonhyperbolic two-scale model (the point with maximal α is generated by the nonhyperbolic element of the partition and contour lines increasing in steps of 0.2); b) function $f(\alpha)$: the linear r.h.s. of the function indicates a phase transition

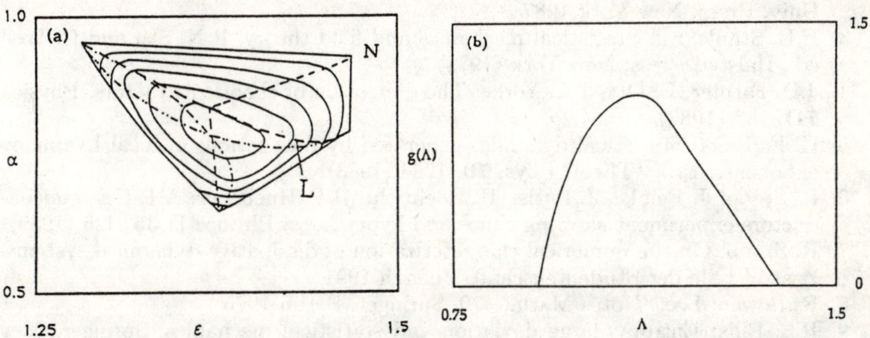

Fig. 18. Nonhyperbolic three-scale Cantor set, asymptotic behavior: a) support of $S_{\mathrm{G}}(\alpha,\varepsilon)$, with contour lines increasing in steps of 0.1; b) function $g(\Lambda)$, preserving the phase transition

only $g(\Lambda)$ undergoes a phase transition (see Fig. 18 a, b). The other specific scaling functions remain unaffected. As can be easily imagined, a different choice of the properties of the added hyperbolic element can provoke other combinations of phase transitions.

5. Conclusions

Using the generalized thermodynamic formalism, a more refined description of the properties of dynamical systems has been given. With the help of appropriate models, the theoretical tools were outlined which permit one to predict and understand the problems which arise for the numerical characterization of the scaling behavior in dissipative dynamical systems. With the same tools, different combinations of first-order phase transitions which appear for generic dynamical systems were elucidated. The improved understanding could also prove to be a valuable help for the development of realistic models for experimental systems.

References

1. D. Ruelle: Thermodynamic formalism. Vol.5 of Encyclopedia of Mathematics and Its Applications. Addison-Wesley, Reading MA 1978
2. H.E. Stanley: Introduction to phase-transition and critical phenomena. Oxford Univ. Press, New York 1987
3. H.E. Stanley: in Statistical mechanics and field theory. R.N. Sen and C. Weil ed., Halsted Press, New York (1972)
4. J.D. Farmer, E. Ott, J.A. Yorke: The dimension of chaotic attractors. Physica **7D**, 153 (1983)
5. H. Fujisaka: Statistical dynamics generated by fluctuations of local Lyapunov exponents. Prog. Theor. Phys. **70**, 1264 (1983)
6. R. Stoop, J. Peinke, J. Parisi, B. Roehricht, R.P. Huebener: A P-GE semiconductors experiment showing chaos and hyperchaos. Physica D **35**, 425 (1989)
7. R. Stoop: On the numerical characterization of dissipative dynamical systems. Zentralstelle der Studentenschaft, Zuerich 1991
8. R. Bowen: Lect. Notes Math., 470, Springer, Berlin 1975
9. R.S. Ellis: Entropy, large deviation and statistical mechanics. Springer, New York 1985
10. Y. Oono: Large deviation and statistical physics. Progr. Theor. Phys. Suppl. **99**, 165 (1989)
11. A.O. Lopez: Entropy and large deviation. Nonlinearity **3**, 527 (1990)
12. A. Muenster: Statistical thermodynamics. Springer, Berlin 1969
13. T.C. Halsey, M.H. Jensen, L.P. Kadanoff, I. Procaccia, B. Shraiman: Fractal measures and their singularities - the characterization of strange sets. Phys. Rev. **A 33**, 1141 (1986)
14. R. Stoop, J. Parisi: Phys. Lett., in press
15. M. Kohmoto: Entropy function for multifractals. Phys. Rev. **A 37**, 1345 (1988)
16. Z. Kovács, T. Tél: to be published
17. Y. Oono, Y. Takahashi: Chaos, external noise and Fredholm theory. Prog. Theor. Phys. **63**, 1804 (1980)
18. B.B. Mandelbrot: The fractal geometry of nature. Freeman, New York 1982
19. J. Peinke, J. Parisi, R. Stoop, O.E. Roessler: An encounter with chaos. Springer, in press (1992)
20. R.L. Devaney: An introduction to chaotic dynamical systems. Benjamin, Menlo Park CAL 1986

21. V.N. Shtern: The dimension of turbulent motion attractors. Dokl. Akad. Nauk. SSSR **270**, 582 (1983)
22. F. Hausdorff: Math. Ann. **79**, 157 (1919)
23. A. Renyi: Probability theory. North-Holland, Amsterdam 1970
24. d'Alessandro, A. Politi: priv. comm.
25. T. Bohr, D. Rand: The entropy function for characteristic exponents. Physica **25D**, 387 (1987)
26. T. Tél, T. Bohr: in Directions in Chaos, vol.2. ed. Hao Bai-Lin, World Scientific 1988
27. T. Tél: Transient chaos. Preprint
28. C. Gunaratne, M.H. Jensen, I. Procaccia: Universal strange attractors on wrinkled tori. Nonlinearity **1**, 157 (1988)
29. R. Artuso, E. Aurell, P. Cvitanović: Recycling of strange sets. Nonlinearity **3**, 325 (1990)
30. R. Stoop, J. Parisi: Calculation of Lyapunov exponents avoiding spurious elements. Phys. Lett. **A 161**, 67 (1991)
31. R. Stoop, J. Parisi: to be published (Physica D)
32. R. Stoop, J. Parisi: Dynamical phase transitions in a parametrically modulated radio-frequency laser. Phys. Rev. **A 43**, 1802 (1991)
33. R. Stoop, J. Peinke, J. Parisi: Phase transitions in experimental systems. Physica **50D**, 405 (1991)
34. P. Szépfalusy, T. Tél, A. Cordás, Z. Kovács: Phase-transitions associated with dynamical properties of chaotic systems. Phys. Rev. **A 36**, 3525 (1987)
35. M.J. Feigenbaum, I. Procaccia, T. Tél: Scaling properties of multifractals as an eigenvalue problem. Phys. Rev. **A 39**, 5359 (1989)
36. R. Stoop, Z. Naturforsch: 46 a, 1117 (1991)
37. R. Stoop: to be published (Phys. Rev. A)
38. R. Stoop, J. Parisi: submitted

New Numerical Methods for High Dimensional Hopf Bifurcation Problems

J. Wu and K. Zhou

Department of Mechanics,
Peking University, Beijing 100871, PRC

Abstract

In the paper a new numerical method for the analysis of both static and Hopf bifurcation is presented. The study of the problem of solution bifurcation is reduced to the study of the local behavior near the singular points of the vector field. In the case of the Hopf bifurcation, the problem is reduced to searching for a pair of complex conjugate eigenvalues with maximum norm of a matrix. Two examples support the introduced theory.

1. Introduction

It is trivial to say that bifurcation phenomena play a fundamental role in nonlinear dynamics. They divide the parameters space into subspaces related to the qualitatively different dynamical behavior. One of the most important bifurcations opening the door between equlibria and movement, is the celebrated Hopf bifurcation. Although many papers have been contributed to the theory and numerical methods of the Hopf bifurcation problem, there are still numerous questions of numerical adaptation which need to be answered. Especially all numerical methods now available for Hopf bifurcation are only effective for low dimensional problems. When the dimension becomes larger, the computation work load becomes tremendously high. Therefore, we propose some new numerical methods for solving the above mentioned problem.

Let us consider an autonomous system of ordinary differential equations

$$\frac{dx}{dt} = F(x, \mu) \tag{1}$$

where $x \in X \subset \mathbb{R}^n$, $\mu \in \Lambda \subset \mathbb{R}$, μ is a real valued parameter and $F : \mathbb{R}^n \times \mathbb{R} \to \mathbb{R}^n$ is a vector valued function.

Let system (1) have an isolated stationary point $x = x^*(\mu)$, i.e.,

$$F(x^*(\mu), \mu) = 0 \,. \tag{2}$$

The Jacobian matrix of F at a stationary point is defined as follows:

$$A(\mu) = D_x F(x^*(\mu), \mu) = \left(\frac{\partial F_i}{\partial x_j}(x^*(\mu), \mu) \ , \ i, j = 1, 2, \ldots, n \right) . \tag{3}$$

Assume that $A(\mu)$ has eigenvalues $\lambda_i(\mu)$ $(i = 1, 2, \ldots, n)$.

If $\mathrm{Re}(\lambda_i(\mu)) \neq 0$, $(i = 1, 2, \ldots, n)$, according to the Implicit Function Theorem, it is obvious that the stationary point of system (1) is unique, and that the nonlinear system (1) is equivalent to its linearized system

$$\frac{dx}{dt} = A(\mu)x \, . \tag{4}$$

The bifurcation phenomenon will occur at some μ^*, where $\mathrm{Re}(\lambda_i(\mu^*)) = 0$.

If $A(\mu^*)$ is singular, and there is a stationary point $x^*(\mu^*)$ such that, for any neighborhood U of (x^*, μ^*) in $\mathbb{R}^n \times \mathbb{R}$, there is more than one independent solution $(x_1(\mu), \mu)$, $(x_2(\mu), \mu), \ldots, (x_m(\mu), \mu) \in U$; that is, there is a $\mu \in \Lambda$, $x_1(\mu)$, $x_2(\mu), \ldots, x_m(\mu) \in X$, $x_i \neq x_j$, $(i, j = 1, 2, \ldots, m)$, such that $(x_1(\mu), \mu)$, $(x_2(\mu), \mu), \ldots, (x_m(\mu), \mu) \in U$ satisfy equation (1), which corresponds to what is called static bifurcation. If $A(\mu)$ has a pair of complex conjugate eigenvalues λ_1, and λ_2

$$\lambda_1(\mu) = \overline{\lambda_2(\mu)} = \alpha(\mu) + i\omega(\mu) \, , \tag{5}$$

such that at $\mu = \mu^*$

$$\alpha(\mu^*) = 0 \, , \quad \omega(\mu^*) = \omega_0 > 0 \, , \quad \alpha'(\mu^*) \neq 0 \, , \tag{6}$$

and the eigenvalues of $A(\mu^*)$, other than $\pm i\omega_0$, all have strictly negative real parts, then system (1) has a family of periodic solutions. This appearance of periodic solutions out of equilibrium state is called Hopf bifurcation.

In this contribution, a new numerical method with low computational complexity is presented for the computation of static bifurcation, Hopf bifurcation, and tracking the closed orbit on a large-scale.

Section 2 deals with problems of the static bifurcation and the pseudo-arclength method. By means of a constructed vector field, an equivalent relationship between invariant manifolds of the vector fields and the solution manifolds of nonlinear equation system is established. Consequently, the study of the problem of finding the bifurcation is turned into that of the local behavior near the singular points of the vector field, and a new numerical method for tracking the solution manifold is provided.

Section 3 deals with the methods for Hopf bifurcation. In Section 3.1, two important theorems are introduced. In Section 3, by taking a transformation, the problem of seeking the Hopf bifurcation points of system (1) is turned into finding a pair of complex conjugate eigenvalues with the maximum norm of a matrix. The problem of tracking the closed orbits on a large-scale is equivalent to tracking the solution manifold of the nonlinear equation system in higher dimensional space.

In Section 4, by using a program suitable for the large-scale computation of nonlinear dynamics in high dimensional space, several examples are calculated to show the efficiency of the introduced method.

2. Static Bifurcation and Pseudo-Arclength Method

The solutions $x(\mu)$ of equations

$$F(x, \mu) = 0 \tag{7}$$

are called the stationary points of system (1). Hereafter F is assumed to be smooth enough. $x(\mu)$ is a one-dimensional manifold except some special points, and we call it solution manifold. As the parameter μ changes, in order to understand the development process of the solution of (7), the continuation method with predictor-corrector [1, 2] is usually used for tracking the solution curve $x(\mu)$.

In setting out to compute the static bifurcation of nonlinear equations (7) by using the continuation method, there are three issues with which one has to deal:

1. Tracking the solution curve.
2. Determining and finding the singular points.
3. Searching for the bifurcation directions starting, from the bifurcation point.

Problem 1 is to solve the equations (7). From (7) one obtains:

$$\begin{aligned} \frac{dx}{d\mu} &= -\left[D_x F(x\mu)\right]^{-1} D_\mu F(x, \mu) , \\ x(\mu_0) &= x_0 , \end{aligned} \tag{8}$$

where $D_x F(x, \mu)$ is a real $n \times n$ matrix, which is the Jacobian of $F(x, \mu)$ about x, and (x_0, μ_0) is the original stationary point.

The system (8) can be solved by using the predictor-corrector method. This method requires that the Jacobian $D_x F(x, \mu)$ is non-degenerate, otherwise it will faile.

In problem 2, the singular points of equations (7) are the stationary points (x^*, μ^*), where $\det D_x F(x^*, \mu^*) = 0$. The turning points and static bifurcation points are all singular points. Let $y = (x, \mu) \in \mathbb{R}^{n+1}$, the Frechet's derivative $DF(y) = (\frac{\partial F_i}{\partial y_j})$ of F is a real $n \times (n+1)$ matrix. At $y^* = (x^*, \mu^*)$, if there exists at least one $n \times n$ sub-matrix of $DF(y^*)$, which is non-degenerate, then y^* is called a turning point; and if all the $n \times n$ sub-matrices of $DF(y^*)$ are degenerate, then y^* is called a static bifurcation point.

At present, there are two methods which allow one to find the singular points. First, a test function $\Delta = \det D_x F(x, \mu)$ can be used for singular points $\Delta = 0$. As the predictor-corrector process is executed, if there are two points near each other with opposite sign of Δ, then there must exist a singular point between them. The singular point is determined by using the half interval search method or other interpolation methods. This method is effective for some simple bifurcation points. Secondly, one can form a more complicated system of equations [3, 4] which is solved by using the Newtonian iterated method. This method is effective only for some specific problems, and it is not a general method.

For problem 3, searching for the bifurcation directions that start from the bifurcation point might be the most difficult problem in the computation of static bifurcation. Until now, the Algebraic Bifurcation Equation [5] has been used widely. However, this method needs to compute all the secondary derivatives of F, and two coupled systems of nonlinear equations should be solved. It is so complicated that it is suitable only for some simple problems, for example, for problems with two bifurcation branches.

In [6], the authors introduced a numerical scheme that is effective for large-scale computations of static bifurcation, determining all the singular points of the solution curve, and finding all the bifurcation directions. This method is called the pseudo-arclength method [7–9]. As compared with other schemes, the pseudo-arclength method has lower computational complexity. Here, we would like to give the main idea of the paper [6].

The nonlinear equations (7) are written as

$$F(y) = 0\,, \tag{9}$$

where $y = (x, \mu) \in \mathbb{R}^{n+1}$ and F is smooth enough.

Now, define a vector field

$$\mathbf{v}(y) = [v^1, v^2, \ldots, v^{n+1}]^T\,, \tag{10}$$

every component of which consists of the cofactors of j-column of DF, i.e.,

$$v^j = (-1)^{j+1} \det\left(\frac{\partial F}{\partial y_1}, \frac{\partial F}{\partial y_2}, \ldots, \frac{\partial F}{\partial y_{j-1}}, \widehat{\frac{\partial F}{\partial y_1}}, \frac{\partial F}{\partial y_{j+1}}, \ldots, \frac{\partial F}{\partial y_{n+1}}\right), \quad j = 1, 2, \ldots, n+1\,. \tag{11}$$

Where "$\widehat{\ }$" denotes "deleted".

This vector field is smooth and possesses the following property

$$DF \cdot v(y) = 0\,, \qquad \forall y \in \mathbb{R}^{n+1}\,. \tag{12}$$

It can be obtained from the equation

$$\det\begin{bmatrix} DF \\ \frac{\partial F_i}{\partial y_1}, \ldots, \frac{\partial F_i}{\partial y_{n+1}} \end{bmatrix} = 0\,, \qquad i = 1, 2, \ldots, n \tag{13}$$

by expanding the determinant by the last row.

The vector field v is accompanied by a dynamic system

$$\frac{dy}{dt} = v(y) . \tag{14}$$

At the singular point y^*, it is obvious that $v(y^*) = 0$ for a static bifurcation point and $v^{n+1}(y^*) = 0$, $v(y^*) \neq 0$ for a turning point.

In paper [6], the authors proved the equivalence between the solution manifold of $F(y)$ =const., and the invariant manifold of v. Thus, solving the zero valued manifold of equations (7) is reduced to solving the invariant manifold of v, i.e.,

$$\frac{dy}{ds} = \tau(s) := \frac{v(y)}{|v(y)|} , \tag{15}$$

$$y(0) = y_0 , \tag{16}$$

where $v(y_0) \neq 0$, and s is the arclength along the solution curve. The predictor-corrector can be described as follows below.

With the aid of the induced vector field the Eulerian prediction $y_k^{(0)}$ can be obtained from the previous point y_{k-1} by moving it along the direction $\tau(y_{k-1})$, i.e.,

$$y_k^{(0)} = y_{k-1} + \tau(y_{k-1}) \cdot \Delta s_{k-1} , \quad k = 1, 2, \dots , \tag{17}$$

where Δs_{k-1} is the step length.

After several steps, the Newtonian iterative correction is executed if necessary, and the iterative directions perpendicular to $v(y_k^{(0)})$ are given by the algorithm

$$y_k^{(l)} = y_k^{(l-1)} - \begin{bmatrix} DF(y_k^{(0)}) \\ v^T(y_k^{(0)}) \end{bmatrix} \cdot \begin{bmatrix} F(y_k^{(l-1)}) \\ 0 \end{bmatrix} , \quad l = 1, 2, \dots . \tag{18}$$

If $v(y_k)$ is not too small, and $y_k^{(0)}$ is close to the solution manifold, it is expected that the sequence $y_k^{(1)}$, $y_k^{(2)}$, ..., converges to a point y_k^* on the solution manifold $F(y_k^*) = 0$.

It is proved that this process succeeds in passing through the turning points. For determining the static bifurcation points, paper [6] introduced two methods: the sign method and the limit value method. For seeking the bifurcation directions, the same paper also introduced two methods. In the case of simple bifurcation (with two bifurcation branches), the algebraic bifurcation equation is used, and for the case of multiple bifurcation, the simplex method is used.

3. The Numerical Methods for Hopf Bifurcation

Consider the following nonlinear dynamic system

$$\frac{dx}{dt} = F(x, \mu) . \tag{19}$$

In order to deal numerically with the Hopf bifurcation of the system (19), three problems arise, as in the case of static bifurcation:

1. Tracking the solution curve.
2. Determining and finding the Hopf bifurcation point.
3. Following the closed orbits starting from the Hopf bifurcation point.

To solve problem 1, the pseudo-arclength method is used, as mentioned above.

To successfully proceed further, it is necessary to introduce the Hopf bifurcation Theorem and the Center Manifold Theorem. They are important for the computation of Hopf bifurcation.

Hopf Bifurcation Theorem *([10–13]). Let the stationary point of the system (19) be $x^* = 0$, and the Jacobian*

$$A(\mu) = \left(\frac{\partial F_i}{\partial y_j}(0, \mu) \ , \ i, j = 1, 2, \ldots, n \right) , \tag{20}$$

if

(1) $F(0, \mu) = 0$ for μ in an open interval containing 0, and $0 \in \mathbb{R}^n$ is an isolated stationary point of F;
(2) F is analytic in x and μ in a neighborhood of (0,0) in $\mathbb{R}^n \times \mathbb{R}$,
(3) $A(\mu)$ has a pair of complex conjugate eigenvalues λ and $\overline{\lambda}$, such that

$$\lambda(\mu) = \alpha(\mu) + i\omega(\mu) \tag{21}$$

where:

$$\alpha(0) = 0 \, , \quad \omega(0) = \omega_0 > 0 \, , \quad \alpha'(0) \neq 0 \, ; \tag{22}$$

(4) The remaining $n-2$ eigenvalues of $A(0)$ have strictly negative real parts.

Then the system (23) has a family of periodic solutions: there exists an $\varepsilon_{\mathrm{H}} > 0$ and an analytic function

$$\mu(\varepsilon) = \sum_{i=2}^{\infty} \mu_i \varepsilon^i \, , \qquad (0 < \varepsilon < \varepsilon_{\mathrm{H}}) \, , \tag{23}$$

such that for each $\varepsilon \in (0, \varepsilon_{\mathrm{H}})$, there exits a periodic solution $P_\varepsilon(t)$ occurring for $\mu = \mu(\varepsilon)$. There is a neighborhood η of $x = 0$ and an open interval Λ containing 0 such that for any $\mu \in \Lambda$ the only nonconstant periodic solutions *of (19) that lie in η are members of the family $P_\varepsilon(t)$ for values of ε satisfying $\mu(\varepsilon) = \mu$, $\varepsilon \in (0, \varepsilon_H)$. The period $T(\varepsilon)$ of $P_\varepsilon(t)$ is an analytic function*

$$T(\varepsilon) = \frac{2\pi}{\omega_0}\left[1 + \sum_{i=2}^{\infty} T_i \varepsilon^i\right], \qquad (0 < \varepsilon < \varepsilon_H). \tag{24}$$

Exactly two of the Floquet exponents of $P_\varepsilon(t)$ approach 0 as $\varepsilon \to 0$. One is 0 for $\varepsilon \in (0, \varepsilon_H)$, and the other is described by an analytic function

$$\beta(\varepsilon) = \sum_{i=2}^{\infty} \beta_i \varepsilon^i, \qquad (0 < \varepsilon < \varepsilon_H). \tag{25}$$

The periodic solution $P_\varepsilon(t)$ is orbitally asymptotically stable with an asymptotic phase if $\beta(\varepsilon) < 0$ but is unstable, if $\beta(\varepsilon) > 0$.

For the proof of the Hopf Bifurcation Theorem, the reader is encouraged to study the reference [11] pp. 22–24.

Center Manifold Theorem *([11, 14, 15]). Consider an abstract differential equation*

$$\frac{dx}{dt} = f(x) \equiv Lx + h(x) \tag{26}$$

with x in some open set U in a real Hilbert space H containing the origin. Here, L is the infinitesimal generator of a C^0 semigroup e^{Lt} and $h(\cdot)$ is an H-valued c^r $(r \geq 1)$ function on U, such that $h(0) = 0$ and $Dh(0) = 0$. A submanifold M in U with $0 \in M$ is said to be locally invariant, if to each x in M the solution $\varphi_t(x)$ of (26) with the initial condition $\varphi_0(x) = x$ remains in M for a certain interval $0 \leq t \leq T$, where $T = T(x) > 0$. The tangent space T_0M of M at 0 is invariant under the linear operator L.

Assume now that L satisfies the following spectral conditions ;

1. *H is the direct sum of two closed invariant spaces V_s, V_c, where $\mathrm{Re}\sigma(L|V_s) < \alpha < 0$ for some α, and V_c has finite dimension with $\mathrm{Re}\sigma(L|V_c) = 0$.*
2. *$\sigma(e^{Lt}) = e^{\sigma(L)t} \cup \{0\}$ for all $t > 0$.*

Then the equation (26) possesses a c^{r-1} center manifold M such that:

(a) M is locally invariant.
(b) M is locally attractive in the following sense. There exists an open neighborhood V of 0 such that whenever $\varphi_t(x) \in V$ for all $t \geq 0$, $\varphi_t(x)$ approaches M as $t \to \infty$.

If $H = \mathbb{R}^n$, it is obvious that the equation (26) is an ordinary differential equation as (19). For ordinary differential equations with $\sigma(A) \leq 0$, the spectral conditions of the center manifold theorem are always met (for the proof of the center manifold theorem, see reference [11], pp. 268).

The theorem of center manifolds introduced here is sufficient for the ordinary differential equations examined in this paper (more details on invariant manifolds and center manifold theories, can be found in references [14, 15]). Now numerical methods to solve the mentioned above problems will be introduced [16].

Let's consider the nonlinear dynamic system

$$\frac{dx}{dt} = F(x, \mu) , \tag{27}$$

for which the conditions of Hopf bifurcation theorem are satisfied.

At present, there are two methods for determining and finding the Hopf bifurcation points on the solution curve $x(\mu)$:

1. Calculation of all the eigenvalues of the Jacobian $A(\mu)$. At certain $\mu = \mu^*$, if $A(\mu)$ has a pair of complex conjugate eigenvalues which cross the imaginary axis, then the stationary point $x(\mu^*)$ is a Hopf bifurcation point ([11, 15]).
2. The classical Hurwitz theorem [17].

By the second method, instead of calculating the roots of a characteristic polynomial, only the sign of a series of Hurwitz determinants is needed, according to which the stability of system (27) can be determined. But this method is based on elimination method, which will diminish computational precision seriously when it deals with the high-dimensional problems. Also, calculation of the eigenvalues of $A(\mu)$ is unsuitable for high-dimensional problems. In fact, once the degree of $A(\mu)$ exceeds 50, the results got from QR method are dubious. Furthermore, what we are concerned with is only the pair of complex conjugate eigenvalues, which will cross the imaginary axis. Thus, calculating all the eigenvalues of $A(\mu)$ obviously greatly increases the computational complexity.

After obtaining the Hopf bifurcation point, there are three methods mainly addressed to numerical calculation of the closed orbits.

1. The method of using the Hopf bifurcation theorem and the center manifold theorem to reduce the n-dimensional system to the Poincaré normal form. By this method, the coefficients of (23)–(25) as well as the stability of the closed orbits, can be gained [11, 15]. However, this method requires the calculation of all the third derivatives of F. Therefore, the work load is too much even for low-dimensional problems, and will be much more for high dimension.
2. The method of averaging [18]. This method is also effective for low-dimensional problems. For problems in high dimension, reducing the system (27) to the averaged equations is very complicated.
3. The direct method. In setting out system (27), numerical methods, such as Runge-Kutta method etc., are used to solve ordinary differential equations

to shoot for closed orbits. By this method, the selection of the original point is so blind that it is only used as a method for some specific problems and not for general problems.

Furthermore, these three methods all have the further defect that they are only used to calculate the closed orbits which are situated near the Hopf bifurcation point. As the parameter μ is far from μ^* (corresponding to the Hopf bifurcation point), the computational precision is not ensured. Thus, an accurate and efficient numerical scheme for the large-scale computation of Hopf bifurcation in high-dimensional space is still waiting to be developed.

In this section we would like to introduce a numerical scheme, which is simple but perfect, and which fulfills the purpose mentioned above.

In calculating the Hopf bifurcation of the system (27), we always assume that one starts from a stable stationary point or from the results of the former computation. Therefore all the eigenvalues of $A(\mu)$ have strictly negative real parts at the original points. Let

$$B = (A + I)(A - I)^{-1} , \tag{28}$$

where I is a real $n \times n$ unit matrix.

Let us assume that the eigenvalues of A and B are λ_A^i and λ_B^i respectively, $(i = 1, 2, \ldots, n)$. Then we have

$$\lambda_B^i = \frac{\lambda_A^i + 1}{\lambda_A^i - 1} , \qquad i = 1, 2, \ldots, n . \tag{29}$$

It is obvious that the transformation (29) turns the imaginary axis into the complex plane, which is presented in Fig. 1.

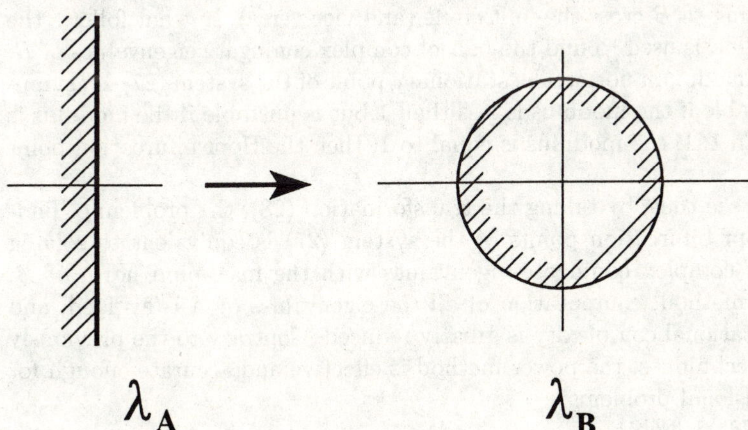

Fig. 1. Illustration of the transformation (29)

Thus, at the original point, the system (27) is a stable one and all eigenvalues of A lie in the left half plane of the complex plane. Likewise, the eigenvalues of B are all inside the unit circle. Additionally, when A has a pair of complex conjugate eigenvalues near the imaginary axis then the following formula holds:

$$\lambda_A = -\varepsilon \pm i\beta \, , \tag{30}$$

where $\varepsilon > 0$ is small and β is arbitrary. Correspondingly, B has a pair of complex conjugate eigenvalues defined below:

$$\lambda_B = \frac{\lambda_A + 1}{\lambda_A - 1} = 1 - \frac{2(1+\varepsilon)}{(\varepsilon+1)^2 + \beta^2} \pm \frac{a\beta}{(\varepsilon+1)^2 + \beta^2} i \, , \tag{31}$$

which has the following modulus:

$$|\lambda_B| = \sqrt{1 - \frac{4\varepsilon}{(\varepsilon+1)^2 + \beta^2}} \approx 1 - \frac{2\varepsilon}{(\varepsilon+1)^2 + \beta^2} \tag{32}$$

and

$$1 - |\lambda_B| \approx \frac{2\varepsilon}{(\varepsilon+1)^2 + \beta^2} > 0 \tag{33}$$

is small.

When A possesses a pair of complex conjugate eigenvalues near the imaginary axis, then B has a pair of complex conjugate eigenvalues near the unit circle. Furthermore, as the bifurcation parameter μ changes, no matter how λ_A^i $(i = 1, 2, \ldots, n)$ are changed, once a pair of complex conjugate eigenvalues of A crosses the imaginary axis, the complex conjugate eigenvalues corresponding to B cross the unit circle (and vice versa). In what follows the power method is used to find the pair of complex conjugate eigenvalues of B, which has maximum norm. The stationary point of the system (27) is asymptotically stable if the modulus is less than 1 but is unstable if the modulus is greater than 1. If the modulus is equal to 1, then the Hopf bifurcation point is found.

We can see that, by taking the transformation (28), the problem of finding the Hopf bifurcation points of the system (27) is equivalent to solving the pair of complex conjugate eigenvalues with the maximum norm of B. Using this method, computation of all the eigenvalues of A is avoided, and the computational complexity is greatly reduced. Contrary to the previously described techniques, the power method is effective and accurate enough for high-dimensional problems.

From (28) one obtains

$$B = (A + I)(A - I)^{-1} = I + 2(A - I)^{-1} \, , \tag{34}$$

which leads to

$$z_{k+1} = Bz_k = z_k + 2(A - I)^{-1} z_k \,, \tag{35}$$

i. e.

$$(A - I)(z_{k+1} - z_k) = 2z_k \,. \tag{36}$$

Let z_0 be an arbitrary n-dimensional vector ($z_0 \neq 0$). The iterative algorithm of the power method is as follows:

$$\begin{cases} 1)\ (A - I)y_k = 2z_{k-1} \,, & \text{find } y_k \\ 2)\ \tilde{y}_k = y_k + 2z_{k-1} & \\ 3)\ m_k = \max(\tilde{y}_k) & \\ 4)\ z_k = \tilde{y}_k / m_k & \\ 5)\ k = k + 1 \,, & \text{goto 1),} \end{cases} \tag{37}$$

where $\max(\tilde{y}_k)$ is the maximum component of $\tilde{y}_k$.

Assume a pair of complex conjugate eigenvalues with maximum norm of B to be

$$\lambda_B^1 = \overline{\lambda}_B^2 \,, \quad |\lambda_B^1| = |\lambda_B^2| > |\lambda_B^i| \,, \qquad i = 3, 4, \ldots, n \,, \tag{38}$$

so we have

$$B^{k+2} z_0 + p_1 B^{k+1} z_0 + p_2 B^k z_0 \to 0 \,, \text{ as } k \to \infty \,, \tag{39}$$

where:

$$\lambda_B^1 + \lambda_B^2 = -p_1 \,, \tag{40}$$

$$\lambda_B^1 \cdot \lambda_B^2 = p_2 \,. \tag{41}$$

From (37) we obtain

$$m_{k+1} m_{k+2} z_{k+2} + p_1 m_{k+1} z_{k+1} + p_2 z_k \to 0 \,, \text{ as } k \to \infty \,. \tag{42}$$

For k big enough we assume that

$$m_{k+1} m_{k+2} z_{k+2} + p_1 m_{k+1} z_{k+1} + p_2 z_k = 0 \,. \tag{43}$$

Equations (43) are solved by using the least square method, i.e.,

$$\begin{aligned} &\begin{bmatrix} m_{k+1}^2 (z_{k+1}^T z_{k+1}) & m_{k+1}(z_k^T z_{k+1}) \\ m_{k+1}(z_k^T z_{k+1}) & z_k^T z_k \end{bmatrix} \begin{bmatrix} P_1^{(k)} \\ P_2^{(k)} \end{bmatrix} = \\ &\qquad = -m_{k+1} m_{k+2} \begin{bmatrix} m_{k+1} z_{k+1}^T z_{k+2} \\ z_k^T z_{k+2} \end{bmatrix} , \end{aligned} \tag{44}$$

so

$$p_1^{(k)} = -m_{k+2} \frac{(z_{k+1}^T z_{k+2})(z_k^T z_k) - (z_k^T z_{k+1})(z_k^T z_{k+2})}{\Delta} \,, \tag{45}$$

$$p_2^{(k)} = -m_{k+1}m_{k+2}\frac{(z_{k+1}^T z_{k+1})(z_k^T z_k) - (z_k^T z_{k+1})(z_{k+1}^T z_{k+2})}{\Delta}, \tag{46}$$

where $\Delta = (z_{k+1}^T z_{k+1})(z_k^T z_k) - (z_k^T z_{k+1})^2$.

In the iterated process, if the sequences $p_1^{(1)}, p_1^{(2)}, \ldots$, and $p_2^{(1)}, p_2^{(2)}, \ldots$ converge to p_1^* and p_2^* respectively, then the approximate value of the eigenvalue λ_B^1 is written as

$$\mathrm{Re}(\lambda_B^1) = -\frac{p_1^*}{2}, \qquad \mathrm{Im}(\lambda_B^1) = \frac{1}{2}\sqrt{4p_2^* - p_1^{*2}}\,. \tag{47}$$

The eigenvector which corresponds to λ_B^1 is equal to

$$z_k + i(\zeta z_k - \tilde{y}_{k+1})/\eta\,, \tag{48}$$

where $\zeta = \mathrm{Re}(\lambda_B^1)$, and $\eta = \mathrm{Im}(\lambda_B^1)$.

Now we will discuss the question of determining and finding the Hopf bifurcation points. As the pseudo-arclength method is used for solving the Hopf bifurcation of the system (27) for each step, the matrix A is formed and then λ_B^1 and u are obtained according to (37)–(48). Usually, at the Hopf bifurcation point the Jacobian A is nonsingular, thus the procedure goes on successfully. Furthermore, near the Hopf bifurcation point, the iterated process (37)–(48) is always convergent.

Now we want to propose two very economical methods for finding the Hopf bifurcation points.

(1) The sign method. In the predictor-corrector procedure, let $y_1 = (x_1, \mu_1)$, $y_2 = (x_2, \mu_2)$ be two points which are near each other. $\lambda_{\max}^1$ and $\lambda_{\max}^2$ are the complex conjugate eigenvalues with maximum norm of B respectively. If $(1 - |\lambda_{\max}^1|)(1 - |\lambda_{\max}^2|) < 0$ then there must exist a Hopf bifurcation point between y_1, and y_2. Taking parabolic interpolation of the solution curve, we have

$$y(s) = \left(1 - \frac{s^2}{l^2}\right)y_1 + \frac{s^2}{l^2}y_2 + s\left(1 - \frac{s}{l}\right)\tau_1\,, \quad s \in [0, l]\,, \tag{49}$$

where l is the arclength of solution curve between y_1 and y_2, and l can be substituted by $\|y_2 - y_1\|$. $\tau_1 = \frac{v_1}{\|v_1\|}\mathrm{sign}(v_1^T \cdot (y_2 - y_1))$, $v_1 = v(y_1)$ mentioned in (10) and (11), and additionally

$$y(0) = y_1\,, \qquad y(l) = y_2\,, \qquad y'(0) = \tau_1\,. \tag{50}$$

From (29), the complex conjugate eigenvalues near the imaginary axis of A corresponding to $\lambda_{\max}^1$ and $\lambda_{\max}^2$ are defined as follows

$$\overline{\lambda}_1 = \frac{\lambda_{\max}^1 + 1}{\lambda_{\max}^1 - 1}, \quad \text{and} \quad \overline{\lambda}_2 = \frac{\lambda_{\max}^2 + 1}{\lambda_{\max}^2 - 1}. \tag{51}$$

By taking linear interpolation for $\mathrm{Re}(\overline{\lambda})$, we have

$$\mathrm{Re}(\overline{\lambda}) = \left(1 - \frac{s}{l}\right)\mathrm{Re}(\overline{\lambda}_1) + \frac{s}{l}\mathrm{Re}(\overline{\lambda}_2)\,, \quad s \in [0, l]\,. \tag{52}$$

Thus the approximate zero-valued point is

$$s^* = \frac{\mathrm{Re}(\overline{\lambda}_1)}{\mathrm{Re}(\overline{\lambda}_1) - \mathrm{Re}(\overline{\lambda}_2)} l\,, \tag{53}$$

and the first-degree approximation of the Hopf bifurcation point is obtained as $y(s^*)$ from (49). The result is corrected by use of Newtonian iterated method and a new interval (y_1^*, y_2^*) is obtained such that $(1-|\lambda_{\max}(y_1^*)|)(1-|\lambda_{\max}(y_2^*)|) < 0$. By repeating this procedure, the Hopf bifurcation point is computed with the necessary accuracy.

(2) The limit value method. Let $y_0 = (x_0, \mu_0)$, $y_1 = (x_1, \mu_1)$ and $y_2 = (x_2, \mu_2)$ be three successive points yielded by the predictor-corrector procedure, and $\lambda^i_{\max} = \lambda_{\max}(y_i)$, $i = 1, 2, 3$. Let $E_i = |\lambda^i_{\max}|$, $i = 1, 2, 3$. If $E_1 < E_2$ and $E_2 > E_3$, then the function $\lambda_{\max}(y)$ might have a maximum point between y_0 and y_2.

Let $s_1 = \widetilde{y_0 y_1}$, $s_2 = \widetilde{y_0 y_2}$ and

$$y(s) = \frac{(s-s_1)(s-s_2)}{s_1 s_2} y_0 + \frac{s(s_2 - s)}{s_1(s_2 - s_1)} y_1 + \frac{s(s - s_1)}{s_2(s_2 - s_1)} y_2\,,\ s \in [0, s_2]\,. \tag{54}$$

For $s' = s_1/2$, $s'' = (s_1 + s_2)/2$, $y(s')$ and $y(s'')$ are computed from (54), and the Newtonian iterated method is used to correct the results. Thus, a new triple (y_0^*, y_1^*, y_2^*) is obtained such that $E(y_0^*) < E(y_1^*)$, and $E(y_1^*) > E(y_2^*)$. By repeating this procedure, the maximum point y^* is gained. If $E(y^*)$ approaches 1 then y^* is a Hopf bifurcation point. But it is not a Hopf bifurcation point if $E(y^*) < 1$.

Once the Hopf bifurcation point of the system (27) has been obtained, the problem of studying the characteristic features of the closed orbits emerges immediately, such as the shape, stability and period of the closed orbit and so on. There are some open problems listed below. How do the changes of the bifurcation parameter μ affect the shape changes of the closed orbit in $\mathbb{R}^n$? And when is there a singular point at the closed orbit such that the bifurcation from closed orbit to tori or the homoclinic orbit and inhomoclinic orbit might occur? To deal with these problems, one should solve the following problem first: How can one trace the closed orbits on a large-scale by using numerical methods?

The closed orbit of the system (27) is a one-dimensional manifold in $\mathbb{R}^n$. Assume that s is the arclength on the closed orbit. Let

$$ds = \sqrt{dx^T dx} = \sqrt{F^T F} dt\,. \tag{55}$$

Taking into account the above equation, the system (27) is equivalent to

$$\frac{dx}{ds} = \tau(x, \mu) := \frac{1}{\sqrt{F^T F}} F(x, \mu)\,, \tag{56}$$

$$\frac{ds}{dt} = \sqrt{F^T F} \,. \tag{57}$$

In our computation, the equi-distance Hermite interpolation is used, which allows the transformation of the equation (6) into nonlinear equations that are similar to (7).

Assuming that $x = x(s)$ is a closed orbit in $\mathbb{R}^n$, and s is the arclength, $x^i(s)$ is a component of $x(s)$, for $i = 1, 2, \ldots, n$ and x^i_j are the interpolation points of $x^i(s)$, for $j = 1, 2, \ldots, m$. These m points divide the closed curve into m segments with same arclength $2l$. Let the arclength of the closed orbit be L; then we have

$$L = 2ml \,. \tag{58}$$

Let additionally

$$s = (1 + z)l + s_i \,, \tag{59}$$

where s_i is the arclength corresponding to x^i_j. Thus, the transformation (59) turns x^i_j and x^i_{j+1} into $z = -1$ and $z = 1$, respectively.

For $x^i \in [x^i_j, x^i_{j+1}]$ we have

$$x^i(z) = [z][G] \left\{ \begin{matrix} x^i_j \\ x^{i'}_j \\ x^i_{j+1} \\ x^{i'}_{j+1} \end{matrix} \right\} , \tag{60}$$

where $x^{i'}_j = \frac{dx^i_j}{dz}$ and the following matrices are defined:

$$[z] = [\,1 \quad z \quad z^2 \quad z^3\,] , \tag{61}$$

$$[G] = \frac{1}{4} \begin{bmatrix} 2 & 1 & 2 & 1 \\ -3 & -1 & 3 & -1 \\ 0 & -1 & 0 & 1 \\ 1 & 1 & -1 & 1 \end{bmatrix} . \tag{62}$$

The term $(d^2 x^i_j / ds^2)$ in $x^i \in [x^i_j, x^i_{j+1}]$ is calculated as

$$\begin{aligned} \frac{d^2 x^i_{j+1}}{ds^2} &= \frac{1}{l^2} \frac{d^2 x^i_{j+1}(z)}{dz^2} = \frac{1}{l^2} [0 \quad 0 \quad 2 \quad 6][G] \left\{ \begin{matrix} x^i_j \\ x^{i'}_j \\ x^i_{j+1} \\ x^{i'}_{j+1} \end{matrix} \right\} = \\ &= \frac{1}{l^2} \left[(x^{i'}_j + 2x^{i'}_{j+1}) + \frac{3}{2}(x^i_j - x^i_{j+1}) \right] . \end{aligned} \tag{63}$$

On the other hand $(d^2 x^i_{j+1} / ds^2)$ in $x^i \in [x^i_{j+1}, x^i_{j+2}]$ is calculated as

$$\frac{d^2x^i_{j+1}}{ds^2} = \frac{1}{l^2}\frac{d^2x^i_{j+1}(z)}{dz^2} = \frac{1}{l^2}[0 \quad 0 \quad 2 \quad -6][G]\begin{Bmatrix} x^i_{j+1} \\ x^{i'}_{j+1} \\ x^i_{j+2} \\ x^{i'}_{j+2} \end{Bmatrix} = \\ = \frac{1}{l^2}\left[-2(x^{i'}_{j+1} - x^{i'}_{j+2}) + \frac{3}{2}(x^i_{j+2} - x^i_{j+1})\right] . \tag{64}$$

Because $(d^2x^i_j/ds^2)$ should be continuous we obtain

$$x^{i'}_j + 4x^{i'}_{j+1} + x^{i'}_{j+2} = \frac{3}{2}(x^{i'}_{j+2} - x^{i'}_j)\,, \qquad j = 1, 2, \ldots, m\,, \tag{65}$$

where: $x^i_{m+1} = x^i_1$, $x^i_{m+2} = x^i_2$ and $x^{i'}_{m+1} = x^{i'}_1$, $x^{i'}_{m+2} = x^{i'}_2$. This leads to the equation

$$\begin{bmatrix} 1 & 4 & 1 & \cdots & \\ 1 & 1 & 4 & \cdots & \\ \vdots & & & \ddots & \\ 1 & & & 1 & 4 \\ 4 & 1 & & \cdots & 1 \end{bmatrix}\begin{Bmatrix} x^{i'}_1 \\ x^{i'}_2 \\ \vdots \\ x^{i'}_m \end{Bmatrix} = \frac{3}{2}\begin{bmatrix} -1 & 0 & 1 & \cdots & \\ 1 & -1 & 0 & \cdots & \\ \vdots & & & \ddots & \\ 1 & & & -1 & 0 \\ 0 & 1 & & \cdots & -1 \end{bmatrix}\begin{Bmatrix} x^i_1 \\ x^i_2 \\ \vdots \\ x^i_m \end{Bmatrix} . \tag{66}$$

Note that all the matrices in (66) are circulant. Equation (66) can be (see [19]) solved using the transformation

$$\begin{Bmatrix} x^{i'}_1 \\ x^{i'}_2 \\ \vdots \\ x^{i'}_m \end{Bmatrix} = [R]\begin{Bmatrix} x^i_1 \\ x^i_2 \\ \vdots \\ x^i_m \end{Bmatrix} , \tag{67}$$

where R is a real $m \times m$ matrix with the elements:

$$R_{ij} = -\frac{3}{2}\sum_{k=1}^{m}\frac{2}{c_k}\sin\frac{2\pi k(i-j)}{m}\sin\frac{2\pi k}{m}\,, \qquad i, j = 1, 2, \ldots, m\,, \tag{68}$$

where:

$$c_k = m(4 + 2\cos\frac{2\pi k}{m})\,, \qquad k = 1, 2, \ldots, m\,. \tag{69}$$

It is obvious that R is a skew-symmetric circulant matrix. If m is odd, let $r = (m+1)/2$. It is proved that

$$R_{11} = 0\,, \quad R_{1,r+1} = -R_{1,r}, \quad R_{1,r+2} = -R_{1,r-1}, \ldots, R_{1,m} = -R_{12}. \tag{70}$$

If m is even, we have for $r = m/2$:

$$R_{11} = 0\,, \quad R_{1,r+1} = 0\,, \quad R_{1,r+2} = -R_{1,r}\,, \\ R_{1,r+3} = -R_{1,r-1}\,, \ldots, R_{1,m} = -R_{12}\,. \tag{71}$$

To conclude, regardless of whether m is odd or even, only r elements of R need to be computed and stored.

Substituting (67) into (60), we have

$$x^i(z) = [z][G][T(j, j+1)]\{X^i\} , \tag{72}$$

where $\{X^i\} = (x_1^i, x_2^i, \ldots, x_m^i)^T$, and

$$[T(j, j+1)] = \left\{ \begin{array}{c} e_j \\ R_j \\ e_{j+1} \\ R_{j+1} \end{array} \right\} , \tag{73}$$

is a $4 \times m$ matrix, e_j is the unit $\mathbb{R}^m$-vector, whose jth element is 1 and the others are 0, and R_j is the jth row vector of R.

Thus on the closed curve $x(s)$ for $\forall x \in [x_j, x_{j+1}]$, the interpolation expression is as follows

$$x(z) = \ulcorner [z][G][T(j,j+1)], [z][G][T(j,j+1)], \ldots, [z][G][T(j,j+1)] \lrcorner \{X\} \tag{74}$$

where $\ulcorner \; \lrcorner$ represents the generalized diagonal matrix and

$$\begin{aligned} \{X\} &= (X^{1^T}, X^{2^T}, \ldots, X^{n^T})^T = \\ &= (x_1^1, \ldots, x_m^1, x_1^2, \ldots, x_m^2, x_1^n, \ldots, x_m^n)^T . \end{aligned} \tag{75}$$

Once the m interpolation points are known, the closed orbit is determined by (74).

From (59) we have

$$ds = l dz = \frac{L}{2m} dz . \tag{76}$$

Thus, by taking this interpolation, the system (27) is equivalent to the following one

$$\frac{dx}{dz} = \frac{L}{2m} \tau(x, \mu) , \tag{77}$$

$$L = \oint \sqrt{dx^T dx} = \sum_{j=1}^{m} \int_{x_j}^{x_{j+1}} \sqrt{dx^T dx} = \frac{1}{2m} \sum_{j=1}^{m} \int_{x_j}^{x_{j+1}} \sqrt{\left(\frac{dx}{dz}\right)^T} \tau dz , \tag{78}$$

and

$$\frac{dz}{dt} = \frac{2m}{L} \sqrt{F^T F} . \tag{79}$$

Substituting (74) into (77) and (78), we have for $x \in [x_j, x_{j+1}]$

$$\begin{aligned}&\ulcorner[0\quad 1\quad 2z\quad 3z^2][G][T(j,j+1)],\ldots,\\&\qquad[0\quad 1\quad 2z\quad 3z^2][G][T(j,j+1)]\lrcorner\{X\}=\\&=\frac{L}{2m}\{G^j(x_k^i,\mu)\}\,,\qquad i=1,2,\ldots,n;\ k=1,2,\ldots,m\,,\end{aligned} \tag{80}$$

where $\{G^j(x_k^i,\mu)\} = (G_1^j(x_k^i,\mu), G_2^j(x_k^i,\mu),\ldots,G_n^j(x_k^i,\mu))^T$ is obtained from the discretization of τ.

Equations (80) should, in particular, be accurately satisfied for $z=1$, i.e.,

$$[W(j,j+1)]\{X\} = \frac{L}{2m}\{G^j(x_k^i,\mu)\}\,, \tag{81}$$

where $[W(j,j+1)] = \ulcorner[0\quad 1\quad 2\quad 3][G][T(j,j+1)],\ldots,[0\quad 1\quad 2\quad 3][G]$ $[T(j,j+1)]\lrcorner$.

Similarly, by using this interpolation for $x\in[x_k,x_{k+1}]$, $k=1,2,\ldots,m$, one can obtain the following series of nonlinear equations:

$$\begin{Bmatrix} W(1,2)\\ W(2,3)\\ \vdots\\ W(m,1)\end{Bmatrix}\{X\} = \frac{L}{2m}\begin{Bmatrix} G^{1T}\\ G^{2T}\\ \vdots\\ G^{mT}\end{Bmatrix} \tag{82}$$

$$L = \frac{1}{2m}\sum_{j=1}^{m}\int_{x_j}^{x_{j+1}}\sqrt{\left(\frac{dx}{dz}\right)^T}\tau dz\,. \tag{83}$$

Equations (82) are of $(n\times m+1)$ degree and the Gauss integration is used to calculate L.

By using the equidistance Hermite interpolation, the system (27) is reduced to (82). Once equations (18) are solved, one can get the close orbit from (74). So the problem of following the closed orbits is equivalent to solving (82). The pseudo-arclength method that we have introduced in Section 2 is suggested to solve (82). The period of the closed orbit is computed by using numerical integration of the equation

$$\frac{dt}{dz} = \frac{L}{2m}\frac{1}{\sqrt{F^TF}}\,, \tag{84}$$

and the Jacobian of (81) can be formed from A.

For the computation of closed orbits of system (27), m must be selected appropriately to ensure the accuracy. In many cases, as the closed orbit is a one-dimensional manifold, selecting m as $10-20$ is enough, and m as $20-30$ is suggested for some complicated cases. Comparing with other numerical schemes, this scheme does not require calculation of the third derivatives of F. Furthermore, it has lower computational complexity and higher computational efficiency. This numerical method is suitable for the computation of

Hopf bifurcations in nonlinear dynamical systems, especially for a system in high dimensional space.

In order to solve equations (81) and (82) using the pseudo-arclength method, an original point (original closed orbit) is needed.

Let's assume that μ is a Hopf bifurcation value, and $x^*(\mu)$ is the stationary point of (78), and perform the transformation of variables

$$x = x^*(\mu_c) + Py \,, \tag{85}$$

where P is a real $n \times n$ matrix whose first column is Reu, second column is Imu, and last $n-2$ columns are any set of real, linearly independent vectors r_j, $(3 \le j \le n)$. This means that

$$P = (\mathrm{Re}u, \mathrm{Im}u, r_3, \ldots, r_n) \,, \tag{86}$$

where u is the eigenvector corresponding to the eigenvalues $i\omega_0$ gained from (48). r_j should be selected such that under the transformation (85), the system (27) will have the following structure:

$$\dot{y} = \begin{bmatrix} 0 & -\omega_0 & 0 \\ \omega_0 & 0 & 0 \\ 0 & 0 & 0 \end{bmatrix} + G(y;\mu) \,, \tag{87}$$

where $G(y;\mu) = o(|(y,\mu)|^2)$, D is a real $(n-2) \times (n-2)$ matrix, and all of the eigenvalues of D have strictly negative parts.

It is obvious that the spectral conditions of the center manifold theorem are satisfied for the system (87). Therefore at $\mu = \mu_c + \varepsilon$ (where $|\varepsilon| \ne 0$ is small) the closed orbit of the system (87) is approximated by a circle with radius b

$$\begin{cases} y_1 = b\cos\omega_0 t \,, \\ y_2 = b\sin\omega_0 t \,, \\ y_3 = 0 \,, \\ \vdots \\ y_n = 0 \,. \end{cases} \tag{88}$$

Taking $t = 2\pi k/m\omega_0$, for $k = 1, 2, \ldots, m$ and using the same Hermite interpolation for (88), the coordinate values of the m interpolation points are obtained from (85), which are expressed by a series of functions of b. Substituting the results into (82) and (83); we get equations with the unknown value b, which are solved using the least square method. While $|\mu - \mu_c| = |\varepsilon|$ is small, b and then the coordinate values of the m interpolation points $\{X_0\}$ are obtained; we get $L_0 = 2\pi b$. The Newtonian iterated method is used with the original point $\{X_0\}$ and $L_0 = 2\pi b$ to obtain a more accurate original point $\{X_0^*\}$ and L_0^*, if necessary.

Note that in (88) $y_i = 0$ for $3 \le i \le n$. This implies that in the matrix P of (86) r_j must not be computed at all.

For each step in the predictor-corrector process, one has to compute the pair of complex conjugate eigenvalues with maximum norm $\lambda_{\max}$. After obtaining the Hopf bifurcation point μ_c, $\lambda_{\max}$ on the both sides of $x^*(\mu_c)$ is computed. If $|\lambda_{\max}| < 1$, then the stationary point is asymptotically stable but is unstable if $|\lambda_{\max}| > 1$. The stability of the closed orbit associated with the characteristics of the original closed orbit can be determined intuitively. For example, y_1, and y_2 are two stationary points, where $y_1 = (x_1, \mu_1)$, $y_2 = (x_2, \mu_2)$, and $\mu_1 < \mu_c$, $\mu_2 > \mu_c$. If $|\lambda_{\max}(y_1)| < 1$, $|\lambda_{\max}(y_2)| > 1$ and the original closed orbit occurs at $\varepsilon < 0$ (> 0), i.e., $\mu - \mu_c < 0$ (> 0) then the closed orbit is orbitally asymptotically stable (unstable), and the bifurcation is supercritical (subcritical), and so on.

In the procedure of tracking the closed orbits continuously, the following problems arise. How to determine the singular points which emerge at the closed orbit? How to judge that a new bifurcation occurs, by which the closed orbits become tori or other sorts of orbits? The approximate period T of the closed orbit is obtained from (84), and if T is large enough, a homoclinic orbit or an inhomoclinic orbit might occur.

4. Examples

According to the new methods introduced above for the computation of static bifurcation, Hopf bifurcation, and tracking the closed orbits on a large scale, several examples, such as Van der Pol's equation and Langford's system are now calculated numerically in order to prove the efficiency of the method.

Example 1. The Lorenz system.

The system of equations

$$\begin{aligned} \dot{x} &= -\sigma x + \sigma y \, , \\ \dot{y} &= -xz + rx - y \, , \\ \dot{z} &= xy - bz \, , \end{aligned} \tag{89}$$

was studied by Lorenz [20] as a model for fluid dynamic turbulence. Here σ, r and b are three positive parameters.

The original derivation of the Lorenz equations can be described as follows. A two-dimensional fluid cell is warmed from below and cooled from above, and the resulting convective motion is modelled by a partial differential equation. Roughly speaking, the variable x measures the rate of convective circulation, the variable y measures the horizontal temperature variation, and the variable z measures the vertical temperature variation. The three parameters σ, r and b are proportional to the Prandtl number, the Rayleigh number, and some physical proportions of the region under consideration, respectively.

Let $\sigma = 10$, $b = 8/3$, and r be the bifurcation parameter. The stationary solutions are as follows

$$
\begin{aligned}
x_*^0 &= (0,0,0)^T\,, \qquad \text{for all } r\,, \\
x_*^1 &= \left(\sqrt{b(r-1)}, \sqrt{b(r-1)}, r-1\right)^T \quad \text{and} \\
x_*^2 &= \left(-\sqrt{b(r-1)}, -\sqrt{b(r-1)}, r-1\right)^T \qquad \text{for } r > 1\,.
\end{aligned}
\tag{90}
$$

The system linearized about x_*^0 has the coefficient matrix

$$
\begin{bmatrix} -\sigma & \sigma & 0 \\ r & -1 & 0 \\ 0 & 0 & -b \end{bmatrix}, \tag{91}
$$

whose eigenvalues are

$$
\lambda_{1,2}^0 = \frac{1}{2}\left\{-(\sigma+1) \pm ((\sigma-1)^2 + 4\sigma r)^{1/2}\right\}, \qquad \lambda_3^0 = -b\,. \tag{92}
$$

Thus, for $0 < r < 1$, $\lambda_{1,2}^0$ and λ_3^0 are real and negative. For $r > 1$, λ_2^0 and λ_3^0 are negative, while λ_1^0 is positive. Thus x_*^0 is asymptotically stable for $0 < r < 1$ and unstable for $r > 1$.

It is proved that all three eigenvalues of the linearized coefficient matrix about x_*^1 and x_*^2 have negative real parts for $1 \le r \le 24.737$ ($\sigma = 10, b = 8/3$), and have a negative real value and a complex conjugate pair of eigenvalues with positive real parts for $r > 24.737$.

As r is increased passing through $r = 1$, the stability lost by x_*^0 is gained by x_*^1 and x_*^2 in a stationary bifurcation. And as r is increased passing through $r = 24.737$ the stationary points x_*^1 and x_*^2 lose stability due to a complex conjugate pair of eigenvalues which cross the imaginary axis with non-zero speed (the other eigenvalue is real and negative). It results in a Hopf bifurcation from stationary to periodic solutions. In reference [13] it is shown that the Hopf bifurcation is subcritical and the closed orbits are unstable. At $r = 13.926$ a homoclinic orbit occurs.

The program BF is used for numerical computation of equation (89), and the following bifurcation points are obtained: $r = 1.0$ corresponds to a static bifurcation point, and $r = 24.737$ corresponds to a Hopf bifurcation point. The characters of the closed orbits bifurcating from x_*^1 are shown in Fig. 2 (m is selected as $m = 16$).

For $r = 13.938$, the period of the closed orbit is calculated as $T = 1.13 \times 10^5$, and one can expect the occurrence of a homoclinic orbit. For a more accurate computation, we take $m = 32$, which leads for $r = 13.930$ to a period of $T = 1.09 \times 10^5$.

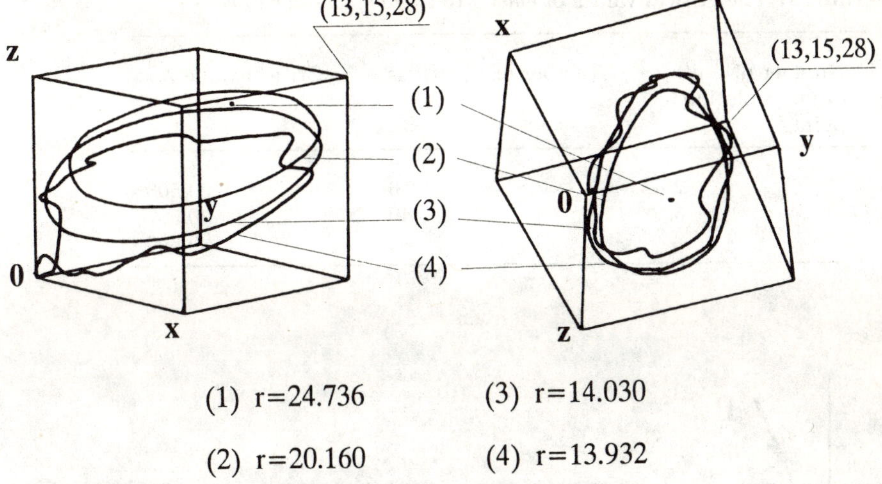

Fig. 2. Characters of closed orbits for the Lorenz equations

Example 2. Elastic rod under a follower force.

Let us consider an elastic rod under the follower force P. We assume that the rod has length l, and is fixed at one end, while free at the other end (see Fig. 3).

We have the governing equation:

$$\rho \ddot{w} + EI \frac{d^4 w}{ds^4} + P \frac{d^2 w}{ds^2} = 0 \,, \tag{93}$$

where ρ denotes the density of the rod, EI is stiffness, and s is the arc length.

The boundary conditions are as follows:

$$s = 0 \,; \qquad w = 0 \,, \qquad \frac{dw}{ds} = 0 \,, \tag{94}$$

$$s = l \,; \qquad \frac{d^2 w}{ds^2} = 0 \,, \qquad \frac{d^3 w}{ds^3} = 0 \,, \tag{95}$$

An elastic rod under follower force is a classic example in elasticity. It is well known that there is a Hopf bifurcation point at the original solutions $w \equiv 0$, and the critical force equals $P_{\text{cr}} = 2\pi^2 (EI/l^2)$.

Hermite interpolation is used to discretize equations (93–95), and the critical value of the bifurcation parameter $\lambda = Pl^2/EI$ is computed. The results are listed in Table 1.

Table 1. The critical values of elastic rod under follower force

Number of units of interpolation points	Number of equations	Critical value λ_{cr}
2	16	20.508
5	40	20.507
8	64	20.506

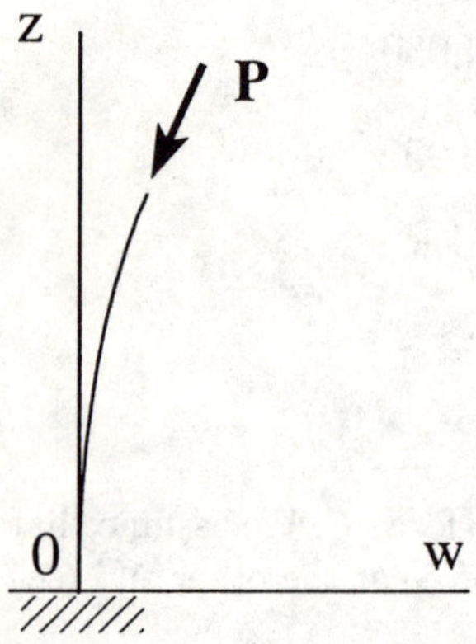

Fig. 3. Elastic rod under the follower force

Acknowledgements. This work is supported by the National Basic Research project "Nonlinear Science" and the National foundation of Natural Science.

References

1. H.J. Wacker(ed.): Continuation methods. Academic Press, New York 1978
2. E. Wasserstrom: Numerical solutions by the continuation method. SIAM Rev. **15**, 89–119 (1973)
3. R. Menzel: Numerical determination of multiple bifurcation points. In [1], 310–318
4. G. Moore, A. Spence: The calculation of turning points of nonlinear equations. SIAM J. Numer. Anal. **17**, 567–576 (1980)
5. H.B. Keller: Numerical solution of bifurcation and nonlinear eigenvalue problems. In applications of bifurcation theory. Edited by P.H. Rabinowitz, Academic Press, Inc., 1977
6. Ji Hai-po, Wu Ji-ke, Hu Hai-chang: Geometric description and computational methods for bifurcation. Science in China(A) **9**, 597–607 (1990)
7. Su Xian-yue: The nonlinear analysis of rotationary shells. Ph. D. Thesis, Peking University (1985)

8. Su Xian-yue, Wang Ying-jian, Wu Ji-ke, Hu Hai-chang: A numerical method for the nonlinear systems with parameters. Computational Structural Mechanics and Applications **3**, 1–9 (1988)
9. Wu Ji-ke, Teng Ning-jun and Yuang Yong: Bifurcation problems and their numerical methods. Mechanics and Practice **4**, 1–7 (1987)
10. A.A. Andronov and A. Witt: Sur la theórie mathematiques des autooscillations. C.R.A-cad. Sci. Paris **190**, 256–258 (1930)
11. B.D. Hassard, N.D. Kazarinoff and Y. H. Wan: Theory and applications of Hopf bifurcation. Cambridge University Press 1981
12. E. Hopf: Abzweigung einer periodischen Lösung von einer stationären Lösung eines differential Systems. Ber. Math-Phys. Sachsische Adademie der Wissenschaften Leipzig **94**, 1–22 (1942)
13. H. Poincaré: Les méthodes nouvelles de la mécanique céleste. Vol.I Paris (1892)
14. A. Kelley: The stable, center-stable, center, center-unstable and unstable manifolds. J. Diff. Eqns. **3**, 546–570 (1967)
15. I.E. Marsden and M. McCracken: The Hopf bifurcation and its applications. Springer-Verlag, New York 1976
16. Wu Ji-ke and Zhou Kun: Numerical computation for high dimension Hopf bifurcation. Acta Scientiarum Naturalium Universitatis Pekinesis (to appear)
17. W. Hahn: Theory and application of Lyapunov's direct method. Prentice-Hall 1963
18. S.N. Chow and J.K. Hale: Methods of bifurcation theory. Springer-Verlag, New York 1982
19. Wu Ji-ke and Shao Xiu-ming: The circulant matrix and its applications in the computation of structures. Mathematicae Numericae Sinica **2**, 144–153 (1979)
20. C. Sparrow: The Lorenz equations: bifurcations, chaos, and strange attractors. Springer-Verlag, New York 1982

Catastrophe Theory and the Vibro-Impact Dynamics of Autonomous Oscillators

G.S. Whiston

National Power PLC, Bilton Centre,
Cleeve Road, Leatherhead, Surrey,
KT22 7SE United Kingdom

Abstract

Non-differentiability in vibro-impact dynamics can lead to breakdown of the global stable manifold theorem, the shredding of stable manifolds and to the generation of homoclinic tangles by the translation of segments of stable and unstable manifolds across each other. This article discusses the non-differentiability of autonomous vibro-impact systems using the methods of catastrophe theory. The singularities are characterised by a hypersurface unfolding the classic singularities such as the cusp and the swallowtail.

1. Introduction

Vibro-impact response, possibly allied with sliding, is a potent damage producing mechanism for many types of industrial plant. For example, many types of nuclear reactors are particularly prone to heat exchanger tube fretting wear through vibro-impact/slide induced by fluid-structural interaction. Vibro-impact can also be a nuisance in many other industrial areas though perhaps in less spectacular ways.

The fluid-structure interactions in heat exchangers are only partially understood, but the dynamics can be divided into two areas: *forced* or non-autonomous motion and *instabilities* or autonomous motion. For example, structural response to turbulent or two-phase flows or to vortex shedding typify 'forced' response. Whirling instabilities due to axial flow in annular gaps or to cross flows over tube bundles typify 'autonomous' response. However, it is arguable that all the above mechanisms derive from an autonomous coupled fluid-structure system.

Clearly, it is highly desirable to obtain a theoretical understanding of the dynamics of vibro-impact. However, even with a well defined excitation mechanism, it is impossible to obtain a global quantitative predictive theory due to the non-linearity of the problem. One can only ever obtain a qualitative global understanding with local quantitative knowledge. The idea of only local quantitative knowledge can be illustrated by consideration of the case of the vibro-impact response of a 1-dimensional linear oscillator to harmonic excitation [1–3]. In this case, one can calculate the stable, period 1, single impact, sub-harmonic motions, but not domains of attraction or multi-impact periodic or chaotic responses.

There are two routes available to obtain a 'global' qualitative understanding of vibro-impact dynamics: performing many computer simulations or by mathematical analysis. The first route can only provide isolated examples of dynamics for specific initial conditions and parameter values. The second route, if feasible, can provide much more complete information pertinent to whole classes of systems. Of course, one is limited to analysis of the simplest models in the latter area. However, the simplicity of the models does not lead to simplicity of the dynamics and many of the phenomena expected to occur in the high dimensional complicated systems will occur in relatively simple low dimensional systems.

This article is dedicated to an expository analysis of the generic discontinuity of the vibro-impact dynamics of autonomous systems. Such discontinuities arise from the lack of 'structural stability' of zero velocity or 'grazing' impacts. That is, there may be impact trajectories arbitrarily close to a grazing trajectory which do not lead to impacts at times close to the graze. This leads to discontinuity of the impact successor map and will occur in any vibro-impact system. These discontinuities can be thought of as leading to a 'loss of predictability' where arbitrarily close initial conditions diverge in discrete steps across a discontinuity.

Much progress has been made in understanding the nature of the discontinuities for forced systems [4], particularly the harmonically forced, undamped, one dimensional linear oscillator. For example, one of the topics investigated in the latter paper was the shredding of stable manifolds which could lead to transversal intersection by translation of filaments of stable and unstable manifolds and thus to the generation of homoclinic tangles and Smale horseshoes. A great deal of the analysis is applicable to non-linear, arbitrarily periodically forced, one dimensional linear oscillators. The analysis reported above made use of singularity (or catastrophe) theory [5–7].

In the present article, the methods developed in [4] are extended to describe the singularity structure of linear autonomous systems. Much of the work is applicable to systems of arbitrary dimension but it is illustrated by reference to a specific 4-dimensional example of two linear oscillators coupled by a stiffness term. The oscillators are assumed identical except that one has positive damping and the other has negative damping.

2. Generalities on Vibro-Impact Dynamics

The vibro-impact dynamical system is derived from the free response of an n-dimensional non-linear oscillator with an arbitrary periodic forcing:

$$\frac{d^2Y}{dt^2} + f\left(Y, \frac{dY}{dt}\right) = g(z\tau) , \tag{1}$$

where g is assumed to have period $2\pi/z$. In later sections, it will be assumed that g is identically zero. The above second order equation on $\mathbb{R}^n$ is transformed into a first order equation on $\mathbb{R}^{2n+1}$ in the usual way:

$$\frac{dX}{d\tau} = G(X) \quad \text{for} \quad X : \tau \to \left(Y(t), \frac{dY}{d\tau}(\tau), \tau\right) , \tag{2}$$

where

$$G : \ ((q,p),\tau) \to (p, -f(q,p) + g(\tau), 1) .$$

The phase space $\mathbb{R}^{2n+1}$ has the usual (p, q) coordinates:

$$\begin{aligned} &x_i \equiv q_i \text{ (displacement) for } 1 \le i \le n , \\ &x_i \equiv p_{i-n} \text{ (velocity) for } n < i \le 2n , \\ &x_{2n+1} \equiv -\tau \text{ (time)} . \end{aligned} \tag{3}$$

It will be supposed that a single amplitude constraint $q_1 \le c$ is imposed and it is therefore useful to define the following subspaces of $\mathbb{R}^{2n+1}$:

$$\begin{aligned} L_c &\equiv \{ x \in \mathbb{R}^{2n+1} \ : q_1(x) \le c\} , \\ E_c &\equiv \{ x \in L_c \ : q_1(x) = c\} , \\ E_c^\pm &\equiv \{ x \in E_c \ : p_1(x) \ge 0(+) \text{ or } p_1(x) \le O(-)\} , \\ E_c^0 &\equiv \{ x \in E_c \ : p_1(x) = 0\} . \end{aligned} \tag{4}$$

Let F be the flow map of the vector field G:

$$\begin{aligned} &F : \mathbb{R}^{2n+1} \times \mathbb{R} \to \mathbb{R}^{2n+1} , \\ &F : (x, \tau) \to U_\tau(x) , \end{aligned} \tag{5}$$

where U_τ is the local diffeomorphism of $\mathbb{R}^{2n+1}$ onto itself which transforms an initial condition x into its time evolution at time τ along a trajectory.

An initial condition x in L_c is said to lead to an impact if its trajectory first crosses $E_c^\pm$ from L_c. If $y = F(x, \tau_1) \in E_c^\pm$ is the first crossing point, we write $y = P_2(x)$ and say that x leads to an impact y at time τ_1. Clearly, any point of the trajectory from x to y leads to the same impact and y is an impact at time $\tau_0(y)$. Define further subsets:

$$\begin{aligned} I_c &\equiv \{\, x \in L_c \ : \ x \text{ leads to an impact} \,\}\,, \\ N_c &\equiv L_c \setminus I_c \qquad \text{and} \qquad N_c^- \equiv E_c^- \cap N_c\,. \end{aligned} \tag{6}$$

The subset N_c comprises those points of L_c not leading to impact. Clearly, I_c is the part of the backwards flow of E_c^+ in L_c. Any point of E_c^+ with $p_1 \neq 0$ is an impact and, in general, the impact event is represented by a map:

$$P_1 : E_c^+ \setminus E_c^0 \to E_c^- \setminus E_c^0\,. \tag{7}$$

The definition of P_1 depends on the physical model of the impact process. One way of doing this is to model the impact restitution as due to a (possibly non-linear) elastic spring in the q_1 direction. In this case, P_1 is defined via the flow of an allied dynamical system obtained by appropriately modifying the vector field G in the q_1 direction. The simplest model (which will be assumed from now on) is the instantaneous inelastic rebound with coefficient of restitution r:

$$\begin{aligned} q_i(P_1(x)) &= q_i(x) && \text{for } 1 \leq i \leq n\,, \\ p_i(P_1(x)) &= -r \cdot p_1(x) && \text{where } 0 < r < 1\,, \\ p_i(P_1(x)) &= p_i(x) && \text{for } 1 < i \leq n\,. \end{aligned} \tag{8}$$

Having defined P_1, one can try to define a global impact successor map. If x is a point of E_c^+ not in E_c^0, $P_1(x)$ is an initial condition for motion into the interior of L_c. If $P_1(x)$ lies in I_c, the free trajectory from $P_1(x)$ leads to another impact, written $P(x) \equiv P_2(P_1(x))$. We can thus try to define a global map mapping an impact into its successor impact. To this end, define:

$$\begin{aligned} &D_c \equiv E_c^- \setminus N_c^-\,, \qquad D_c^+ \equiv E_c^+ \setminus P_1^{-1}(N_c^-)\,, \\ &\text{and } P \equiv P_2 \circ P_1 : D_c^+ \to E_c^-\,. \end{aligned} \tag{9}$$

Notice that there is no guarantee that $P(x) \in D_c^+$, that is, $P(x)$ leads to another impact. In general, infinite trains of impacts need not occur—impacting can 'stop' if $P(x)$ lies in N_c^-.

In general, N_c^- consists of points whose trajectories either immediately enter and then remain in $q_1 < c$, remain in $q_1 = c$ or immediately enter $q_1 > c$. The latter two types of point must lie in E_c^0, that is, represent 'grazes' on $q_1 = c$. The singularity structure of vibro-impact dynamics depends entirely on such points.

By definition, a point x of E_c^0 is an extremum of the displacement-time trajectory $q_1(x, \tau)$ from x. That is:

$$q_1(x, \tau_0(x)) = c \quad \text{and} \quad \frac{dq_1}{d\tau}(x, \tau_0(x)) = 0\,. \tag{10}$$

In order to decide upon the fate of a degenerate impact, it is necessary to decide on the nature of the extremum or singularity of q_1 as determined by its higher derivatives. We regard $q_1(x, \tau)$ as a family of functions parameterised by initial conditions x.

Thus let $f : \mathbb{R} \to \mathbb{R}$ be a differentiable function on a neighbourhood of a point t_0 and let c be a real number. Let f_c be the function $t \mapsto f(t) - c$. f_c is said to have an $A_k(\pm)$ singularity at t_0 if:

$$f_c^j(t_0) = 0 \quad \text{for} \quad 0 \leq j < k \quad \text{but} \quad f_c^k(t_0) > 0 \text{ or } f_c^k(t_0) < 0 \,, \tag{11}$$

where f_c^j denotes $d^j f_c/dt^j$. The interpretation of these singularities goes as follows.

An $A_0(\pm)$ singularity has $f(t_0) = c$ and $f(t_0) > 0$ or < 0. In the present case, the sets $A(0, \pm)$ of $A_0(\pm)$ singularities of q_1^c are $E_c^\pm \setminus E_c^0$. An $A_1(\pm)$ singularity of f_c is a minimum at level c ($A_1(+)$) or a maximum at level c ($A, (-)$). Similarly, an $A_2(\pm)$ singularity of f_c is a level c ordinary inflection point which is either a crossing from $y = f(t) < c$ to $y > c$ (an $A_2(+)$) or a crossing from $y > c$ to $y < c$ (an $A_2(-)$). Singularities $A_k(\pm)$ for $k > 2$ are called higher inflection points. It is easy to show that at an $A_k(\pm)$ singularity, f_c is locally equivalent to the function $c \pm (t - t_0)^{k+1}$. In this way, we can label the $A_{2k+1}(\pm)$ singularities as generalised minima or maxima since f_c is locally concave or convex at t_0.

Clearly, $A_{2k+1}(+)$ and $A_{2k}(-)$ singularities of q_1^c cannot represent impacts because $q_c^c(x, t)$ crosses E_c from $q_1 < c$ and in the former case, immediately re-enters $q_1 > c$. Singularities of type $A_{2k+1}(-)$, generalised maxima, represent grazing trajectories.

Of most interest are the inflections $A_{2k}(+)$ which represent ‘trapping’ behaviour. At such a singularity, a trajectory crosses $q_1 = c$ from L_c with zero velocity (and thus represents an impact) and immediately crosses into $q_1 > c$. However, vibro-impact trajectories are not allowed to enter $q_1 > c$ and the vibro-impact convention is to confine the dynamics to E_c until the out of plane acceleration $a(\tau)$ defined by:

$$a(\tau) = -f\left(Y^*(x, \tau), \frac{dY^*}{d\tau}(x, \tau)\right) + g(z\tau) \,, \tag{12}$$

(where Y^* is the trajectory restricted to the plane $q_1 = c$ and $q_1 = 0$) changes sign from positive to negative, representing an $A_{2k}(-)$ singularity of q_1^c. The function q_1^* is called the ‘lift-off’ function. Note that lift-off need never occur. Suppose that $A(k, \pm)$ denotes the set of $A_k(\pm)$ singularities on E_c^0 and:

$$\begin{aligned} A^+ &\equiv \bigcup_{k\geq 0} (A(2k+1, +) \cup A(2k, +)) \,, \\ A^- &\equiv \bigcup_{k\geq 0} (A(2k+1, -) \cup A(2k, -)) \,. \end{aligned} \tag{13}$$

Then $A^+ \subset N_c^-$ because if $x \in A^+$, $q_1(x, \tau)$ does not first cross E_c^+ from $q_1 < c$. The points of A^- represent 'allowable' impacts, that is, zero velocity impacts from $q_1 < c$.

Let S_c denote the set of points of E_c which lead to zero velocity impacts, $S_c \equiv P_2^{-1}(A^-)$. Then the map P_2 is differentiable on $D_c^- \setminus S_c$. This follows from the implicit function theorem. For, if x is a point of $D_c^- \setminus S_c$, the impact time $\tau_0(P_2(x))$ is the first solution, τ_1, of $q_1(x,\tau) = c$ and $q_1 \neq 0$. The implicit function theorem therefore yields a unique solution $\tau' = w(x')$ to $q_1(x',\tau') = c$ for x' in a neighbourhood U of x. It follows that if F is the flow of the free dynamical system, P_2 coincides with the map $x' \mapsto F(x', w(x'))$ on $U \cap D_c^-$ and is therefore differentiable. The above argument breaks down on S_c because its points are by definition, non-regular for q_1. Indeed, P_2 is in general, *discontinuous* on S_c, for if x lies in S_c, there may be points x' arbitrary close to x such that $q_1(x',\tau)$ need not cross $q_1 = c$ within a prescribed time interval about $\tau_0(P_2(x))$. That is, if $D^{2n-1} \times D_b^1$ is a rectangular neighbourhood of $P_2(x)$ with width $2b$ about $\tau_0(P_2)$ and $P_2(x)$ is convex (an $A_{2k+1}(-)$ singularity), $q_1(x',\tau)$ need not cross $q_1 = c$ within the time interval D_b^1 whist remaining in $q_1 < c$ for times $\tau_0(x') < \tau < \tau_0(x)$. Equivalently, it may be that there exist neighbourhoods U of $P_2(x)$ in E_c such that for all neighbourhoods V of x, $P_2(V)$ is not a subset of U. This has to be proved by analysis of the 'unfolding' of the singularity by the family of nearby displacement-time trajectories $q_1(x',\tau)$ for x' close to x in E_c^-. This question will be further investigated in section (5).

We now return to the construction of the impact successor map P. Recall that the orbits of points of E_c^+ can terminate, in which case it is not possible to define a *global* discrete dynamical system. However, in many cases of interest, there will always be infinite trains of impacts. The condition for this is that $N_c^- = A^+$ when orbits cannot be trapped forever in E_c^0 or in $L_c \setminus E_c$. For example, this is true for the vibro-impact response of an undamped 1-dimensional linear oscillator under any periodic excitation. The condition is also met by linear systems where the oscillator transient response has at least one eigenvalue with positive real part—the origin of the unforced system is a saddle or an unstable vertex—whose eigen-direction crosses E_c^-.

We now turn to consider the types of singularities that can occur in autonomous systems, regarding these as the case $g \equiv 0$. In this case, we can replace all the subspaces of $\mathbb{R}^{2n+1}$ defined above by subspaces of $\mathbb{R}^{2n}$. It is useful to consider the family of all displacement functions $Y_i(x,\tau) = q_i(F(x,\tau))$. At an $A_{\geq 1}(\pm)$ singularity of $Y_c(x,\tau) = Y(x,\tau) - c$, we must have $Y_c(x,\tau_1) = 0$, $\frac{dY_c}{d\tau}(x,\tau_1) = 0$ and:

$$\frac{d^2Y_c}{d\tau^2}(x,\tau_1) = -f\left(F(x,\tau_1), \frac{\partial F}{\partial \tau}(x,\tau_1)\right). \tag{14}$$

The higher derivatives $d^kY_c/d\tau^k(x,\tau_1)$ for $k > 2$ are related to the lower derivatives. For example, $d^3Y_c/d\tau^3(x,\tau_1)$ is given by:

$$\sum_i \left(\frac{\partial f_1}{\partial q_i}\left(F(x,\tau_1), \frac{\partial F}{\partial \tau}(x,\tau_1)\right)\frac{dY_i}{d\tau} + \right.$$
$$\left. + \frac{\partial f_1}{\partial p_i}\left(F(x,\tau_1), \frac{\partial F}{\partial \tau}(x,\tau_1)\right)\frac{d^2Y_i}{d\tau^2}\right) . \tag{15}$$

Consider the case of a 1-dimensional non-linear autonomous system. In this case, the latter expression implies that at an $A_2(\pm)$ singularity: $Y_c = 0$, $dY_c/d\tau = 0$ and $d^2Y_c/d\tau^2 = 0$, $d^3Y_c/d\tau^3$ also vanishes. Indeed, $d^kY_c/d\tau^k = 0$ for all $k \geq 0$. Such a singularity is called 'flat'. It is easy to see that the singularity must be a fixed point of the free system (which is defined by $dY_c/d\tau = 0$ and $d^2Y_c/d\tau^2 = 0$) at $(0, c)$. Note that the fixed points of oscillator systems always occur on $v_0 = 0$. In the trapping interpretation, $A_2(\pm)$ singularities lead to permanent trapping!

Next consider the $A_1(\pm)$ singularities at level c, where the acceleration is given by:

$$\frac{d^2Y_c}{d\tau^2}(x, \tau_0(x)) = -f_1(c, 0) \equiv m(c) . \tag{16}$$

Clearly, these singularities are maxima if $m(c) > 0$ or minima if $m(c) < 0$, the zeros, or $A_{>1}(\pm)$ singularities corresponding to the fixed points. The nature of the singularity is a function of the clearance. Consider the case of the oscillator with a non-linear spring defined by:

$$f_1(y, y') = y(a^2 - y^2) , \qquad m(c) = c(a^2 - c^2) . \tag{17}$$

In this case, $m(c) = 0$ at $c = 0$ and at $c = \pm a$, which are fixed points. For $0 < c < a$, $m(c) > 0$ and the point $(0, c)$ is a maximum, whilst for $c > a$, $m(c) < 0$ and $(0, c)$ is a minimum. The change in nature of $(0, c)$ as c varies represents a global bifurcation in the nature of vibro-impacting. The unconstrained system is Hamiltonian and the free integral curves correspond with the level surfaces of the Hamiltonian $H(p, q) = \frac{p^2}{2} + q^2 \frac{(a^2 - \frac{q^2}{2})}{2}$. These are sketched in the phase portrait shown in Fig. 1.

Consider the two clearance levels d and e where $0 < d < a < e$. At clearance d, the homoclinic orbit from $(0, a)$ to $(0, -a)$ crosses $q = c$ at a point $(c, -v_*)$ of E_c and all points (c, p) of E_c^- with $p < -v^*$ do not lead to further impacts. N_c^- coincides with this set. The point $(d, 0)$ is a maximum and corresponds to a fixed point of the impact successor map. Indeed, the Poincaré map of the system is defined as follows: $N_d^- = \{d\} \times [-v_*, -\infty[$, $P_1^{-1}(N_d^-) = \{d\} \times [v_*/r, \infty[$ so that $D_d^- = \{d1\} \times [0, -v_*[$ and $D_d^+ = \{d\} \times [0, v_*/r[$. In this case, the map $P_2 : D_d^- \to D_d^+$ is given by $(d, -p) \mapsto (d, p)$ by energy conservation and it follows that the Poincaré map is given by $P : (d, p) \mapsto (d, rp)$. Thus for inelastic rebound, $0 < r < 1$, the only fixed point (or periodic vibro-impact response) of P is the graze $(d, 0)$.

At clearance e, the singularity situation is different in that $(e, 0)$ is a maximum and $A(2k+1, +) = \{(e, 0)\}$. The stable manifold of the saddle

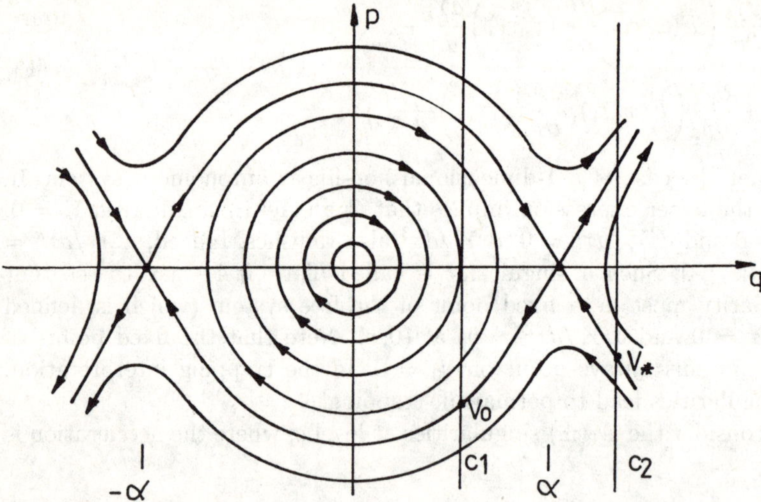

Fig. 1. A global bifurcation in the nature of the 1-dimensional non-linear autonomous system

point $(0, \alpha)$ crosses E_c^- at velocity $-w_*$ and N_c^- consists of the union of $(e, 0)$ and $\{e\} \times [-w_* - \infty[$. The Poincaré map now has domain $D_d^+ = \{d\} \times]0, w_*/r[$ and is once again given by $P : (e, p) \mapsto (e, rp)$, but this time it has no fixed points.

At clearance d, the subset A^- is empty, but for $0 < d < \alpha$, A^- comprises $A(1, -) = \{(d, 0)\}$. In the latter case, the singularity subspace $S_c = P_2^{-1}\{(d, 0)\} = \{(d, 0)\}$, because $(d, 0)$ is a fixed point of P. The question now arises of the continuity of P on S_c. To answer this, we consider the general 1-dimensional case.

Clearly, for any 1-dimensional oscillator, S_c is either empty or it is a singleton set, $\{(c, v_*/r)\}$ where the trajectory $Y_1((c, -v_*), \tau)$ passes through $(c, 0)$ at some time τ_1. To establish the continuity of P at $(c, v_*/r)$, it is sufficient to establish the continuity of P_2 at $(c, -v_*)$. It is necessary to consider the trajectories of points $(c, -v)$ for v close to v_* in E_c^- and for times τ close to τ_1. For $x' \equiv (c, -v)$ and v close to v_*, one can approximate the trajectory $Y_1(x', \tau)$ by:

$$Y_1(x', \tau) = Y_1(x, \tau) + (v - v_*)\frac{\partial Y_1}{\partial p}(x, \tau), \tag{18}$$

where $x \equiv (c, v^*)$. By definition, $Y_1, (x, \tau)$ crosses $q_1 = c$ at time τ_1 with zero velocity and negative acceleration. In this case, we can use Hademard's lemma to replace $Y_1(x, \tau)$ for τ close to τ_1 by $(c - (\tau - \tau_1)^2 \cdot h(\tau_1)$ where h is positive. Therefore:

$$Y_1(x', \tau) \sim c - (\tau - \tau_1)^2 + (v - v_*)\frac{\partial Y_1}{\partial p}(x, \tau_1). \tag{19}$$

The behaviour of $Y_1(x', \tau) - c$ for x' close to x and τ close to τ_1 is therefore governed by the polynomial:

$$\phi(\tau) \equiv (\tau - \tau_1)^2 - d_0 \text{ for } d_0 \equiv \frac{\frac{\partial Y_1}{\partial p}(x, \tau_1)(v - v_*)}{h(\tau_1)} . \tag{20}$$

If d_0 is positive, ϕ always has a real root, but if d_0 is negative, it has no real roots. In the former case, the initial condition x' leads to a crossing of E_c whilst in the latter case, it does not. It is easy to establish that $(\partial Y_1/\partial p)(x, \tau_1) > 0$. For, if $v < v_*$ the trajectory $Y((c, -v), \tau)$ starts above $Y(x, \tau)$ and because it cannot cross $Y(x, \tau)$, it must cross $p_1 = 0$ at $q_1 < c$, i.e.: ϕ does not have real roots so that $d_0 < 0$. If $v > v_*$, $Y((c, -v), \tau)$ starts below $Y(x, \tau)$ and must therefore cross $p_1 = 0$ for $q_1 > c$ and therefore cross E_c^+.

It follows that if one can choose $v < v_*$, P is not continuous on S_c. The discontinuity corresponds to an end point of the domain of P_2 because the orbit of $(c, -v_*)$ separates trajectories that can never cross E_c^+ from those that can. It must therefore be true that if $v < v_*$, (c, v) does not lead to impact and that if v is close to and greater than v_*, (c, v) leads to impact. If $v_* = 0$ we cannot choose $v < v_*$ in E_c^- so that if S_c is non-empty, it is a fixed point of P. Such a fixed point must correspond to a closed orbit of the free system, either a limit cycle or part of a system of centres—as in the Hamiltonian example discussed above.

3. The Geometry of Singularity Subspaces

This section presents a discussion of singularity structure in autonomous vibro-impact dynamics from a general point of view. Later, we specialise to general linear systems. It is shown that the singularity surface has a 'typical' geometry of a smooth manifold except at singularities which unfold the geometry of the higher elementary catastrophes. For example, in dimension 4, the 'surface' S_c is locally a two-manifold of maxima except at cusp ridges arising from $A_2(+)$ singularities or swallowtail singularities arising from $A_3(\pm)$ singularities. Such singularities are typical of any linear 4-dimensional system. The appearance of the geometry of the classical singularities depends upon the property of 'versality' of the unfolding of the singularities by the free flow and much of the following material is concerned with resolving this question.

It is useful to consider the global behaviour of all the time developed displacement-time trajectories of the free flow of the dynamical system:

$$\frac{d^2Y}{d\tau^2} + f\left(Y, \frac{dY}{d\tau}\right) = 0 \text{ on } \mathbb{R}^n . \tag{21}$$

To this end, consider the following embedding:

$$
\begin{aligned}
&i : \mathbb{R}^{2n} \times \mathbb{R} \to \mathbb{R}^{2n+2} \equiv \mathbb{R} \times (\mathbb{R}^{2n} \times \mathbb{R}) \,, \\
&i : (x, \tau) \mapsto \big(Y_1(x, \tau), (x, \tau)\big) \,,
\end{aligned}
\tag{22}
$$

and let the hypersurface $\mathrm{Im}(i)$ be labelled Γ. It is easy to show that Γ is an injectively immersed hypersurface of $\mathbb{R}^{2n+2}$ and it is to be interpreted as a $(2n+1)$-dimensional graph comprising the totality of the q_1-displacements of the flow of the system. A more useful but equivalent way to visualise Γ is to define a related family Y_1' of trajectories by adding displacement as a parameter. Thus define:

$$
\begin{aligned}
&Y_1' : \mathbb{R}^{2n+2} \equiv \mathbb{R} \times (\mathbb{R}^{2n} \times \mathbb{R}) \to \mathbb{R} \,, \\
&Y_1' : \big(a, (x, \tau)\big) \mapsto Y_1(x, \tau) - a \,.
\end{aligned}
\tag{23}
$$

The function Y_1' is obviously a submersion so that the fibre $Y_1^{-1}\{0\} = \Gamma$ is a smooth, closed, codimension one submanifold of $\mathbb{R}^{2n+2}$. If the space of initial conditions R^{2n} is replaced by a one-dimensional subspace, $\mathbb{R}^{2n+2}$ is replaced by $\mathbb{R}^3$ and Γ is replaced by a two-surface whose 'wrinkles' in the displacement direction correspond to oscillations in the flow off the one-dimensional subspace (see Fig. 2.).

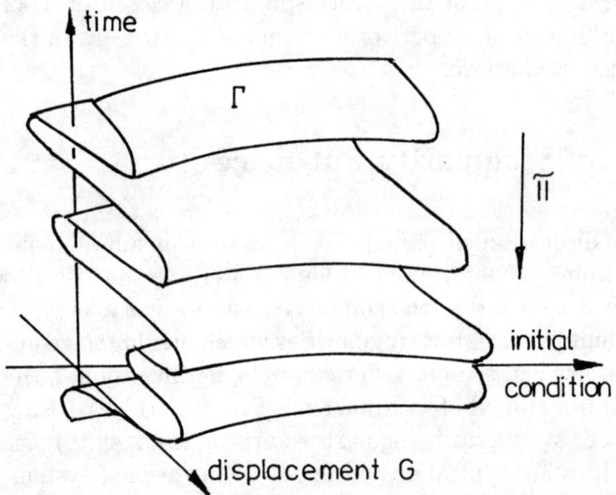

Fig. 2. A two two-surface Γ with 'wrinkles'

The loci of the values of the extrema of the family Y_1' is represented as a hypergraph, or as the 'visible contour' of Γ (in the direction of time increasing) by projection onto the hypersurface $G \equiv \tau^1\{0\}$:

$$
\begin{aligned}
&\pi : \mathbb{R}^{2n+2} \equiv \mathbb{R} \times (\mathbb{R}^{2n} \times \mathbb{R}) \to \mathbb{R}^{2n+1} \equiv \mathbb{R} \times (\mathbb{R}^{2n} \times \{0\}) \equiv G \,, \\
&\pi : \big(a, (x, \tau)\big) \mapsto (a, x) \,.
\end{aligned}
\tag{24}
$$

The set of singularities of the family of curves Y_1' coincides with the singularity set of $\mu \equiv \pi|\Gamma$ (π restricted to Γ). The latter set is defined as the 'fold' set of μ, the set of points γ having $\mathrm{Ker}(T_\gamma\mu) \neq 0$, where $T_\gamma\mu$ denotes the Jacobian of μ at γ. The tangent space $T(\Gamma,\gamma)$ to Γ at γ is the orthogonal complement of the subspace of $T(\mathbb{R}^{2n+2},\gamma)$ spanned by $\nabla Y_1'$. Also, $\mathrm{Ker}(T_\gamma\pi)$ is the subspace of $T(\mathbb{R}^{2n+2},\gamma)$ parallel to $\frac{\partial}{\partial\tau}$ at γ. It follows that $\mathrm{Ker}(T_\gamma\mu) = 0$ unless $\nabla Y_1'$ is horizontal:

$$\langle \nabla Y_1'(\gamma), \frac{\partial}{\partial\tau}(\gamma)\rangle = \frac{\partial Y_1'}{\partial\tau}(\gamma) = \frac{\partial Y_1}{\partial\tau} = 0\,. \tag{25}$$

The set of all $A_{\geq 1}(+)$ singularities plus values therefore coincides with the fold set S_c of Y_1'.

Away from its bifurcation subset, B_*, S_* is a smooth $2n$-manifold. Our main interest is in its projection $\pi(S_*)$ into G, where there will be self-intersections and singularities $\pi(B_*)$. The singularity subspace S_c of the corresponding vibro-impact system is part of the locally $(2n-1)$-dimensional cross section of $\pi(S_*)$ at constant displacement c. More precisely, the initial condition parameter space $\mathbb{R}^{2n}$ is replaced by its $(2n-1)$-dimensional subspace $\{q_1 = c\}$, the above surface is constructed (it now has dimension $2n$) and its $(2n-1)$-dimensional fold set S^c is projected into G. The singularity subspace S_c of the vibro-impact system is the part corresponding to the $(2n-2)$-dimensional cross section of $\pi(S^c)$ by the plane $q_1 = c$. In order to understand the geometry of S_c we have to understand that of S^c.

The set of all trajectories close to a singularity constitutes an example of an 'unfolding' of the singularity. Certain unfoldings are known as 'versal' and unfold all possible behaviour which reduces to the singularity. The geometry of versal unfoldings is well understood (at least in low dimensions), corresponding to the discriminants of polynomials.

A standard method of deciding upon the versality of an m-parameter unfolding of an A_k singularity γ of a family F of functions from $\mathbb{R}$ to $\mathbb{R}$ is to compute the rank of the matrix of coefficients of the $(k-1)$-jets of the parameter gradient $\nabla F(\gamma)$. The latter matrix, $M_{k-1}(\gamma)$ has elements:

$$(M_{k-1})_{ij} \equiv \frac{\frac{\partial^{i+1}F}{\partial\tau^i\partial x_j}(\gamma)}{i!} \qquad 0 \leq i < k\,; \quad 1 \leq j \leq m\,. \tag{26}$$

The following paragraphs present some propositions concerning the versality of flow unfoldings of coupled oscillator systems. Unless otherwise stated, we shall be dealing with the family Y_1 rather than Y_1'.

Consider a system of $n \geq 2$ coupled linear oscillators described by the ordinary differential equation:

$$\frac{d^2\underline{Y}}{d\tau^2} + D(\underline{Y}) + K(\underline{Y}) = 0 \text{ for } \underline{Y} : \mathbb{R} \to \mathbb{R}^n\,. \tag{27}$$

This is equivalent to a first order ODE on $\mathbb{R}^{2n}$ which is coordinatised by $x_i \equiv q_i$, $1 \leq i \leq n$ (displacement) and $x_i \equiv p_{i-n}$, $n < i \leq 2n$ (velocity). If

$\underline{X} : \mathbb{R} \to \mathbb{R}^n$ is $\tau \mapsto (\underline{Y}(\tau), \frac{dY}{d\tau}(\tau))$, the linear vector field $L : \mathbb{R}^{2n} \to \mathbb{R}^{2n}$ is defined by the matrix:

$$L = \begin{bmatrix} 0 & 1 \\ -K & -D \end{bmatrix} , \tag{28}$$

and the flow of the free dynamical system $\underline{F} : \mathbb{R}^{2n} \times \mathbb{R} \to \mathbb{R}^{2n}$ satisfies $\frac{\partial \underline{F}}{\partial \tau} = L \circ \underline{F}$. We write $\underline{F}(x, \tau) = U_\tau(x)$ where $U_\tau : \mathbb{R}^{2n} \to \mathbb{R}^{2n}$ is the corresponding group of automorphism $\exp(\tau L)$. We seek versality conditions for the family $F_1 \equiv q_1 \circ \underline{F}$ of displacement-time curves and therefore need conditions on the linear dependence of the derivatives $(\partial^k \nabla F_1 / \partial \tau^k)$, where ∇F_1 is the parameter gradient of F_1 with components $(\partial F_1 / \partial x_i)$, for $1 \le i \le 2n$.

Define:

$$F_j \equiv x_j \circ \underline{F} = \begin{cases} q_j \circ \underline{F} & 1 \le i \le n \\ p_{j-n} \circ \underline{F} & n < j \le 2n \end{cases} \quad \text{or } F_j \equiv \langle \beta_j, \underline{F} \rangle , \tag{29}$$

where $\{\beta_j\}$ is the linear basis for $\mathbb{R}^{2n}$. The definition of $\underline{F}(x, \tau) = U_\tau(x)$ then yields:

$$\frac{\partial \underline{F}}{\partial x_i}(x, \tau) = U_\tau\left(\frac{\partial x}{\partial x_i}\right) = U_\tau(\beta_i) , \tag{30}$$

and it follows that:

$$\langle \beta_j, \frac{\partial \underline{F}}{\partial x_i} \rangle = \langle \beta_j, U_\tau(\beta_i) \rangle \quad \text{or} \quad \frac{\partial \underline{F}_j}{\partial x_i}(x, \tau) = U_{\tau ij} , \tag{31}$$

which means that the vector $\nabla F_j(x, \tau)$ is the j-th row of the matrix of the linear automorphism U_τ, in the basis $\{\beta_j\}$. This implies:

Proposition 1. *The vectors $\nabla F_j(x, \tau)$ are linearly independent.*

This has the following immediate corollary:

Proposition 2. *The vectors $\nabla F_1(x, \tau)$ and $(\partial F_1 / \partial \tau)(x, \tau)$ are linearly independent.*

Proof : By definition, $\frac{\partial \underline{F}}{\partial \tau} = L \circ \underline{F}$ which implies:

$$\frac{\partial}{\partial x_i} \frac{\partial F_1}{\partial \tau} = \frac{\partial}{\partial \tau} \frac{\partial \underline{F}}{\partial x_i} = L \circ \frac{\partial \underline{F}}{\partial x_i} . \tag{32}$$

Therefore:

$$\frac{\partial}{\partial \tau} \frac{\partial F_1}{\partial x_i} = \sum_j L_{ij} \left(\frac{\partial F_j}{\partial x_i} \right) = \frac{\partial F_{n+i}}{\partial x_i} . \tag{33}$$

The last equation implies that $(\partial \nabla F_1 / \partial \tau) = \nabla F_{n+1}$ and Proposition 1 therefore implies the result.

We next consider the linear dependence of ∇F_1, $(\partial \nabla F_1/\partial \tau)$ and $(\partial^2 \nabla F_1/\partial \tau^2)$.

Proposition 3. *The vectors* ∇F_1, $(\partial \nabla F_1/\partial \tau)$ *and* $(\partial^2 \nabla F_1/\partial \tau^2)$ *are linearly independent if and only if the first oscillator is coupled to a least one other oscillator.*

Proof : Using the result $(\partial \nabla F_1/\partial \tau) = \nabla F_{n+1}$ it follows that:

$$\frac{\partial^2 \nabla F_1}{\partial \tau^2} = \frac{\partial \nabla F_{n+1}}{\partial \tau} = \sum_{j=1}^{2n} L_{n+1,j} \nabla F_j \tag{34}$$

that is:

$$\frac{\partial^2 \nabla F_1}{\partial \tau^2} = -\sum_{j=1}^{n} K_{1j} \nabla F_j - \sum_{j=1}^{n} D_{1j} \nabla F_{n+j} , \tag{35}$$

or

$$\frac{\partial^2 \nabla F_1}{\partial \tau^2} = -K_{11} \nabla F_1 - D_{11} \frac{\partial \nabla F_1}{\partial \tau} - \sum_{j>1} K_{1j} \nabla F_j - \sum_{j>1} \nabla F_{j+n} . \tag{36}$$

Clearly, if $K_{1j} = 0$ and $D_{1j} = 0$ for all $j > 1$ (the first oscillator is uncoupled from all the rest), the three vectors are linearly independent. Conversely, suppose that the three vectors are linearly dependent. Then, because ∇F_1, and $(\partial \nabla F_1/\partial \tau)$ are linearly independent, there must exist scalars a and b such that:

$$\frac{\partial^2 \nabla F_1}{\partial \tau^2} = a \nabla F_1 + b \frac{\partial \nabla F_1}{\partial \tau} = a \nabla F_1 + b \nabla F_{n+1} . \tag{37}$$

It therefore follows that if K^* and D^* are defined by $K^*_{1j} = K_{1j}$, $D^*_{1j} = D_{1j}$ for $j > 1$ and $K^*_{11} = K_{11} - a$, $D^*_{11} = D_{11} - b$, we have an expression:

$$\sum_{j=1}^{n} K^*_{1j} \nabla F_j + \sum_{j=1}^{n} D^*_{1j} \nabla F_{n+j} = 0 . \tag{38}$$

By proposition 1, the vectors ∇F_j, $1 \leq j \leq 2n$ are linearly independent. This implies that K^*_{1j} and D^*_{1j} are zero for all $1 \leq j \leq n$. Hence linear dependence implies no coupling. It has therefore been established that the three vectors are linearly dependent if and only if the first oscillator is uncoupled to the rest. The converse of this is the statement of the proposition.

The situation regarding singularities of type $A_{>3}$ is much more complicated so at this stage we gather the above propositions into a theorem formulated in terms of the versality of the unfolding of singularities of type $A_{\leq 3}$ by the flow for systems of linear oscillators.

Theorem 1. *If the j-th in a family of $n \geq 2$ linear oscillators is coupled to at least one other oscillator, the family F_j of displacement-time trajectories is versal for singularities of type $A_{\leq 3}$.*

In order to examine versality at $A_{>3}$ singularities (which can only occur for systems of at least 3 oscillators), we have to start by examining the linear dependence of $\{(\partial^r \nabla F_1/\partial \tau^r)\}$, $r \geq 3$.

$$\frac{\partial^2 \underline{F}}{\partial \tau^2} = \frac{\partial (L \circ \underline{F})}{\partial \tau} = L \circ \frac{\partial \underline{F}}{\partial \tau} = L^2 \circ \underline{F}\,, \tag{39}$$

where:

$$L^2 = \begin{bmatrix} -K & -D \\ D \circ K & D^2 - K \end{bmatrix}, \tag{40}$$

$$\frac{\partial^3}{\partial \tau^3}\frac{\partial F_1}{\partial x_i} = \frac{\partial^2}{\partial \tau^2}\frac{\partial F_{n+1}}{\partial x_i} = \sum_{j=1}^{2n} L^2_{n+1,j}\frac{\partial F_j}{\partial x_i}\,, \tag{41}$$

or

$$\frac{\partial^3}{\partial \tau^3}\frac{\partial F_1}{\partial x_i} = \sum_{j=1}^{n}(D \circ K)_{1j}\nabla F_j + \sum_{j=1}^{n}(D^2 - K)_{1j}\nabla F_{n+j}\,. \tag{42}$$

Suppose that oscillator 1 is coupled to some other. Then by proposition 3, the vectors ∇F_1, $(\partial \nabla F_1/\partial \tau)$ and $(\partial^2 \nabla F_1/\partial \tau^2)$ are linearly independent. It follows that if the addition of $(\partial^3 \nabla F_1/\partial \tau^3)$ yields a dependent set, there exist real numbers a, b and c such that:

$$\frac{\partial^3 \nabla F_1}{\partial \tau^3} = a\frac{\partial^2 \nabla F_1}{\partial \tau^2} + b\frac{\partial \nabla F_1}{\partial \tau} + c\nabla F_1\,. \tag{43}$$

Substituting the expression for $(\partial^2 \nabla F_1/\partial \tau^2)$ in terms of ∇F_1 in the (43) expression yields:

$$\frac{\partial^3 \nabla F_1}{\partial \tau^3} = -\sum_{j=1}^{n} K^*_{1j}\nabla F_j - \sum_{j=1}^{n} D^*_{1j}\nabla F_{n+j}\,, \tag{44}$$

where

$$\begin{aligned} D^*_{1j} &\equiv aD_{1j} \text{ for } j \neq 1, \qquad D^*_{11} = aD_{11} - b\,, \\ K^*_{1j} &\equiv aK_{1j} \text{ for } j \neq 1, \qquad K^*_{11} = aK_{11} - c\,. \end{aligned} \tag{45}$$

Equating the above expression for $(\partial^3 \nabla F_1/\partial \tau^3)$ to the expression derived from the L^2 representation leads to:

$$\sum_{j=1}^{n}(D \circ K + K^*)_{1j}\nabla F_j + \sum_{j=1}^{n}(D^2 - K + D^*)_{1j}\nabla F_{n+j} = 0\,. \tag{46}$$

The linear independence of $\{\nabla F_j\}$ $1 \leq j \leq 2n$ therefore implies the following results:

$$\begin{aligned} 0 &= (D \circ K)_{1j} + aK_{1j} \quad \text{and} \quad 0 = (D^2 - K)_{1j} + aD_{1j} \qquad \text{for } j \neq 1 \,, \\ 0 &= (D \circ K)_{11} + aK_{11} - c \quad \text{and} \quad 0 = (D^2 - K)_{11} + aD_{11} - b \,, \end{aligned} \tag{47}$$

which are difficult to interpret in the general case. However, for stiffness coupled systems D and D^2 are diagonal and the relationship $(D^2 - K)_{1j} + aD_{1j} = 0$ for $j \neq 1$ reduces to $K_{1j} = 0$ for $j \neq 1$, which is not true by hypothesis. It therefore follows that the assumption of linear dependence made above is false. The following proposition is therefore true:

Proposition 4. *In a stiffness coupled family of $n \geq 3$ oscillators, if the j-th oscillator is coupled to at least one other oscillator, the family of displacement-time trajectories is versal for $A_{\leq 4}(\pm)$ singularities.*

Having established the above general results, we return to the system of a pair of stiffness coupled linear oscillators, where only singularities of type $A_{\leq 3}(\pm)$ are permitted. It has been established that the *full*, 4 parameter family of displacement-time trajectories versally unfolds its singularities. These unfoldings are:

a) $A_1(\pm)$ singularities unfold to three-surfaces of maxima/minima;
b) $A_2(\pm)$ singularities unfold to two-surfaces of cusps;
c) $A_3(\pm)$ singularities unfold to lines of swallowtails.

In order to interpret these geometrical structures, we note that the $A_2(\pm)$ unfoldings can be regarded as the Cartesian product of a cusp ridge with a line. The former structure is a tangential intersection of a sheet of minima with a sheet of maxima:

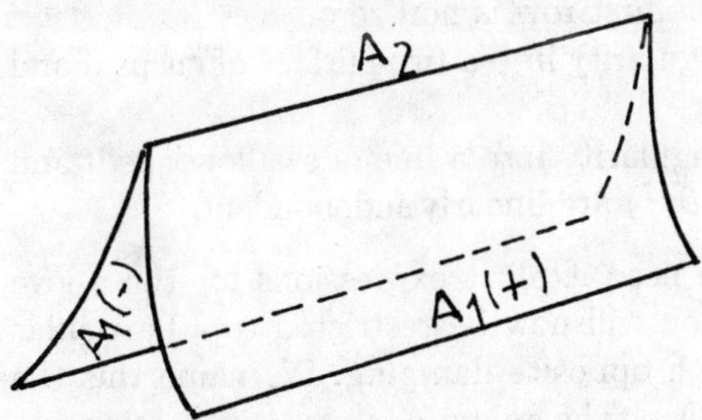

Fig. 3. The cusp ridge

The $A_3(\pm)$ unfolding is the product of an interval with the famous swallowtail singularity of catastrophe theory.

Our results concerning the versality of the full, four-parameter unfolding are useful, but not exactly the right ones for vibro-impact. There are two

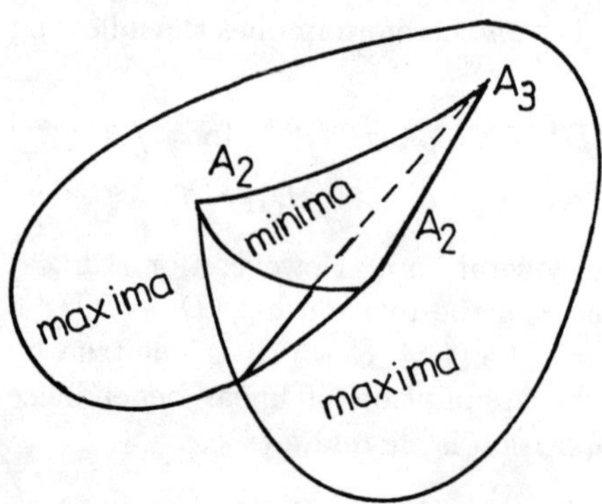

Fig. 4. The swallowtail formed from two cusp ridges with intersections

more relevant families. The first is denoted Y_1^5, the 5 parameter family obtained from the 4 parameter unfolding (Y_1^4) discussed above by the additional displacement parameter. The versality of Y_1^4 trivially implies versality for Y_1^5 because ∇Y_1^5 is the vector $(-1, \nabla Y_1^4)$. The other family, denoted by Z_1^4 is obtained from Y_1^5 by restricting to the subspace $q_1 = c$.

The versality of Y_1^4 for $A_{\leq 3}(\pm)$ singularities is governed by the matrix M_3^4:

$$\begin{bmatrix} -1 & (\partial Y_1/\partial q_2) & (\partial Y_1/\partial p_1) & (\partial Y_1/\partial p_2) \\ 0 & (\partial^2 Y_1/\partial\tau\partial q_2) & (\partial^2 Y_1/\partial\tau\partial p_1) & (\partial^2 Y_1/\partial\tau\partial p_2) \\ 0 & (\partial^3 Y_1/\partial\tau^2\partial q_2) & (\partial^3 Y_1/\partial\tau^2\partial p_1) & (\partial^3 Y_1/\partial\tau^2\partial p_2) \end{bmatrix} . \tag{48}$$

Denote the three-gradient of Y_1 with respect to q_2, p_1 and p_2 by DY_1. We make the following observations about the above unfolding matrix:

1) The family Y_1^4 versally unfolds any $A_1(\pm)$ singularity into a three-surface of maxima or minima because the first row is non zero.
2) Y_1^4 versally unfolds an $A_2(\pm)$ singularity into a two-surface of cusps if and only if $(\partial DY_1/\partial\tau) \neq 0$.
3) Y_1^4 versally unfolds an $A_3(\pm)$ singularity into a line of swallowtails if and only if $(\partial DY_1/\partial\tau)$ and $(\partial^2 DY_1/\partial\tau^2)$ are linearly independent.

In order to proceed further, we need explicit expressions for the above derivatives. Because of this, attention will now be restricted to a particular example—the pair of oscillators with opposite damping. We name this the S/U oscillator for obvious reasons. Using the analytic integrals of motion one can prove the following result after a lot of tedious but elementary algebra:

Theorem 2. *Suppose that Y_1^4 is the family of displacement-time trajectories of the S/U oscillator. Then:*

1) Y_1^4 versally unfolds all $A_1(\pm)$ and $A_2(\pm)$ singularities.

2) Y_1^4 *versally unfolds an* $A_3(\pm)$ *singularity if and only if it is not an initial* $(\tau = 0)$ *singularity.*

Theorem 2 describes the geometry of the projection $\pi(S_4)$ of the fold set of Y_1^4 into G. Thus $\pi(S_4)$ is locally a three-manifold in the neighbourhood of an $A_1(+)$ singularity; modulo intersections at the same level of the local three-surfaces of minima corresponding to equal amplitudes at different times. $\pi(S_4)$ is not a three-manifold at self intersections or at points of the bifurcation set $\pi(B_4)$ comprising projections of the higher $(A_{>1}(\pm))$ singularities. However, by theorem 2, if (a, x, τ) is an A_2 singularity, $\pi(a, x, \tau) = (a, x)$ lies in a two-plane of cusp singularities. If (a, x, τ) is an A_3 singularity and $\tau \neq 0$, $\pi(a, x, \tau) \equiv (a, x)$ lies on a line of swallowtail singularities. If $\tau = 0$, the A_3 is not versally unfolded. However, we do have some idea of the geometry of A_3 singularities at level c and $\tau = 0$. Indeed, an A_3 singularity at level c and $\tau = 0$ comprises a point, $(c, (kc/K, 0, 0))$ of $\pi(S_4)$. As c varies, we obtain a line of swallowtails in $\pi(S_4)$, but at constant c, we do not get the characteristic tangent cusp ridges at the A_3. Instead, we have antilinear cusp ridges, a non-versal unfolding.

$\pi(S_4)$ contains all information on the changes in structure of S_c as the clearance c varies. Indeed, if P_c denotes the three-plane of constant clearance c in G, S_c is part of the cross section $P_c \cap \pi(S_4)$. It is important to note that if P_c transversally intersects the versal unfoldings of the singularities, we obtain a versal unfolding in $P_c \cap \pi(S_4)$. Thus modulo self intersections, we obtain a model of the geometry of $P_c \cap \pi(S_4)$—which must be a two-manifold in the neighbourhood of A_1 singularities. An A_2 singularity must lie on a cusp ridge and an A_3 must lie at the terminal intersection of two cusp ridges. We therefore obtain a global picture of S_c as part of a 'pleated' surface with self intersections (Fig. 5.). The part of $P_c \cap \pi(S_4)$ corresponding to S_c will be described in detail later. For example, as we are only interested in A_1 maxima at level c, we have to excise sheets of level c minima.

It is useful to introduce local coordinates in $\pi(S_4 \backslash B_4)$ at suitable points as follows. If (a, x, τ) lies in $S_4 \setminus B_4$, $Y_1(x, \tau) = a$ and $(\partial Y_1/\partial\tau)(x, \tau) = 0$ but $(\partial^2 Y_1/\partial\tau^2)(x, \tau) \neq 0$. In this case, we can use the implicit function theorem to obtain a map $t : U \to \mathbb{R}$ from a neighbourhood U of x in $\mathbb{R}^3$ which solves $(\partial Y_1/\partial\tau)(x', \tau') = 0$ for x' in U: $\tau' = t(x')$. Let $y : U \to \mathbb{R}$ be the associated extremal value function: $y : x' \mapsto Y_1(x', t(x'))$. Also, let $\{\tau_k\}$ denote the set of successive maxima and minima of $Y_1(x, \tau)$ as τ increases. Applying the above construction, we can define corresponding local functions t_k and y_k and therefore also define local maps:

$$\begin{aligned} &\phi_k : U \to S_4 \subset \mathbb{R}^5 \, , \\ &\phi_k : x \mapsto (y_k(x), x, t_k(x)) \, . \end{aligned} \tag{49}$$

The images $\phi_k(U)$ of U in S_4, are 1-1 immersed three-planes of maxima or minima and are all disjoint. However, the corresponding planes in $\pi(S_4)$,

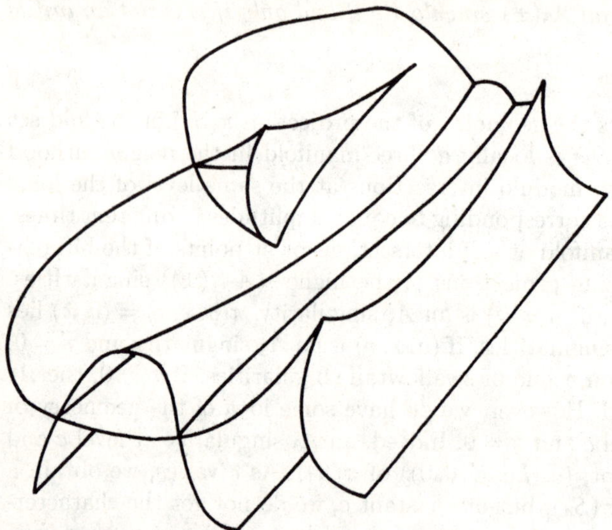

Fig. 5. Pleated surface with self intersections

defined by functions:

$$\begin{aligned} \theta_k &: U_k \to \pi(S_4)\,, \\ \theta_k &: x \mapsto \pi \circ \phi_k(x) \equiv (y_k(x), x)\,, \end{aligned} \tag{50}$$

need not be disjoint since the amplitudes might coincide: $y_k(x) = y_j(x)$ but $t_k(x) \neq t_j(x)$. It is easy to show that the Jacobian of θ_k is given by:

$$\begin{bmatrix} (\partial y_k/\partial q_2) & (\partial y_k/\partial p_1) & (\partial y_k/\partial p_2) \\ 1 & 0 & 0 \\ 0 & 1 & 0 \\ 0 & 0 & 1 \end{bmatrix}, \tag{51}$$

and therefore the sheets $\theta_k(U)$ are transversal to planes of constant q_1, unless $Dy_k(x) = 0$. Thus if $Dy_k(x) \neq 0$, the three-plane $\theta_k(U)$ is transversal to P_c: $P_c \cap \theta_k(U)$ is a two-plane. Indeed, $P_c \cap \theta_k(U)$ is given by $\{x\} \times y_k^{-1}\{c\}$, where $y_k^{-1}\{c\}$ is a regular two-surface in U because $Dy_k(x) \neq 0$. Now $Dy_k(x) = Dy_1(x, t_k(x))$ since, for example:

$$\begin{aligned} \frac{\partial y_k}{\partial x_i}(x) \equiv & \frac{\partial}{\partial x_i}\Big(Y_1(x, t_k(x))\Big) = \\ & \frac{\partial Y_1}{\partial x_i}(x, t_k(x)) + \frac{\partial Y_1}{\partial \tau}(x, t_k(x))\frac{\partial t_k}{\partial x_i}\,, \end{aligned} \tag{52}$$

and $(\partial Y_1/\partial \tau)(x, t_k(x)) = 0$. It follows that $DY_k(x) = 0$ if and only if $t_k(x) = 0$. One can analyse $P_c \cap \theta_k(U)$ for $Dy_k(x) = 0$ using Morse theory.

Next, suppose that x corresponds to a point of $\theta_k(U) \cap \theta_j(U)$ where $Dy_k(x)$ and $Dy_j(x)$ are non zero and non co-linear so that the intersection is transversal. Equivalently, x is a point of $y_k^{-1}\{x\} \cap y_j^{-1}\{x\}$ and we may suppose that $t_k(x) < t_j(x)$. Suppose that both extrema are minima. Then if $Y_1(x,\tau)$ does not cross $q_1 = c$ for $t_k(x) < \tau < t_j(x)$, it follows that x is a point of $P_2^{-1}(Z_c) \cap P_2^{-2}(Z_c)$, or $Y_1(x,\tau)$ has two successive grazes. This is true for all points along the linear intersection.

If $Dy_k(x) \neq 0$ and B is a small enough open ball about x in Σ_c^+, the plane $y_k^{-1}\{c\}$ divides B into hemispheres B_+ and B_-:

$$\begin{aligned} B_+ &\equiv \{x' \in B \mid y_k(x') \geq c\}, \\ B_- &\equiv \{x' \in B \mid y_k(x') \leq c\}. \end{aligned} \tag{53}$$

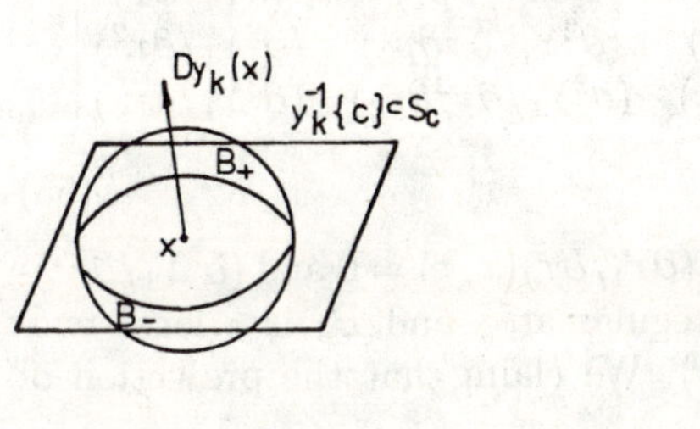

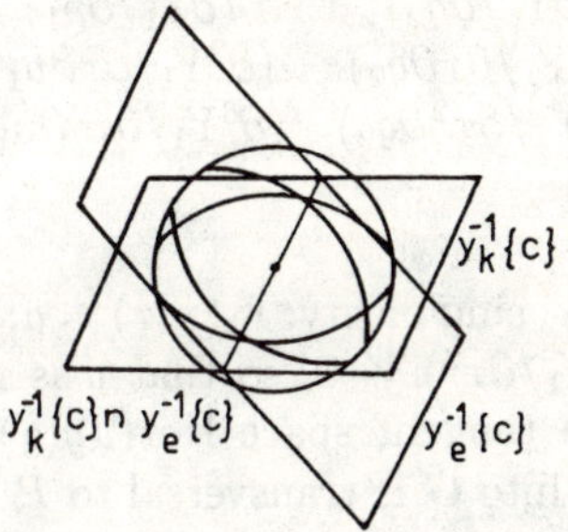

Fig. 6. A schematic of location of the hemispheres B_+ and B_- (a) and transversal intersection of $y_k^{-1}\{c\}$ and $y_j^{-1}\{c\}$ (b)

Points in B_+ lead to impacts at times close to $t_k(x)$, but points in B_- do not, because if $x' \in B_-$, $Y_1(x',\tau)$ need not cross $q_1 = c$ until a time τ_1 close to a subsequent crossing by $Y_1(x,\tau)$ for $\tau > t_k(x)$. Equivalently, P_2 need not be continuous across S_c.

Suppose that x is a point of transversal intersection of $y_k^{-1}\{c\}$ and $y_j^{-1}\{c\}$, both maxima

B is typically divided into 4 sectors by the local two-surfaces. Still supposing that $t_k(x) < t_j(x)$, points in $B_+ \cap y_j^{-1}\{c\}$ cannot correspond to impacts because such points have first crossings of E_c^+ close to time $t_k(x)$. Thus whilst $\pi(S_4)$ has two intersecting sheets of maxima, the surface $B_+ \cap y_j^{-1}\{c\}$ has to be excised in order to obtain S_c. In the present case, S_c locally consists of two abutting sheets. Similar considerations apply to intersections of more than two sheets of maxima.

The procedure for obtaining S_c from $\pi(S_4)$ is:

1) delete all surfaces of minima;
2) delete surfaces of maxima preceded by earlier maxima at the same level.

For example, if the above process is applied to the pleated surface of Fig. 5, one might obtain Fig. 7. Note that cusp ridges in $P_c \cap \pi(S_4)$ correspond to 'edges' in S_c and that an edge merges smoothly into a surface of maxima at a point corresponding to a swallowtail singularity.

The question of non-transversal intersections of sheets of maxima in $\pi(S_c)$ will not be considered here. However, it remains to consider the transversality of versal unfoldings in $\pi(S_4)$ and the planes P_c in G. We first consider A_2 singularities. Define a map:

$$\begin{aligned} &g : \mathbb{R}^5 \equiv \mathbb{R} \times (E_c \times \mathbb{R}) \to \mathbb{R}^3 \, , \\ &g : \big(a, (x, \tau)\big) \mapsto \big(Y_1(x,\tau) - a, Y_1^1(x,\tau), Y_1^2(x,\tau)\big) \, . \end{aligned} \tag{54}$$

Clearly, the fibre $g^{-1}\{0\}$ corresponds to the bifurcation set B_4 of the surface Γ. The Jacobian $Jg(\gamma)$ of g at a singularity γ is given by:

$$\begin{vmatrix} -1 & (\partial Y_1/\partial q_2) & (\partial Y_1/\partial p_1) & (\partial Y_1/\partial p_2) & (\partial Y_1/\partial \tau) \\ 0 & (\partial^2 Y_1/\partial \tau \partial q_2) & (\partial^2 Y_1/\partial \tau \partial p_1) & (\partial^2 Y_1/\partial \tau \partial p_2) & (\partial^2 Y_1/\partial \tau^2) \\ 0 & (\partial^3 Y_1/\partial \tau^2 \partial q_2) & (\partial^3 Y_1/\partial \tau^2 \partial p_1) & (\partial^3 Y_1/\partial \tau^2 \partial p_2) & (\partial^3 Y_1/\partial \tau^3) \end{vmatrix} . \tag{55}$$

If γ is an A_2 singularity: $Y_1(x,\tau) = a$, $(\partial Y_1/\partial \tau)(x,\tau) = 0$ and $(\partial^2 Y_1/\partial \tau^2)$-$= 0$ but $(\partial^3 Y_1/\partial \tau^3) \neq 0$, so that g is regular at γ and B_4 is a local two-manifold with tangent space $\mathrm{Ker}\big(Jg(\gamma)\big)$. We claim that the projection of this subspace into G is transversal to P_c.

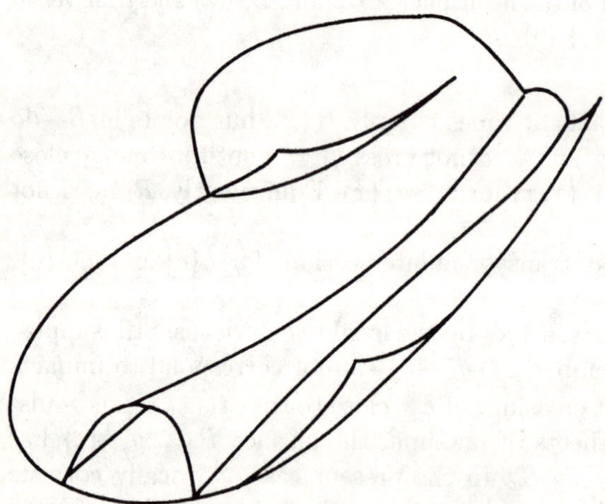

Fig. 7. Singularity surface

Proposition 5. *If $\tau \neq 0$ the tangent subspace $J\pi(\gamma)\big(\mathrm{Ker}(Jg(\gamma))\big)$ is transverse to the fibre P_a for the S/U oscillator.*

Proof : Suppose that $w = (w_1, w_2, w_3)$ is tangent to $\mathbb{R} \times (E_c \times \mathbb{R})$ at γ and that $Jg(\gamma)(w) = 0$. Then w is defined by the three equations:

$$w_1 = \langle DY_1, w \rangle \, , \quad \left\langle \frac{\partial DY_1}{\partial \tau}, w \right\rangle = 0 \quad \text{and} \quad w_5 = -\frac{\left\langle \frac{\partial^2 DY_1}{\partial \tau^2}, w \right\rangle}{\frac{\partial^3 Y_1}{\partial \tau^3}} \, . \tag{56}$$

The second equation states that w is orthogonal to $(\partial DY_1/\partial\tau)$ which is non-zero. Let u_1 and u_2 be two vectors tangent to E_c at γ such that the set $\{(\partial DY_1/\partial\tau), u_1, u_2\}$ is an orthonormal basis to the tangent space. We claim that $\langle DY_1, u_1\rangle = 0$ and $\langle DY_1, u_2\rangle = 0$ is impossible if $\tau \neq 0$. In this case, the corresponding component in the direction of $\frac{\partial}{\partial a}(\pi(\gamma))$ of at least one vector (w_1, u_i) spanning $J\pi(\gamma)(\mathrm{Ker}(Jg(\gamma)))$ is non-zero and the subspace is transverse to P_a. Suppose that $DY_1 \neq 0$ $(\tau \neq 0)$ and that *both* u_1 and u_2 are orthogonal to DY_1. Then DY_1 and $(\partial DY_1/\partial\tau)$ are colinear because:

$$DY_1 = \left\langle DY_1, \frac{\partial DY_1}{\partial \tau} \right\rangle \frac{\partial DY_1}{\partial \tau} + \langle DY_1, u_1\rangle u_1 + \langle DY_1, u_2\rangle u_2 \, . \tag{57}$$

It can be demonstrated that DY_1 and $(\partial DY_1/\partial\tau)$ cannot be colinear if $\tau \neq 0$. If these vectors *are* colinear, the following equation is true:

$$\frac{\partial y_1}{\partial p_1}\frac{\partial^2 y_1}{\partial \tau \partial p_1} = \frac{\partial y_1}{\partial p_2}\frac{\partial^2 y_1}{\partial \tau \partial p_2} \, . \tag{58}$$

Using the analytic solutions, it is possible to show that the latter equation cannot be true if $\tau \neq 0$.

Proposition 5 establishes that for the S/U oscillator, the two-surface of cusps versally unfolded by Y_1 in $\pi(S_4)$ intersects P_a in a line of cusps (a cusp ridge) through γ if $\tau \neq 0$.

If $\tau = 0$, $DY_1 = 0$ and this implies that both vectors u_i spanning $\mathrm{Ker}(Jg(\gamma))$ are tangent to E_c. However, we already know that the A_2 singularities intersect E_c in the line $q_2 = kc/K$ in E_c^0. Thus one of the vectors, u_1 say, must project onto the above line and u_2 projects onto an orthonormal direction tangent to E_c.

Next consider the A_3 singularities using a map f analogous to the map g used above. Thus define :

$$\begin{aligned} &f : \mathbb{R}^5 \equiv \mathbb{R} \times (E_c \times \mathbb{R}) \to \mathbb{R}^4 \, , \\ &f : (a, (x, \tau)) \mapsto (Y_1(x,\tau) - a, Y_1^1(x,\tau), Y_1^2(x,\tau), Y_1^3(x,\tau)) \, . \end{aligned} \tag{59}$$

The fibre $f^{-1}\{0\}$ coincides with the set of $A_{\geq 3}$ singularities of the family Y_1 in Γ. The Jacobian of f at a singularity γ is obtained from the Jacobian of g by adding the obvious fourth row vector, noting that the first three time derivatives are zero at the singularity. If the singularity is an A_3 (so that the fourth time derivative is non-zero and $(\partial DY_1/\partial\tau)$ and $(\partial^2 DY_1/\partial\tau^2)$ are linearly independent), $Jf(\gamma)$ is surjective and $\mathrm{Ker}(Jf(\gamma))$ is a one dimensional subspace which projects onto the line of swallowtails versally unfolded by Y_1 in $\pi(S_4)$. Suppose that w lies in $\mathrm{Ker}(Jf(\gamma))$ and $w = (w_1, \underline{w}, w_5)$. Then:

$$w_1 = \langle DY_1, \underline{w}\rangle \ , \qquad \left\langle \frac{\partial DY_1}{\partial \tau}, \underline{w}\right\rangle = 0 \ , \tag{60}$$

$$\left\langle \frac{\partial^2 DY_1}{\partial \tau^2}, \underline{w}\right\rangle = 0 \text{ and } w_5 = \frac{\left\langle \frac{\partial^3 DY_1}{\partial \tau^3}, \underline{w}\right\rangle}{\frac{\partial^4 Y_1}{\partial \tau^4}} \ . \tag{61}$$

Proposition 6. *If $\tau \neq 0$, the tangent subspace $J\pi(\gamma)(\mathrm{Ker}(Jf(\gamma)))$ is transverse to the fibre P_c for the S/U oscillator.*

Proof : If $\tau \neq 0$, $w_1 = \langle DY_1, \underline{w}\rangle \neq 0$, $DY_1 \neq 0$ and $(\partial DY_1/\partial\tau)$ and $(\partial^2 DY_1/\partial\tau^2)$ are linearly independent. It was established in proposition 5 that $\tau \neq 0$ implies that DY_1 and $(\partial DY_1/\partial\tau)$ are linearly independent. Thus either DY_1 lies in the plane spanned by $(\partial DY_1/\partial\tau)$ and $(\partial^2 DY_1/\partial\tau^2)$ or DY_1, $(\partial DY_1/\partial\tau)$ and $(\partial^2 DY_1/\partial\tau^2)$ are linearly independent. In the latter case, given that w lies in the orthogonal complement of the subspace spanned by the last two vectors, $\langle DY_1, \underline{w}\rangle \neq 0$. The vectors DY_1, $(\partial DY_1/\partial\tau)$ and $(\partial^2 DY_1/\partial\tau^2)$ are linearly independent if and only if the product:

$$\left\langle DY_1, \left[\frac{\partial DY_1}{\partial \tau} \wedge \frac{\partial^2 DY_1}{\partial \tau^2}\right]\right\rangle = 0 \ , \tag{62}$$

where '$\wedge$' denotes the vector cross product in $\mathbb{R}^3$. Note that for the S/U oscillator:

$$\frac{\partial^2 DY_1}{\partial \tau^2} = -KDY_1 - 2\zeta\frac{\partial^2 DY_1}{\partial \tau} + KDY_2 \ , \tag{63}$$

which implies that:

$$\left\langle DY_1, \left[\frac{\partial DY_1}{\partial \tau} \wedge \frac{\partial^2 DY_1}{\partial \tau^2}\right]\right\rangle = K\frac{\partial Y_1}{\partial \tau}[DY_1 \wedge DY_2] \ . \tag{64}$$

A tedious direct calculation of the latter expression establishes that the product is non-zero. Thus if $\tau \neq 0$, $w_1 \neq 0$.

The above proposition establishes that for the S/U oscillator, the line of swallowtail singularities with $\tau \neq 0$ versally unfolded by Y_1 in $\pi(S_4)$ is transversal to a plane of constant clearance. This means that the cross-section $P_c \cap \pi(S_4)$ contains a swallowtail singularity. Similarly, the preceding proposition established that the cross section will contain a cusp ridge corresponding to a line of A_2 singularities with $\tau \neq 0$ versally unfolded by Y_1.

4. Continuity of the Poincaré Map of the S/U Oscillator

The analysis of Sec. 3 has established the generic geometry of the singularity subspace S_c of the S/U oscillator. Moreover, this geometry ought be stable to small perturbations away from the symmetry of the S/U oscillator to more general systems. The present section considers continuity in more detail.

Suppose that $x_0 \in P_2^{-1}(Z_c)$ and that $P_2(x_0)$ is an $A_k(\pm)$ singularity of $Y_1(x_0, \tau)$ at time τ_a. P_2 is continuous at x_0 if, given any neighbourhood V of $P_2(x_0)$, there is a neighbourhood U of x_0 with $P_2(U) \subset V$. That is, for all x in U, the trajectory $Y_1(x, \tau)$ first crosses $q_1 = c$ at time τ_1 with $(Y_1^1(x, \tau_1), Y_2(x, \tau_1), Y_2^1(x, \tau_1))$ in V. If we can establish that for some x, $\tau_1(x)$ is not close to τ_a, $P_2(x)$ is not in V for a suitably chosen V.

Because $Y_1(x_0, \tau) < c$ for $0 < \tau < \tau_a$, we can always ensure that $Y(x, \tau) < c$ except when τ is sufficiently close to τ_a by choosing x sufficiently close to x_0. Thus we only have to examine $Y_1(x, \tau)$ for $\|x - x_0\|$ and $|\tau - \tau_a|$, small. One can decide the continuity of P_2 by counting the roots of $Y_1(x, \tau) = c$ of the polynomial approximations of the above unfolding. Thus if all polynomials have at least one root, P_2 is continuous at x_0. Otherwise, if there exist polynomials with no roots, P_2 cannot be continuous at x_0.

Given that $Y_1(x_0, \tau)$ has an A_k singularity at time τ_0 Hademard's lemma states that we can find a positive smooth function f on a neighbourhood V of τ_a such that:

$$Y_1(x_0, \tau) = c \pm (\tau - \tau_a)^{k+1} f(\tau_a) . \tag{65}$$

We want to examine the analogous expressions for x close to x_0. Consider the first approximation:

$$Y_1(x, \tau) \sim Y_1(x_0, \tau) + \langle DY_1(x_0, \tau), \underline{u} \rangle \text{ for } \underline{u} \equiv (x - x_0) . \tag{66}$$

It is already known that $Y_1(x_0, \tau)$ can be approximated by a polynomial of degree $(k+1)$ for τ close to τ_a so the difference term can be approximated by its $(k-1)$-jet or Taylor expansion truncated at degree $(k-1)$. It is therefore sufficient to consider the $(k-1)$-jet of $DY_1(x_0, \tau)$ at $\tau = \tau_a$. The difference function is therefore approximated by a polynomial of degree $(k-1)$ with coefficients:

$$\frac{\left\langle \frac{\partial^r DY_1}{\partial \tau^r}(x_0, \tau_a), \underline{u} \right\rangle}{r!} . \tag{67}$$

Note that if the unfolding Y_1^3 is versal, all possible polynomials of degree $(k-1)$ are obtained for the difference term. We now consider the allowable $A_k(\pm)$ singularities for $k \leq 3$.

1) **Maxima—$A_1(-)$ Singularities.** $A_1(-)$ singularities are unfolded into polynomials $Y_1(x, \tau) - c \sim -(\tau - \tau_a)^2 f(\tau_a) + d$ where $d \equiv \langle (\partial DY_1/\partial \tau), \underline{u} \rangle$.

If $DY_1(x_0,\tau_a) \neq 0$, i.e.: $\tau_a \neq 0$, one can always find a vector $\underline{u}$ such that $d < 0$ and in this case, $-(\tau-\tau_a)^2 f(\tau_a) + d$ has no real roots. It follows that P_2 cannot be continuous at x_0 if $x_0 \in E_c^0$. The case $DY_1(x_0,\tau_a) = 0$ will be examined later.

2) **Ordinary Inflections—$A_2(+)$ Singularities.** $A_2(+)$ singularities unfold into polynomials $Y_1(x,\tau) - c \sim -(\tau-\tau_a)^3 f(\tau_a) + \langle DY_1(x_0,\tau), \underline{u}\rangle + \langle \partial DY_1(x_0,\tau)/\partial\tau, \underline{u}\rangle(\tau-\tau_a)$, which have the monic form $(\tau-\tau_a)^3 + x_1 + x_2(\tau-\tau_a)$ for $x_1 = \langle DY_1(x_0,\tau), \underline{u}\rangle / f(\tau_a)$ and $x_2 = \langle (\partial DY_1/\partial\tau)(x_0,\tau), \underline{u}\rangle / f(\tau_a)$. If $\tau \neq 0$, $DY_1(x_0,\tau,) \neq 0$ by proposition 5. Also, $(\partial DY_1/\partial\tau)(x_0,\tau_0) \neq 0$ by proposition 6. It follows that for $\tau_2 \neq 0$, Y_1^3 versally unfolds a singularity. For $\tau_a = 0$, the singularity is unfolded into polynomials $(\tau-\tau_a)^3 + (\tau-\tau_a)x_1$. Regarding the versal unfolding as the disciminant of a cubic, $Y_1(x,\tau) - c$ always has at least one root. If $\tau_a = 0$, $x_1 \neq 0$ and there is always at least one real root. It follows that P_2 is continuous on $P_2^{-1}(A(2,+))$.

3) **Higher Inflections—$A_3(\pm)$ Singularities.** $A_3(-)$ singularities are unfolded into polynomials:

$$\begin{aligned} Y_1(x,\tau) - c \sim &- (\tau-\tau_a)^4 f(\tau_a) + \langle DY_1, \underline{u}\rangle + \\ &+ \left\langle \frac{\partial DY_1}{\partial\tau}, \underline{u}\right\rangle (\tau-\tau_a) + \left\langle \frac{\partial^2 DY_1}{\partial\tau^2}, \underline{u}\right\rangle \frac{(\tau-\tau_a)^2}{2}. \end{aligned} \tag{68}$$

If $\tau_a = 0$, $DY_1 = 0$ and $(\partial DY_1/\partial\tau)$ and $(\partial^2 DY_1/\partial\tau^2)$ are linearly independent. Thus A_3 singularities with $\tau_a = 0$ have monic form $(\tau-\tau_a)^4 + x_3(\tau-\tau_a)^3 + x_2(\tau-\tau_a)^2 + x_1$ where $x_1 = 0$ and $x_2 = \mu x_3$ for some real number μ. Regarding the swallowtail singularity as the discriminant set of the general reduced quartic, the line $x_2 = \mu x_3$ through the plane $x_1 = 0$ always has at least one root. If $\tau \neq 0$, the unfolding is versal because DY_1, $(\partial DY_1/\partial\tau)$ and $(\partial^2 DY_1/\partial\tau^2)$ are linearly independent. In this case, one can always find a polynomial with no real roots. It follows that P_2 is continuous at an A_3 singularity in E_c^0 but discontinuous at an A_3 with $\tau_a \neq 0$.

It still remains to discuss the continuity of P_2 at $A_1(-)$ singularities where $DY_1(x_0,\tau_a) = 0$. Recall proposition 5 states that $DY_1(x_0,\tau_a) = 0$ if and only if $\tau_a = 0$. However, singularities with $\tau_a = 0$ cannot lie in S_c because P_2 is defined by the positive time translation $P_2(x) = U_s(x)$ where s solves $Y_1(x_0,s) = c$ for $s > 0$.

Finally, we have to consider the continuity of P_2 at $A_2(+)$ singularities or α-points. By definition, $P_2(\alpha)$ is an ω-point or $A_2(-)$ singularity obtained from α by the trapped flow within E_c^0. The latter flow is smooth so that nearby α-points are mapped to nearby ω-points. However, the $A_2(+)$ singularities form part of the boundary of $A(1,-)$ and it is intuitively clear that maxima close to an α-point need not cross E_c^+ under the global flow at a point close to the ω-point $P_2(\alpha)$. Thus we assert without proof that P_2 is not continuous on $A(2,+)$.

References

1. G.S. Whiston: Impacting under harmonic excitation. Journal of Sound and Vibration **67**, 179–186 (1979)
2. S.W. Shaw, P.J. Holmes: A periodically forced piecewise linear oscillator. Journal of Sound and Vibration **90**, 129–155 (1983)
3. G.S. Whiston: Global dynamics of a vibro-impacting linear oscillator. Journal of Sound and Vibration **118**, 395–429 (1987)
4. G.S. Whiston: Singularities in vibro-impact dynamics. Journal of Sound and Vibration **152**, 427–460 (1992)
5. V.I. Arnold: Catastrophe theory. 2nd edn. Springer Verlag 1986
6. V.I. Arnold, S.M. Guezin-Zade, A.N. Varchenko: Singularities of differentiable maps, Vol.1, Birkhauser 1985
7. J.W. Bruce, P.J. Giblin: Curves and singularities 2nd edn. Cambridge University Press 1987

Codimension Two Bifurcation and Its Computational Algorithm

H. Kawakami [1] and T. Yoshinaga [2]

[1] Department of Electrical and Electronic Engineering,

[2] School of Medical Sciences,
Tokushima University, Tokushima-shi, 770 Japan

Abstract

In this paper we discuss a method for calculating the bifurcation value of parameters of periodic solution in nonlinear ordinary differential equations. The generic bifurcations of the periodic solution are known as codimension one bifurcations: tangent bifurcation, period doubling bifurcation and the Hopf bifurcation. At the parameters for which bifurcation occurs, if a periodic solution satisfies two bifurcation conditions, then the bifurcation is referred to as a codimension two bifurcation. Our method enables us to obtain directly both codimension one and two bifurcation values from the original equations without special coordinate transformation. Hence we can easily trace out various bifurcation sets in an appropriate parameter plane, which correspond to nonlinear phenomena such as jump and hysteresis phenomena, frequency entrainment, etc. Some electrical circuit examples are analyzed and shown to illustrate the validity of our method.

1. Introduction

In nonlinear dynamical systems, variation of system parameters may cause an abrupt change in the qualitative property of their state. The state change is referred to as a bifurcation and the parameter value at which the bifurcation occurs is called the bifurcation value. If we restrict our attention to the bifurcation problem of periodic solution in nonlinear ordinary differential equations, we see three types of generic codimension one bifurcations: the tangent, period doubling and Hopf bifurcation, corresponding to the generation of a pair of periodic solutions, the branching of periodic solution and the appearance of a quasi-periodic solution respectively [1]. On the other hand, if a periodic solution satisfies two bifurcation conditions at a single bifurcation value of parameters, then we class the bifurcation as codimension two.

Such a codimension two bifurcation occurs naturally at a point if system parameters are changed two-dimensionally in a parameter space. Hence the bifurcation value is generally obtained as an isolated point in the planar bifurcation diagram. Note that in the neighbourhood of a codimension two bifurcation value of parameters we may observe complicated features of states. Some types of codimension two bifurcations may relate to the generation of chaotic states. Intuitively, codimension one and two bifurcations appear as follows. Let us consider an n-dimensional periodic nonautonomous system containing an m-dimensional system parameter. Using the Poincaré map defined in the next section, a periodic solution of the system corresponds to a fixed point on the Poincaré map. The set of fixed points consists of an m-dimensional manifold in $(n + m)$-dimensional direct sum space of states and parameters. If a bifurcation condition given by a scalar equation intersects the fixed point manifold transversely, then the intersection becomes an $(m-1)$-dimensional manifold. Projecting the intersection onto the parameter space, we obtain a codimension one bifurcation set [2].

Similarly, if two bifurcation conditions intersect the fixed point manifold simultaneously, then we obtain a codimension two bifurcation set in parameter space. Note that in both cases the transversality property suggests the possibility of calculating the intersection by an appropriate numerical method such as Newton's method. If we consider a two parameter problem (i.e. we fix $(m-2)$ parameters as constant values) then the sets of codimension one and two bifurcation values in the rest of the two dimensional parameter plane are obtained by curves and isolated points, respectively.

In Section 2, we summarize the results of the Poincaré map correlated with a periodic nonautonomous differential equation: classification of a hyperbolic fixed point of the Poincaré map, and codimension one and two bifurcations. In Section 2.3, codimension two bifurcations are classified by the combinatorial intersection of codimension one bifurcation curves in the parameter plane [2, 3].

In Section 3, computational algorithms are used to obtain the codimension one and two bifurcation parameters as well as the bifurcating periodic solutions. Newton's method is used to solve a set of equations appeared in our algorithm. The Jacobian matrix of the equations is obtained from the solutions to the variational equations with respect to the initial condition and the system parameters.

In Section 4, we illustrate some numerical analysis of electrical circuits with a sinusoidal external source. In these examples, we shall see all types of codimension two bifurcations discussed in Section 2. As a global feature of bifurcation diagrams in the parameter plane, we find that types of codimension two bifurcation occur together successively with the period doubling cascade and tangent bifurcations. This bifurcation sequence may offer a new route to the generation of chaotic states.

2. Bifurcations of Fixed Point

This section contains a brief summary of the geometric approach to ordinary differential equations which will be used in this paper. The topological property of periodic solution provides the classification and bifurcation conditions.

2.1 The Poincaré Map and Property of Fixed Points

Let us consider a nonautonomous ordinary differential equation

$$\frac{dx}{dt} = f(t, x, \lambda) , \tag{1}$$

where $x \in \mathbb{R}^n$ is the state and $\lambda \in \mathbb{R}^m$ is the system parameter. Let $f : \mathbb{R} \times \mathbb{R}^n \times \mathbb{R}^m \to R^n$ be sufficiently differentiable for all arguments and periodic in time with period τ, i.e., $f(t+\tau, x, \lambda) = f(t, x, \lambda)$. We also assume that the solution of Eq.(1) with initial condition $u := x(0)$, denoted by $\varphi(t, u, \lambda)$, exists for all t. Since f has the period τ, we can naturally define a diffeomorphism T_λ, called the Poincaré map, from the state space $\mathbb{R}^n$ into itself:

$$T_\lambda : \mathbb{R}^n \to \mathbb{R}^n \; ; \; u \mapsto T_\lambda(u) = \varphi(\tau, u, \lambda) . \tag{2}$$

If a solution $x(t) = \varphi(t, u, \lambda)$ is periodic with period τ, then the point u becomes a fixed point of T_λ:

$$F_\lambda(u) := u - T_\lambda(u) = 0 . \tag{3}$$

Hence the study of a periodic solution with period τ, $x(t) = \varphi(t, u, \lambda)$ of Eq.(1) is topologically equivalent to the study of a fixed point $u \in \mathbb{R}^n$ satisfying Eq.(3). Note that a periodic solution with period $k\tau$, can be studied by replacing T_λ with T_λ^k, k-th iterates of T_λ, in Eq.(3). Therefore in the following we shall consider only the property of a fixed point of T_λ and its bifurcations. A similar argument can be applied to the periodic point of T_λ.

Let $u \in \mathbb{R}^n$ be a fixed point of T_λ. Then the characteristic equation of the fixed point u is defined by

$$\chi(u, \lambda, \mu) = \det(\mu I - DT_\lambda(u)) = \mu^n + a_1\mu^{n-1} + \cdots + a_{n-1}\mu + a_n = 0 , \tag{4}$$

where I is the $n \times n$ identity matrix, and DT_λ denotes the derivative of T_λ. We class u as hyperbolic, if all the absolute values of the eigenvalues of T_λ are different from unity [4]. The topological type of a hyperbolic fixed point is determined by the dim E^u and $\det L^u$, where E^u is the intersection of $\mathbb{R}^n$ and the direct sum of the generalized eigenspaces of $DT_\lambda(u)$ corresponding to the eigenvalues μ such that $|\mu| > 1$, and $L^u = DT_\lambda(u)|_{E^u}$.

A hyperbolic fixed point is called D-type, if $\det L^u > 0$, and I-type if $\det L^u < 0$. By this definition we have $2n$ topologically different types of hyperbolic fixed points. These types are:

$$ {}_kD \quad (k = 0, 1, \ldots, n), \qquad {}_kI \quad (k = 1, \ldots, n-1)\,, $$

where D and I denote the type of the fixed point and the subscript integer indicates the dimension of the unstable subspace: $k = \dim E^u$. This classification is also obtained from the distribution of characteristic multipliers of Eq.(4). That is, D and I correspond to the even and odd number of characteristic multipliers on the real axis $(-\infty, -1)$, and k indicates the number of characteristic multipliers outside the unit circle in the complex plane. The distribution can be checked by the coefficients of Eq.(4). For more detailed information see [2, 5]. Note that for the two dimensional case: $n = 2$, we have four types of hyperbolic fixed points: ${}_0D$ (sink), ${}_1D$, ${}_1I$ (saddle), and ${}_2D$ (source), see [6]. In the following we use the notation ${}_kD^m$ (resp. ${}_kI^m$) denoting a D-type (I-type) hyperbolic m-periodic point of T_λ. If $m = 1$, it will be omitted.

2.2 Codimension One Bifurcations

Bifurcation occurs when the topological type of a fixed point is changed by the variation of its system parameter λ. For the codimension one bifurcation, we have three different types of bifurcations: the tangent, period doubling and Hopf bifurcation. These bifurcations are observed when the hyperbolicity is destroyed, corresponding to the critical distribution of the characteristic multiplier: $\mu = +1$, $\mu = -1$, or $\mu = e^{j\theta}$, where $j = \sqrt{-1}$.

I-a: *Tangent bifurcation (abbr. T-bifurcation)*

Under the change of the parameter λ the generation or extinction of a couple of fixed points occurs at $\lambda = \lambda_0$. The types of bifurcation are

$$ \phi \leftrightarrow {}_{k-1}D + {}_kD\,, \tag{5a} $$

$$ \phi \leftrightarrow {}_{k-1}I + {}_kI\,, \tag{5b} $$

the symbol $\leftrightarrow$ indicating the relation before and after the bifurcation and ϕ denotes the extinction of fixed points. This type of bifurcation is observed if one of the eigenvalues of Eq.(4) satisfies the condition $\mu = 1$, or equivalently

$$ \chi(u, \lambda, 1) = 0\,, \tag{6} $$

and the remainder of the eigenvalues lies off the unit circle. For convenience sake, we shall use the notation G to denote the T-bifurcation set in the bifurcation diagram.

I-b: *Period doubling bifurcation (abbr. P-bifurcation)*

The types of bifurcations are

$$ {}_kD \rightarrow {}_{k+1}I + 2{}_kD^2\,, \tag{7a} $$

$$ {}_kD \rightarrow {}_{k-1}I + 2{}_kD^2\,, \tag{7b} $$

$$_kI \to {}_{k+1}D + 2_kD^2 , \tag{7c}$$

$$_kI \to {}_{k-1}D + 2_kD^2 , \tag{7d}$$

where 2_kD^2 indicates two numbers of 2-periodic point of type $_kD$. This type of bifurcation is observed if $\mu = -1$, or equivalently

$$\chi(u, \lambda, -1) = 0 , \tag{8}$$

and no other eigenvalues are on the unit circle. We use the notation I to denote the P-bifurcation parameter set in the parameter plane.

I-c: *the Hopf bifurcation (abbr. H-bifurcation)*

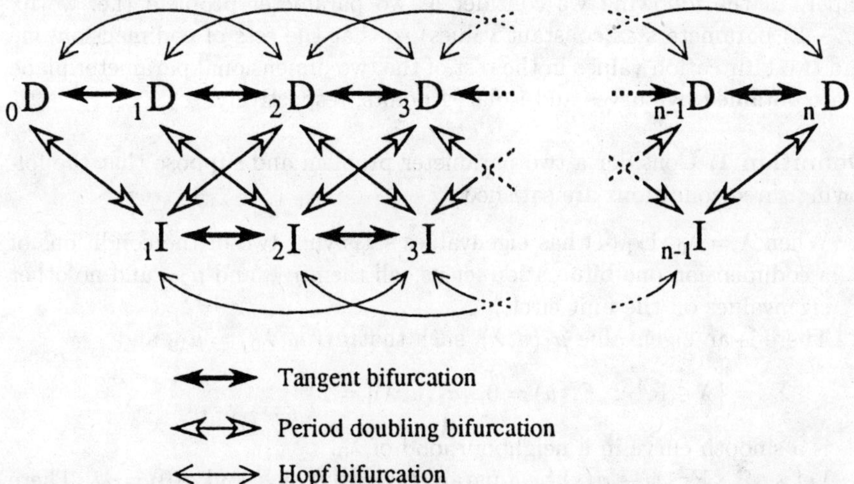

Fig. 1. Schematic diagram of the relation between codimension one bifurcations

The types of bifurcations are

$$_kD \to {}_{k+2}D + ICC , \tag{9a}$$

$$_kD \to {}_{k-2}D + ICC , \tag{9b}$$

$$_kI \to {}_{k+2}I + ICC , \tag{9c}$$

$$_kI \to {}_{k-2}I + ICC , \tag{9d}$$

where ICC indicates an invariant closed curve of T_λ which generally corresponds to quasi-periodic solutions of Eq.(1). This type of bifurcation is observed if a simple pair of complex conjugate roots of Eq.(4) transverses the unit circle in the complex plane. The condition for this type of bifurcation is given by:

$$\chi(u, \lambda, e^{j\theta}) = 0 . \tag{10}$$

Hence, eliminating θ in Eq.(10), we obtain a bifurcation condition

$$\chi_H(u, \lambda) = 0 . \tag{11}$$

Note that in this case we need an additional inequality satisfying the condition: $|\cos\theta| < 1$. We use the notation H to denote the H-bifurcation parameter set.

Fig. 1 shows the relation between the above three generic bifurcations.

2.3 Codimension Two Bifurcations

We give the definition of codimension two bifurcations considered in this paper. In the following we consider a two parameter problem (i.e. we fix $(m-2)$ parameters as constant values) so that the sets of codimension one and two bifurcation values in the rest of the two dimensional parameter plane Λ are obtained by curves and isolated points, respectively.

Definition 1. Consider a two parameter problem and suppose that the following three conditions are satisfied:

1. When $\lambda = \lambda_0$, Eq.(4) has eigenvalues satisfying two of the conditions of a codimension one bifurcation let us call them μ_{10} and μ_{20}, and no other eigenvalues on the unit circle.
2. There is an eigenvalue $\mu_1(u, \lambda)$, such that $\mu_1(u_0, \lambda_0) = \mu_{10}$ and
$$\Sigma := \{\lambda \in \mathbb{R}^2 \ : \ F_\lambda(u) = 0, \ |\mu_1(u, \lambda)| = 1\}$$
is a smooth curve in a neighbourhood of λ_0.
3. Let $s : \mathbb{R} \times \mathbb{R}^2$; $s \to \sigma(s)$ be a parameterization of Σ and $\sigma(0) = \lambda_0$. There exists an eigenvalue $\mu_2(s)$ such that $\mu_2(0) = \mu_{20}$ and $\frac{d}{ds}|\mu_2(0)| \neq 0$.

Then we say that a codimension two bifurcation occurs at the parameter λ_0 and fixed point u_0.

Note that in this paper we do not discuss the degeneracy with respect to state, so the problem of cusp points is not within the framework of our consideration. The conditions 2 and 3 imply that, in the neighbourhood of λ_0, the topological feature of the codimension one bifurcations in the parameter plane is robust for perturbation of states and parameters.

From condition 1 of the definition, there are six types of codimension two bifurcations combining two conditions of codimension one bifurcations. Whilist, in a bifurcation diagram of the Λ plane, we observe codimension two bifurcation at the intersecting point of several curves representing codimension one bifurcations. The possible types of planar bifurcation diagrams including codimension two bifurcation point is sketched in Fig. 2. In each diagram, the condition on codimension two bifurcation is different and the number of types is the same as the number of combinations stated above.

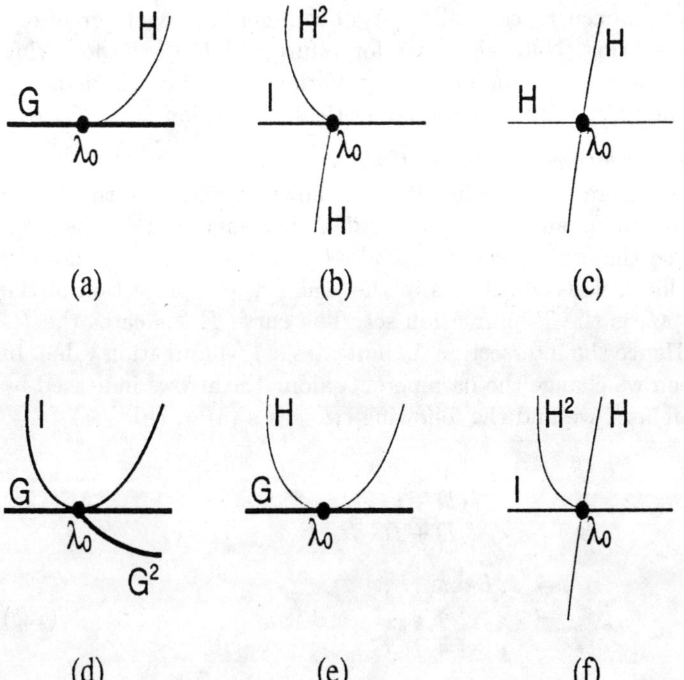

Fig. 2. Classification of codimension two bifurcations by combinatorial intersection of codimension one bifurcation curves in the Λ plane. The point λ_0 indicates the parameter values of (a) T^2-, (b) P^2-, (c) H^2-, (d) TP-, (e) TH- or (f) HP-bifurcation. The curves labelled by the symbols G, I and H denote the T-, P- and H-bifurcation sets for fixed points, respectively. G^2 and H^2 represent the codimension one bifurcation sets for 2-periodic points

According to each diagram shown in Fig. 2 we class the different types of codimension two bifurcation as follows:

II-a: *Double tangent bifurcation (T^2-bifurcation)*

II-b: *Double period doubling bifurcation (P^2-bifurcation)*

II-c: *Double Hopf bifurcation (H^2-bifurcation)*

II-d: *Tangent-Period doubling bifurcation (TP-bifurcation)*

II-e: *Tangent-Hopf bifurcation (TH-bifurcation)*

II-f: *Hopf Period doubling bifurcation (HP-bifurcation)*

Note that the product of all eigenvalues of Eq.(4) becomes positive by the Liouville formula. Hence three eigenvalues such that $\mu_1\mu_2\mu_3 > 0$ can never satisfy the conditions for HP-bifurcation. This means that this type of bifurcation is never observed in a 3-dimensional system. Evidently H^2-bifurcation cannot occur in a 3-dimensional system. A 4-dimensional system has the possibility of occurrence of all types of codimension two bifurcations classified by our setting. Note also that for bifurcated *ICC* the following bifurcation relations are only one possibility. Variants may be considered by changing the combination of super- or sub-critical bifurcation for *ICC*.

II-a: *Double tangent bifurcation (abbr. T^2-bifurcation)*

A schematic diagram of this bifurcation is shown in Fig. 3. In this figure Λ is the parameter plane and the vertical axis indicates a norm of the fixed point of Eq.(3), so that the folded manifold M is a fixed point manifold in $\mathbb{R}^n \times \Lambda$. The bold curve G which lies in the plane Λ, the projection of the folded curve on M, is the T-bifurcation set. The curve H indicates the H-bifurcation set. Hence the intersection λ_0 indicates a T^2-bifurcation value. In this diagram when we change the parameter λ along the arrows indicated by the encircled numbers, we find the following two cases (a) or (b):

$$\begin{array}{llrcl} (a) & ① & \phi & \rightarrow & {}_kD + {}_{k+1}D\,, \\ & ② & \phi & \rightarrow & {}_{k+1}D + {}_{k+2}D\,, \\ & ③ & {}_kD & \rightarrow & {}_{k+2}D + ICC\,, \end{array} \tag{12a}$$

$$\begin{array}{llrcl} (b) & ① & \phi & \rightarrow & {}_kI + {}_{k+1}I\,, \\ & ② & \phi & \rightarrow & {}_{k+1}I + {}_{k+2}I\,, \\ & ③ & {}_kI & \rightarrow & {}_{k+2}I + ICC\,. \end{array} \tag{12b}$$

On the other hand the bifurcation conditions are given by:

$$\chi(u, \lambda, 1) = 0\,, \tag{13a}$$

$$\frac{d}{d\mu}\chi(u, \lambda, 1) = 0 \tag{13b}$$

for $u = u_0$, $\lambda = \lambda_0$.

II-b: *Double period doubling bifurcation (abbr. P^2-bifurcation)*

This bifurcation appears at the intersection of the P- and H-bifurcation curves, see Fig. 4. M^2 indicates the 2-periodic point manifold produced by the P-bifurcation. The H-bifurcation for the 2-periodic point occurs on the curve H^2 which emanates from $\lambda = \lambda_0$. Two kinds of bifurcation patterns are possible:

$$\begin{array}{llrcl} (a) & ① & {}_kD & \rightarrow & {}_{k+1}I + 2{}_kD^2\,, \\ & ② & {}_{k+2}D & \rightarrow & {}_{k+1}I + 2{}_{k+2}D^2\,, \\ & ③ & {}_kD & \rightarrow & {}_{k+2}D + ICC\,, \\ & ④ & {}_kD^2 & \rightarrow & {}_{k+2}D^2 + ICC\,, \end{array} \tag{14a}$$

$$\begin{array}{llrcl} (b) & ① & {}_kI & \rightarrow & {}_{k+1}D + 2{}_kD^2\,, \\ & ② & {}_{k+2}I & \rightarrow & {}_{k+1}D + 2{}_{k+2}D^2\,, \\ & ③ & {}_kI & \rightarrow & {}_{k+2}I + ICC\,, \\ & ④ & {}_kD^2 & \rightarrow & {}_{k+2}D^2 + ICC\,. \end{array} \tag{14b}$$

The bifurcation conditions are given by:

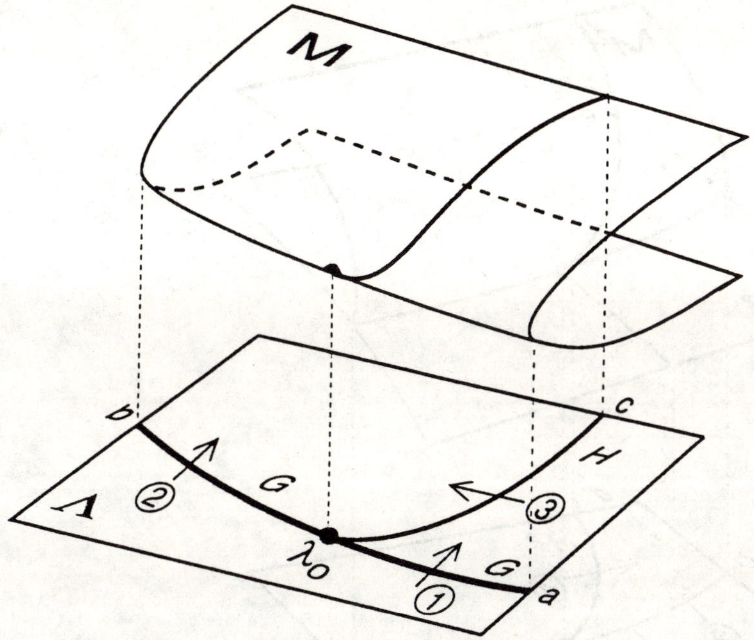

Fig. 3. Schematic diagram of the T^2-bifurcation

$$\chi(u, \lambda, -1) = 0\,, \tag{15a}$$

$$\frac{d}{d\mu}\chi(u, \lambda, -1) = 0 \tag{15b}$$

for $u = u_0$, $\lambda = \lambda_0$.

II-c: *Double Hopf bifurcation (abbr. H^2-bifurcation)*

In this case two H-bifurcation curves transversely intersect at $\lambda = \lambda_0$, see Fig. 5. Two cases are considered when we change the parameters along the arrows:

$$\begin{array}{llrcl} (a) & ① & {}_kD & \rightarrow & {}_{k+2}D + ICC\,, \\ & ② & {}_{k+2}D & \rightarrow & {}_{k+4}D + ICC\,, \\ & ③ & {}_kD & \rightarrow & {}_{k+2}D + ICC\,, \\ & ④ & {}_{k+2}D & \rightarrow & {}_{k+4}D + ICC\,, \end{array} \tag{16a}$$

$$\begin{array}{llrcl} (b) & ① & {}_kI & \rightarrow & {}_{k+2}I + ICC\,, \\ & ② & {}_{k+2}I & \rightarrow & {}_{k+4}I + ICC\,, \\ & ③ & {}_kI & \rightarrow & {}_{k+2}I + ICC\,, \\ & ④ & {}_{k+2}I & \rightarrow & {}_{k+4}I + ICC\,, \end{array} \tag{16b}$$

where the appearance of ICC on the right and left hand side depends on the higher order nonlinearity of the system. The bifurcation conditions are given by:

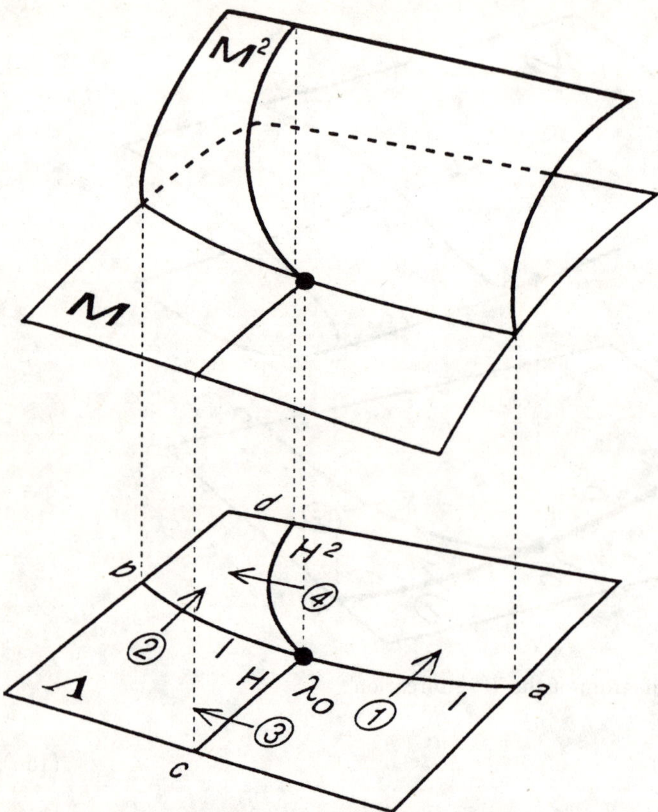

Fig. 4. Schematic diagram of the P^2-bifurcation

$$\chi(u\lambda, \mathrm{e}^{j\theta_1}) = 0\,, \tag{17a}$$

$$\chi(u, \lambda, \mathrm{e}^{j\theta_2}) = 0 \tag{17b}$$

for $u = u_0$, $\lambda = \lambda_0$. By eliminating θ_1 and θ_2 in Eqs. (17) we obtain the conditions

$$\chi_{H_1}(u, \lambda) = 0\,, \tag{18a}$$

$$\chi_{H_2}(u, \lambda) = 0 \tag{18b}$$

with additional inequalities satisfying $|\cos\theta_1| < 1$ and $|\cos\theta_2| < 1$. For example, if we consider a 4-dimensional system, Eqs. (18) are actually written as:

$$\chi_{H_1}(u, \lambda) = a_1 - a_3 = 0\,, \tag{19a}$$

$$\chi_{H_2}(u, \lambda) = a_4 - 1 = 0\,, \tag{19b}$$

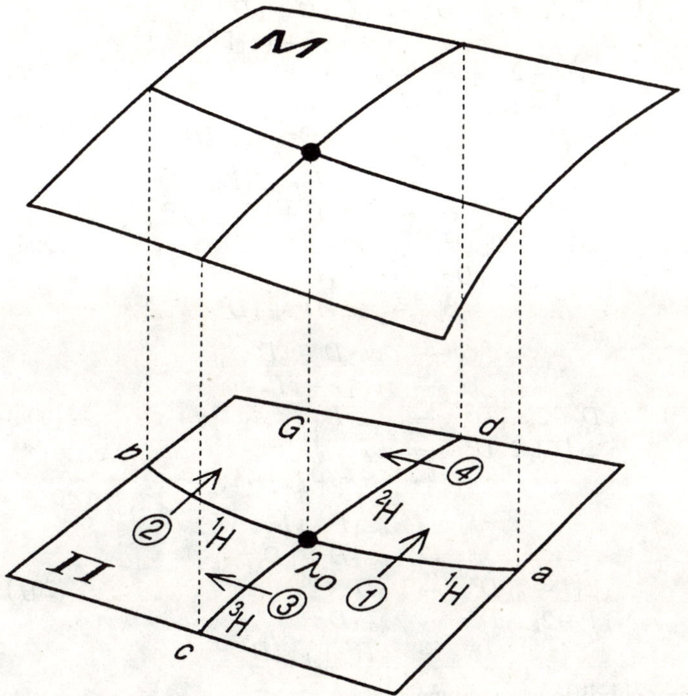

Fig. 5. Schematic diagram of the H^2-bifurcation

and the inequalities to be checked are reduced to the conditions $|a_1| - a_2/2 < 1$ and $a_1^2/4 - a_2 + 2 > 0$, where a_k, $k = 1, \ldots, 4$, is a coefficient of Eq.(4).

II-d: *Tangent-Period doubling bifurcation (abbr. TP-bifurcation)*

Fig. 6 illustrates this bifurcation. Fixed and bifurcated 2-periodic manifolds are both folded so that the T-bifurcation is forced to exist for the 2-periodic point as shown in Fig. 6. Six possible cases are considered:

$$(a)\quad \begin{array}{lrcl} ① & \phi & \to & {}_{k-1}D + {}_kD\,, \\ ② & \phi & \to & {}_kI + {}_{k+1}I\,, \\ ③ & {}_kD + 2{}_{k+1}D^2 & \to & {}_{k+1}I\,, \\ ④ & {}_{k-1}D + 2{}_kD^2 & \to & {}_kI\,, \\ ⑤ & \phi & \to & {}_kD^2 + {}_{k+1}D^2\,, \end{array} \tag{20a}$$

$$(b)\quad \begin{array}{lrcl} ① & \phi & \to & {}_kI + {}_{k+1}I\,, \\ ② & \phi & \to & {}_{k-1}D + {}_kD\,, \\ ③ & {}_{k+1}I + 2{}_kD^2 & \to & {}_kD\,, \\ ④ & {}_kI + 2{}_{k-1}D^2 & \to & {}_{k-1}D\,, \\ ⑤ & \phi & \to & {}_kD^2 + {}_{k+1}D^2\,, \end{array} \tag{20b}$$

$$
(c)\quad
\begin{array}{cr@{\;\rightarrow\;}l}
① & \phi & {}_{k-1}D + {}_kD\,, \\
② & \phi & {}_{k-2}I + {}_{k-1}I\,, \\
③ & {}_kD + 2{}_{k-1}D^2 & {}_{k-1}I\,, \\
④ & {}_{k-1}D + 2{}_{k-2}D^2 & {}_{k-2}I\,, \\
⑤ & \phi & {}_{k-2}D^2 + {}_{k-1}D^2\,,
\end{array}
\tag{20c}
$$

$$
(d)\quad
\begin{array}{cr@{\;\rightarrow\;}l}
① & \phi & {}_{k-2}I + {}_{k-1}I\,, \\
② & \phi & {}_{k-1}D + {}_kD\,, \\
③ & {}_{k-1}I + 2{}_kD^2 & {}_kD\,, \\
④ & {}_{k-2}I + 2{}_{k-1}D^2 & {}_{k-1}D\,, \\
⑤ & \phi & {}_{k-1}D^2 + {}_kD^2\,,
\end{array}
\tag{20d}
$$

$$
(e)\quad
\begin{array}{cr@{\;\rightarrow\;}l}
① & \phi & {}_{k-1}D + {}_kD\,, \\
② & \phi & {}_{k-1}I + {}_kI\,, \\
③ & {}_kD + 2{}_{k-1}D^2 & {}_{k-1}I\,, \\
④ & {}_{k-1}D + 2{}_kD^2 & {}_kI\,, \\
⑤ & \phi & {}_kD^2 + {}_{k-1}D^2\,,
\end{array}
\tag{20e}
$$

$$
(f)\quad
\begin{array}{cr@{\;\rightarrow\;}l}
① & \phi & {}_{k-1}I + {}_kI\,, \\
② & \phi & {}_{k-1}D + {}_kD\,, \\
③ & {}_{k-1}I + 2{}_kD^2 & {}_kD\,, \\
④ & {}_kI + 2{}_{k-1}D^2 & {}_{k-1}D\,, \\
⑤ & \phi & {}_kD^2 + {}_{k-1}D^2\,.
\end{array}
\tag{20f}
$$

The bifurcation conditions are given by:

$$\chi(u, \lambda, 1) = 0\,, \tag{21a}$$

$$\chi(u, \lambda, -1) = 0 \tag{21b}$$

for $u = u_0$, $\lambda = \lambda_0$.

II-e: *Tangent-Hopf bifurcation (abbr. TH-bifurcation)*
This bifurcation occurs at the intersection of the T- and H-bifurcation curves, see Fig. 7. Two possible cases can arise:

$$
(a)\quad
\begin{array}{cr@{\;\rightarrow\;}l}
① & \phi & {}_{k-1}D + {}_kD\,, \\
② & \phi & {}_{k+1}D + {}_{k+2}D\,, \\
③ & {}_kD & {}_{k+2}D + ICC\,, \\
④ & {}_{k-1}D & {}_{k+1}D + ICC\,,
\end{array}
\tag{22a}
$$

$$
(b)\quad
\begin{array}{cr@{\;\rightarrow\;}l}
① & \phi & {}_{k-1}I + {}_kI\,, \\
② & \phi & {}_{k+1}I + {}_{k+2}I\,, \\
③ & {}_kI & {}_{k+2}I + ICC\,, \\
④ & {}_{k-1}I & {}_{k+1}I + ICC\,.
\end{array}
\tag{22b}
$$

The bifurcation conditions are given by:

$$\chi(u, \lambda, 1) = 0\,, \tag{23a}$$

$$\chi(e^{j\theta}) = 0 \tag{23b}$$

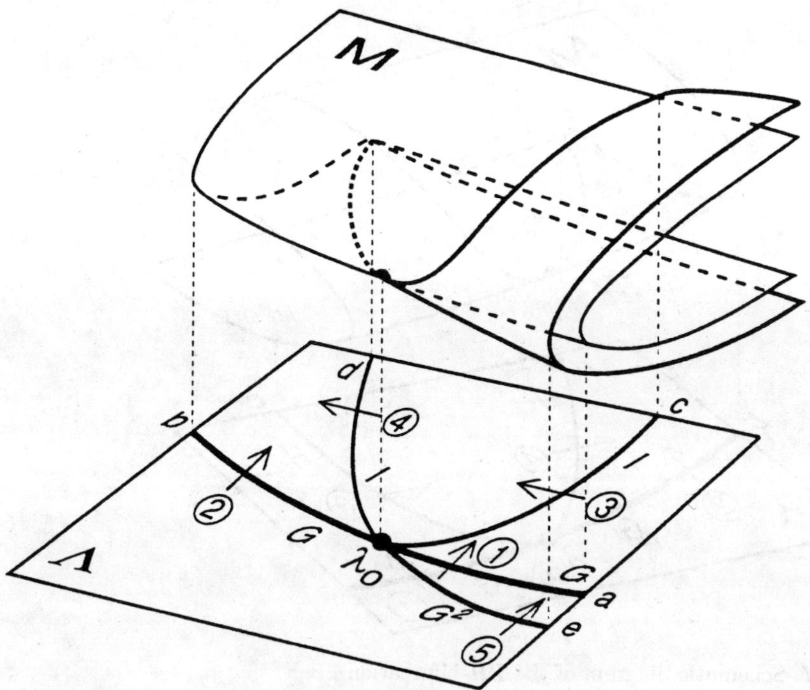

Fig. 6. Schematic diagram of the TP-bifurcation

for $u = u_0$, $\lambda = \lambda_0$.

II-f: *Hopf Period doubling bifurcation (abbr. HP-bifurcation)*
This bifurcation occurs at the intersection of the T- and H-bifurcation curves, see Fig. 8. Four possible cases can arise:

$$
\begin{array}{llrcl}
(a) & ① & {}_kD & \rightarrow & {}_{k+1}I + 2{}_kD^2 \,, \\
 & ② & {}_{k+2}D & \rightarrow & {}_{k+3}I + 2{}_{k+2}D^2 \,, \\
 & ③ & {}_kD & \rightarrow & {}_{k+2}D + ICC \,, \\
 & ④ & {}_{k+1}I & \rightarrow & {}_{k+3}I + ICC \,, \\
 & ⑤ & {}_kD^2 & \rightarrow & {}_{k+2}D^2 + ICC \,,
\end{array}
\tag{24a}
$$

$$
\begin{array}{llrcl}
(b) & ① & {}_{k+1}I & \rightarrow & {}_kD + 2{}_{k+1}D^2 \,, \\
 & ② & {}_{k+3}I & \rightarrow & {}_{k+2}D + 2{}_{k+3}D^2 \,, \\
 & ③ & {}_{k+1}I & \rightarrow & {}_{k+3}I + ICC \,, \\
 & ④ & {}_kD & \rightarrow & {}_{k+2}D + ICC \,, \\
 & ⑤ & {}_{k+1}D^2 & \rightarrow & {}_{k+3}D^2 + ICC \,,
\end{array}
\tag{24b}
$$

$$
\begin{array}{llrcl}
(c) & ① & {}_kD & \rightarrow & {}_{k-1}I + 2{}_kD^2 \,, \\
 & ② & {}_{k+2}D & \rightarrow & {}_{k+1}I + 2{}_{k+2}D^2 \,, \\
 & ③ & {}_kD & \rightarrow & {}_{k+2}D + ICC \,, \\
 & ④ & {}_{k-1}I & \rightarrow & {}_{k+1}I + ICC \,, \\
 & ⑤ & {}_kD^2 & \rightarrow & {}_{k+2}D^2 + ICC \,,
\end{array}
\tag{24c}
$$

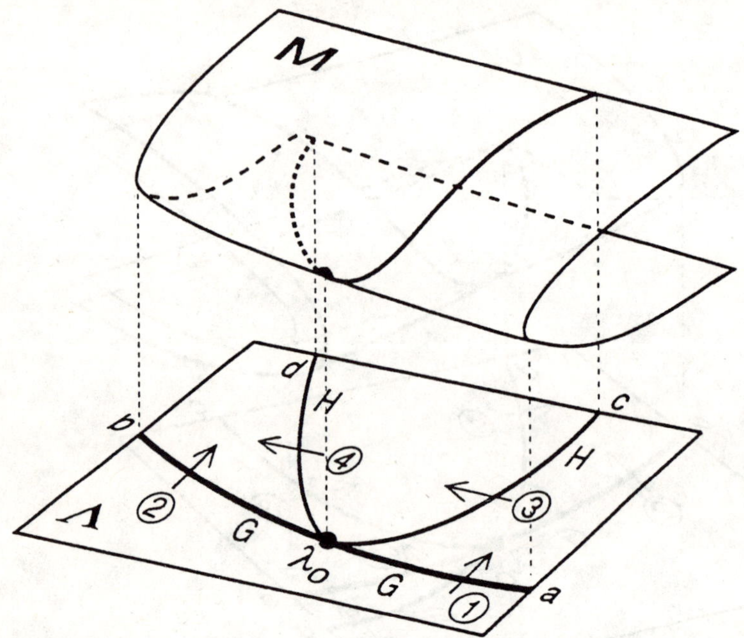

Fig. 7. Schematic diagram of the TH-bifurcation

$$
\begin{array}{llcl}
(d) & ① & {}_{k-1}I \rightarrow {}_{k}D + 2{}_{k-1}D^2 \,, & \\
& ② & {}_{k+1}I \rightarrow {}_{k+2}D + 2{}_{k+1}D^2 \,, & \\
& ③ & {}_{k-1}I \rightarrow {}_{k+1}I + ICC \,, & (24d) \\
& ④ & {}_{k}D \rightarrow {}_{k+2}D + ICC \,, & \\
& ⑤ & {}_{k-1}D^2 \rightarrow {}_{k+1}D^2 + ICC \,. &
\end{array}
$$

The bifurcation conditions are given by:

$$\chi(u, \lambda, -1) = 0 \,, \tag{25a}$$

$$\chi(e^{j\theta}) = 0 \tag{25b}$$

for $u = u_0$, $\lambda = \lambda_0$.

Note finally that a cascade of P-bifurcations is frequently observed in many dynamical systems. If the P-bifurcation occurs successively together with the P^2-, TP- or HP-bifurcation, then we may find the successive occurrences of the codimension two bifurcation.

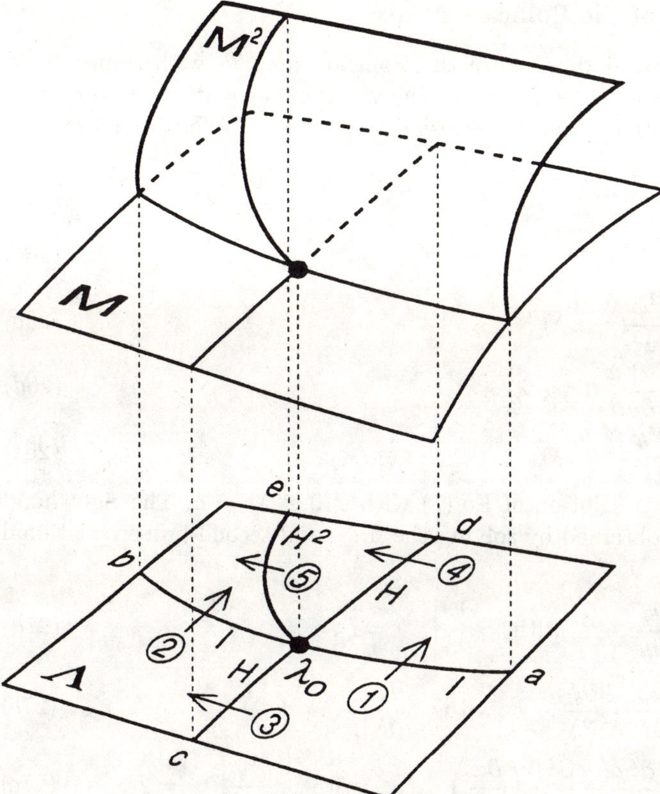

Fig. 8. Schematic diagram of the HP-bifurcation

3. Computational Algorithms

In this section we consider a computational method for finding the bifurcation values of parameters. We use a shooting algorithm combined with Newton's method. The fixed point equation (3) and the bifurcation conditions stated in the preceding section are solved by this method. Note that Eq.(3) becomes singular, that is, the Jacobian matrix becomes singular, at the point of tangent bifurcation. This kind of singularity can be eliminated however by solving Eq.(3) and the bifurcation condition simultaneously with respect to the location of the fixed point and one component of the parameter λ ($(n+1)$-variables). We have already applied this strategy to the codimension one bifurcations and obtain effective algorithms, see [2, 7]. Hence in the following section we shall mainly apply our algorithm to codimension two bifurcations. At first we introduce derivatives of the Poincaré map T_λ, which will be used to calculate Jacobian matrix.

3.1 Derivatives of the Poincaré Map

The first and higher derivatives of the Poincaré map T_λ with respect to u and λ are obtained by the solutions of the variational equations with respect to the initial condition u and the system parameter λ. In fact we have:

$$\frac{\partial}{\partial u}T_\lambda(u) = \frac{\partial \varphi}{\partial u}(\tau, u, \lambda)\,, \tag{26a}$$

$$\frac{\partial}{\partial \lambda}T_\lambda(u) = \frac{\partial \varphi}{\partial \lambda}(\tau, u, \lambda)\,, \tag{26b}$$

$$\frac{\partial^2}{\partial u^2}T_\lambda(u) = \frac{\partial^2 \varphi}{\partial u^2}(\tau, u, \lambda)\,, \tag{26c}$$

$$\frac{\partial^2}{\partial u \partial \lambda}T_\lambda(u) = \frac{\partial^2 \varphi}{\partial u \partial \lambda}(\tau, u, \lambda)\,, \tag{26d}$$

$$\frac{\partial^2}{\partial \lambda^2}T_\lambda(u) = \frac{\partial^2 \varphi}{\partial \lambda^2}(\tau, u, \lambda)\,, \tag{26e}$$

where $\varphi(t, u, \lambda)$ is a solution of Eq.(1) with $\varphi(0, u, \lambda) = u$. The right-hand side of Eq.(26) is obtained by solving the first- and second-order variational equations:

$$\frac{d}{dt}\frac{\partial \varphi}{\partial u} = \frac{\partial f}{\partial x}\frac{\partial \varphi}{\partial u}\,, \qquad \text{with} \quad \left.\frac{\partial \varphi}{\partial u}\right|_{t=0} = I\,, \tag{27a}$$

$$\frac{d}{dt}\frac{\partial \varphi}{\partial \lambda} = \frac{\partial f}{\partial x}\frac{\partial \varphi}{\partial \lambda} + \frac{\partial f}{\partial \lambda}\,, \qquad \text{with} \quad \left.\frac{\partial \varphi}{\partial \lambda}\right|_{t=0} = I\,, \tag{27b}$$

$$\frac{d}{dt}\frac{\partial^2 \varphi}{\partial u^2} = \frac{\partial f}{\partial x}\frac{\partial^2 \varphi}{\partial u^2} + \frac{\partial^2 f}{\partial x^2}\left(\frac{\partial \varphi}{\partial u}\right)^2\,, \qquad \text{with} \quad \left.\frac{\partial^2 \varphi}{\partial u^2}\right|_{t=0} = I\,, \tag{27c}$$

$$\frac{d}{dt}\frac{\partial^2 \varphi}{\partial u \partial \lambda} = \frac{\partial f}{\partial x}\frac{\partial^2 \varphi}{\partial u \partial \lambda} + \frac{\partial^2 f}{\partial x^2}\frac{\partial \varphi}{\partial u}\frac{\partial \varphi}{\partial \lambda} + \frac{\partial^2 f}{\partial x \partial \lambda}\frac{\partial \varphi}{\partial u}\,, \qquad \text{with} \quad \left.\frac{\partial^2 \varphi}{\partial u \partial \lambda}\right|_{t=0} = I\,, \tag{27d}$$

$$\frac{d}{dt}\frac{\partial^2 \varphi}{\partial \lambda^2} = \frac{\partial f}{\partial x}\frac{\partial^2 \varphi}{\partial \lambda^2} + \frac{\partial^2 f}{\partial x^2}\left(\frac{\partial \varphi}{\partial \lambda}\right)^2 + 2\frac{\partial^2 f}{\partial x \partial \lambda}\frac{\partial \varphi}{\partial \lambda} + \frac{\partial^2 f}{\partial \lambda^2}\,, \qquad \text{with} \quad \left.\frac{\partial^2 \varphi}{\partial \lambda^2}\right|_{t=0} = 0\,. \tag{27e}$$

These relations are easily obtained by differentiating the following equations as necessary as:

$$\frac{d}{dt}\varphi(t, u, \lambda) = f(t, \varphi(t, u, \lambda), \lambda) \tag{28}$$

and

$$\varphi(0, u, \lambda) = u\,. \tag{29}$$

3.2 Numerical Method of Analysis

Let us consider a method for finding the bifurcation value $\lambda = \lambda_0$ and the fixed point $u = u_0$ at $\lambda = \lambda_0$. As shown above the method is accomplished by solving Eq.(3) and the bifurcation conditions, simultaneously. To illustrate our method consider the T^2-bifurcation. The fixed point equation and the bifurcation conditions are given by Eqs. (3) and (13):

$$\begin{aligned} F(u,\lambda) &:= u - T_\lambda(u) = 0\,, \\ G_1(u,\lambda) &:= \chi(u,\lambda,1) = 0\,, \\ G_2(u,\lambda) &:= \frac{d}{d\mu}\chi(u,\lambda,1) = 0\,. \end{aligned} \tag{30}$$

The unknown variables are chosen as $u \in \mathbb{R}^n$ and two components of $\lambda = (\lambda_1, \lambda_2, \lambda_3, \ldots, \lambda_n) \in \mathbb{R}^m$, where we fix $(m-2)$ components $\lambda_3, \ldots, \lambda_m$ as constant parameters. Hence the problem is reduced to the determination of $(n+2)$ u-variables and (λ_1, λ_2) for $(n+2)$ number of equations (30). We can prove that if appropriate components λ_1 and λ_2 are selected from λ, then the Jacobian matrix of Eqs. (30) becomes nonsingular. Hence we can apply Newton's method for solving Eqs. (30).

Rewriting $(n+2)$ variables and Eqs. (30) as

$$\begin{aligned} y &= (u, \lambda_1, \lambda_2)\,, \\ F(y) &:= (F(y), G_1(y), G_2(y)) = 0\,, \end{aligned} \tag{31}$$

and assuming y^* is an isolated solution of Eqs. (31), we obtain an iterative algorithm as follows:

1. Guess an initial $y^{(0)}$, i.e., an approximate solution $y^{(0)}$ of y^*.
2. Compute $y^{(k+1)}$ from $y^{(k)}$ using the following relation:

$$\begin{aligned} y^{(k+1)} &= y^{(k)} + h\,, \\ DF(y^{(k)})h + F(y^{(k)}) &= 0\,. \end{aligned}$$

3. Repeat until

$$\|y^{(k+1)} - y^{(k)}\| < \varepsilon_y\,, \text{ or } \|F(y^{(k+1)})\| < \varepsilon_F$$

is satisfied, where ε_y and ε_F are prescribed small positive constants.

Example. As an example for Eqs. (31), we consider the Duffing-van der Pol equation:

$$\frac{d^2x}{dt^2} - \varepsilon(1-x^2)\frac{dx}{dt} + x^3 = B\cos\nu t\,. \tag{32}$$

In this two dimensional case we only observe T^2- or P^2-bifurcation stated in the relations (12) or (14) as codimension two bifurcations. Let us derive Eqs.

(31) in order to calculate the T^2-bifurcation point. We assume that B and ν are parameters.

Rewriting Eq.(32) as

$$\begin{aligned}\frac{dx_1}{dt} &= x_2\,, \\ \frac{dx_2}{dt} &= \frac{\varepsilon}{\nu}(1-x_1^2)x_2 - \frac{1}{\nu^2}x_1^3 + \frac{B}{\nu^2}\cos t\,,\end{aligned} \tag{33}$$

and denoting the solution of Eqs. (33) as

$$\begin{aligned}x_1(t) &= \varphi_1(t, u_1, u_2, B, \nu)\,, \\ x_2(t) &= \varphi_2(t, u_1, u_2, B, \nu)\,,\end{aligned} \tag{34}$$

we have the fixed point equation (3) and the bifurcation conditions (13) as:

$$\begin{aligned}F_1(u_1, u_2, B, \nu) &= u_1 - \varphi_1(2\pi, u_1, u_2, B, \nu) = 0\,, \\ F_2(u_1, u_2, B, \nu) &= u_2 - \varphi_2(2\pi, u_1, u_2, B, \nu) = 0\,,\end{aligned} \tag{35a}$$

$$\begin{aligned}G_1(u_1, u_2, B, \nu) &= 1 - \left(\frac{\partial\varphi_1}{\partial u_1} + \frac{\partial\varphi_2}{\partial u_2}\right) + \frac{\partial\varphi_1}{\partial u_1}\frac{\partial\varphi_2}{\partial u_2} - \frac{\partial\varphi_1}{\partial u_2}\frac{\partial\varphi_2}{\partial u_1} = 0\,, \\ G_2(u_1, u_2, B, \nu) &= 2 - \left(\frac{\partial\varphi_1}{\partial u_1} + \frac{\partial\varphi_2}{\partial u_2}\right) = 0\,.\end{aligned} \tag{35b}$$

To evaluate the Jacobian matrix of Eqs. (35), we must obtain numerical values for Eqs. (34) at $t = 2\pi$ and

$$\begin{aligned}&(x_3, x_4) := \left(\frac{\partial\varphi_1}{\partial u_1}, \frac{\partial\varphi_2}{\partial u_1}\right), \quad (x_5, x_6) := \left(\frac{\partial\varphi_1}{\partial u_2}, \frac{\partial\varphi_2}{\partial u_2}\right), \\ &(x_7, x_8) := \left(\frac{\partial\varphi_1}{\partial B}, \frac{\partial\varphi_2}{\partial B}\right), \quad (x_9, x_{10}) := \left(\frac{\partial\varphi_1}{\partial \nu}, \frac{\partial\varphi_2}{\partial \nu}\right), \\ &(x_{11}, x_{12}) := \left(\frac{\partial^2\varphi_1}{\partial u_1^2}, \frac{\partial^2\varphi_2}{\partial u_1^2}\right), \quad (x_{13}, x_{14}) := \left(\frac{\partial^2\varphi_1}{\partial u_1\partial u_2}, \frac{\partial^2\varphi_2}{\partial u_1\partial u_2}\right), \\ &(x_{15}, x_{16}) := \left(\frac{\partial^2\varphi_1}{\partial u_2^2}, \frac{\partial^2\varphi_2}{\partial u_2^2}\right), \quad (x_{17}, x_{18}) := \left(\frac{\partial^2\varphi_1}{\partial u_1\partial B}, \frac{\partial^2\varphi_2}{\partial u_1\partial B}\right), \\ &(x_{19}, x_{20}) := \left(\frac{\partial^2\varphi_1}{\partial u_2\partial B}, \frac{\partial^2\varphi_2}{\partial u_2\partial B}\right), \quad (x_{21}, x_{22}) := \left(\frac{\partial^2\varphi_1}{\partial u_1\partial \nu}, \frac{\partial^2\varphi_2}{\partial u_1\partial \nu}\right), \\ &(x_{23}, x_{24}) := \left(\frac{\partial^2\varphi_1}{\partial u_2\partial \nu}, \frac{\partial^2\varphi_2}{\partial u_2\partial \nu}\right).\end{aligned} \tag{36}$$

These values are calculated from the solutions of the following variational equations:

$$\frac{dx_{2k+1}}{dt} = x_{2k} , \qquad k = 1, 2, \ldots, 11 ,$$
$$\frac{dx_4}{dt} = P_1 x_3 + P_2 x_4 ,$$
$$\frac{dx_6}{dt} = P_1 x_5 + P_2 x_6 ,$$
$$\frac{dx_8}{dt} = P_1 x_7 + P_2 x_8 + P_3 ,$$
$$\frac{dx_{10}}{dt} = P_1 x_9 + P_2 x_{10} + P_4 ,$$
$$\frac{dx_{12}}{dt} = P_1 x_{11} + P_2 x_{12} + Q_1 x_3^2 + 2Q_2 x_3 x_4 ,$$
$$\frac{dx_{14}}{dt} = P_1 x_{13} + P_2 x_{14} + Q_1 x_3 x_5 + Q_2(x_3 x_6 + x_4 x_5) ,$$
$$\frac{dx_{16}}{dt} = P_1 x_{15} + P_2 x_{16} + Q_1 x_5^2 + 2Q_2 x_5 x_6 ,$$
$$\frac{dx_{18}}{dt} = P_1 x_{17} + P_2 x_{18} + Q_1 x_3 x_7 + Q_2(x_4 x_7 + x_3 x_8) ,$$
$$\frac{dx_{20}}{dt} = P_1 x_{19} + P_2 x_{20} + Q_1 x_5 x_7 + Q_2(x_6 x_7 + x_5 x_8) ,$$
$$\frac{dx_{22}}{dt} = P_1 x_{21} + P_2 x_{22} + Q_1 x_3 x_9 + Q_2(x_4 x_9 + x_3 x_{10}) + Q_3 x_3 + Q_4 x_4 ,$$
$$\frac{dx_{24}}{dt} = P_1 x_{23} + P_2 x_{24} + Q_1 x_5 x_9 + Q_2(x_6 x_9 + x_5 x_{10}) + Q_3 x_5 + Q_4 x_6 ,$$

where

$$P_1 = -\frac{2\varepsilon}{\nu} x_1 x_2 - \frac{3}{\nu^2} x_1^2 , \qquad P_2 = \frac{\varepsilon}{\nu}(1 - x_1^2) , \qquad P_3 = \frac{1}{\nu^2} \cos t ,$$
$$P_4 = -\frac{\varepsilon}{\nu^2}(1 - x_1^2) x_2 + \frac{2}{\nu^3} x_1^3 - \frac{2B}{\nu^3} \cos t , \qquad Q_1 = -\frac{2\varepsilon}{\nu^2} x_2 - \frac{6}{\nu^2} x_1 ,$$
$$Q_2 = -\frac{2\varepsilon}{\nu} x_1 , \qquad Q_3 = \frac{2\varepsilon}{\nu^2} x_1 x_2 + \frac{6}{\nu^3} x_1^2 , \qquad Q_4 = -\frac{\varepsilon}{\nu^2}(1 - x_1^2) .$$

4. Numerical Examples

To illustrate the computational algorithms for codimension two bifurcations discussed in the previous section, we analyze two kinds of oscillatory circuits. We can find all types of codimension two bifurcations defined in Section 2.3, and obtain several new nonlinear phenomena correlated with the codimension two bifurcations.

In the examples below, we illustrate various bifurcation sets, i.e., curves and points in the parameter plane. Therefore, for convenience' sake, we use notations: I_k^m – period doubling bifurcation set, G_k^m – tangent bifurcation set, H_k^m – the Hopf bifurcation set, where $(\ )^m$ indicates a bifurcation set for the m-periodic point of T_λ, and k denotes the number used to distinguish several sets of $(\ )^m$, if they exist. If $m = 1$ or $k = 1$, they will be omitted.

4.1 Circuit Model for Chemical Oscillation at a Water-Oil Interface

We investigate the synchronization phenomena observed in a forced oscillatory circuit whose free oscillator has a limit cycle and the successive period doubling process within the circuit. The illustrating circuit is shown in Fig. 9. Using the notation of the figure, we can set up three-dimensional nonautonomous differential equations:

$$\begin{aligned} L_1 \frac{di_1}{dt} &= E_1 - R_1 i_1 - v \,, \\ L_2 \frac{di_2}{dt} &= E_2 - R_2 i_2 - v \,, \\ C \frac{dv}{dt} &= i_1 + i_2 + v - v^3/3 + j(t) \,, \end{aligned} \tag{37}$$

where $j(t)$ is a sinusoidal current source:

$$j(t) = J \cos(\nu t) \,, \tag{38}$$

and the several system parameters are fixed as

$$\begin{aligned} L_1^{-1} &= 0.4 \,, \quad R_1 = 0.5 \,, \quad E_1 = 0.2 \,, \\ L_2^{-1} &= 0.1 \,, \quad R_2 = 0.5 \,. \end{aligned} \tag{39}$$

The setting of the system parameters is such that the occurrence of an oscillation in the free system is ensured. This oscillation is constructed by removing the forcing term from Eqs. (37). Before considering the forced oscillator, we discuss the autonomous system.

4.1.1 An Autonomous Chemical Oscillator

Let us consider periodic responses observed in a free system such as the circuit shown in Fig. 9, that is, the system equation is described by Eqs. (37) with

$$j(t) \equiv 0 \,. \tag{40}$$

This autonomous system is considered to be a circuit model for a chemical oscillation at a water-oil interface [8, 9]. It is known that the electrical potential across the water-oil interface in the presence of chemical species displays self-oscillatory phenomena. The frequency, amplitude and shape of the oscillations change on addition of various chemicals to the aqueous phase solutions.

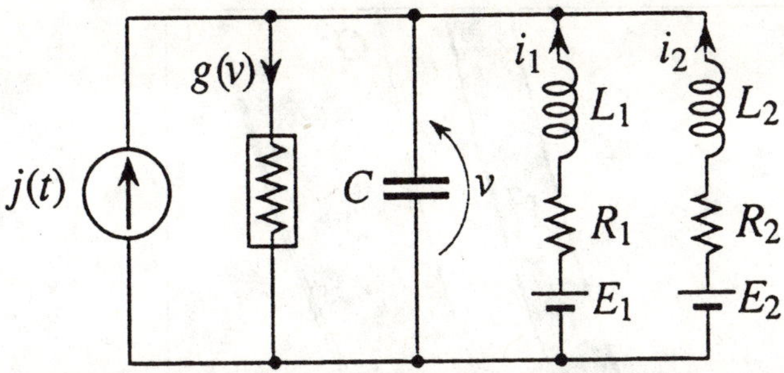

Fig. 9. Oscillatory circuit with nonlinear conductor $g(v) = v^3/3 - v$, and current source $j(t)$

Equations (37) with $j(t) \equiv 0$, $C^{-1} = 3$ and $E_2 = 1.29$, have a unique unstable equilibrium point and a stable limit cycle. By increasing the value of E_2 from 1.29, the period doubling bifurcations proceed until finally chaotic attractors are observed at $E_2 = 1.291$, for example. Table 1 shows a list of the numerical results of the value of E_2 at which the period doubling bifurcation of the limit cycle (with order of period up to 16) occurs.

Table 1. A period doubling process of limit cycles

Order of Period	E_2
1	1.2901772
2	1.2907955
4	1.2909273
8	1.2909554
16	1.2909614

Fig. 10 shows the bifurcation diagram in the (E_2, C^{-1}) plane, including the successive period doubling bifurcations for limit cycles. Examples of stable limit cycles are shown in Fig. 11.

For latter use, we now define an average natural frequency of a limit cycle. First, we choose a local cross section or Poincaré section as a half plane Π:

$$\Pi = \{(i_1, i_2, v) \; : \; v = v_0, \; i_1 + i_2 > i_{10} + i_{20}\} \; , \tag{41}$$

where (i_{10}, i_{20}, v_0) is the unique equilibrium point of Eqs. (37) with $j(t) \equiv 0$. Then we can define the Poincaré map as the first return map from Π to Π by the solution. Note that the return time $\tau(p)$ represents the time in which a trajectory emanating from a point p on Π will hit this section again. The average natural frequency, say AF for convenience' sake, of an m-periodic limit cycle is now defined as

$$AF = \frac{2\pi}{AT} \; , \tag{42}$$

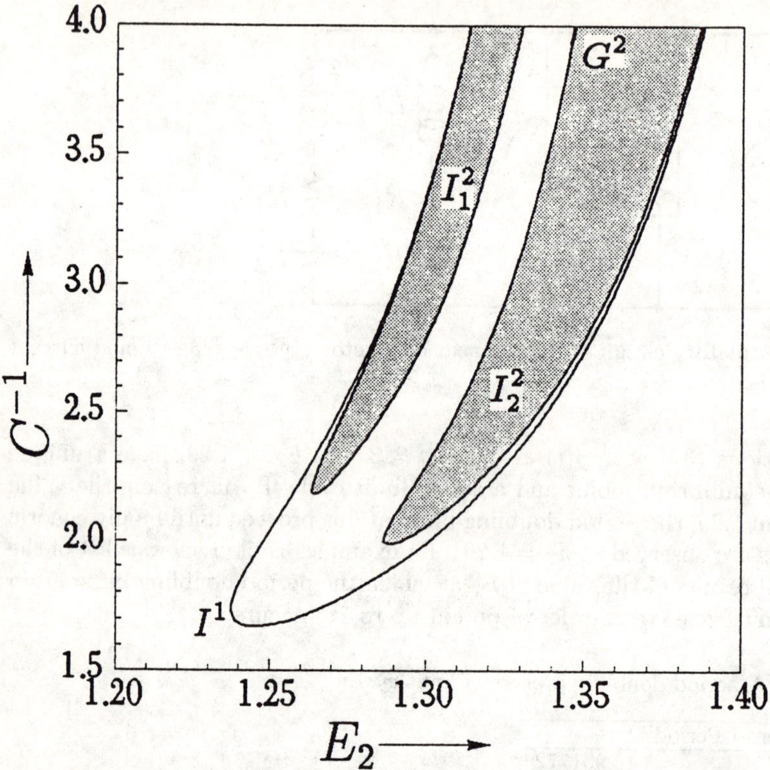

Fig. 10. Bifurcation diagram of limit cycles for the free system $j(0) \equiv 0$ in Eqs. (37). In the shaded regions we observe a chaotic attractor or periodic limit cycles

where AT (average return time) is the time mean of the return time:

$$AT = \frac{1}{m} \sum_{k=0}^{m-1} \tau(T_\lambda^k(p)), \tag{43}$$

for an m-periodic point p. The definition of the average natural frequency can be applied to a chaotic attractor by replacing the upper limit of Eq. (43), i.e., $(m-1)$ with infinity.

Figure 12 shows a graph of AF vs. E_2. In this figure, each curve labelled by a symbol Λ^m represents an m-periodic point on the Poincaré map. Then the branching of a curve represents the period doubling bifurcation. The solid and broken curves indicate a stable and unstable periodic point, respectively. We see that each curve of λ^m forms a tree branch. This pattern is called tree-like pattern. A similar tree-like pattern is found in a mean value defined on a period doubling cascade in one-dimensional dynamical systems. This property can be explained by a renormalization of a function series on the cascade [10].

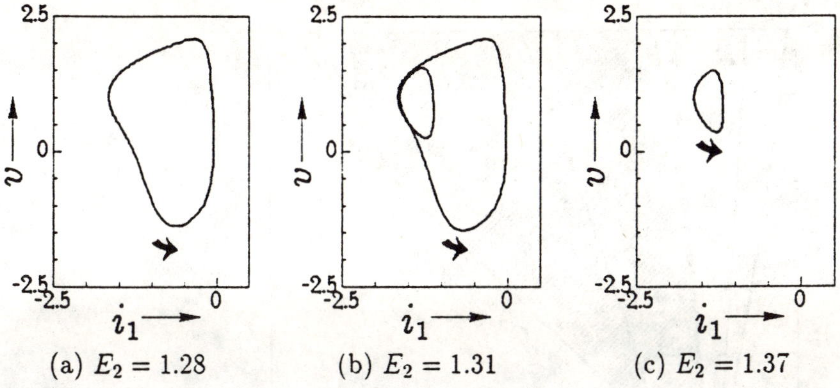

Fig. 11. Limit cycles in the (i_1, v) plane. The arrow indicates the direction of trajectory. $C^{-1} = 3$

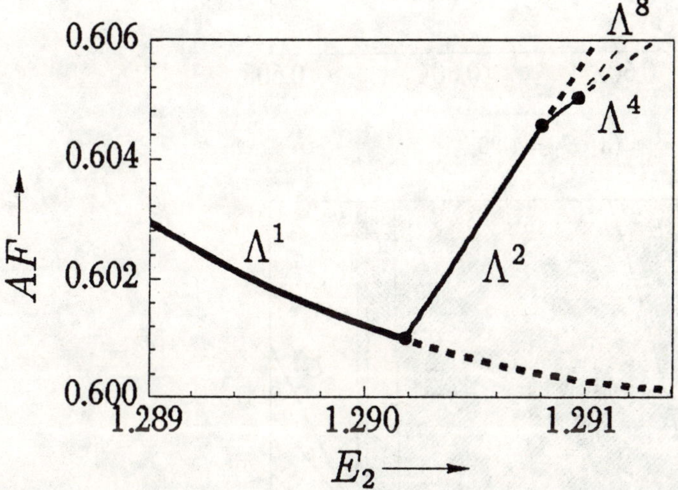

Fig. 12. The average natural frequencies of limit cycles observed in the free system for Eqs. (37)

4.1.2 Bifurcations in a Forced Chemical Oscillator [11]

Now we consider the oscillator with an external forcing term, that is, $j(t)$ is given by Eq.(38). The limit cycle observed in the free system may be entrained to the external force with appropriate parameter values of J and ν. Qualitative analysis of a periodic solution due to the entrainment can be reduced to the study of the periodic point on the Poincaré map. The Poincaré map T_λ is defined by Eq.(2) with $x = (i_1, i_2, v)$. A fixed and an m-periodic point of T_λ correspond to harmonic and $1/m$-subharmonic oscillations in Eqs. (37), respectively.

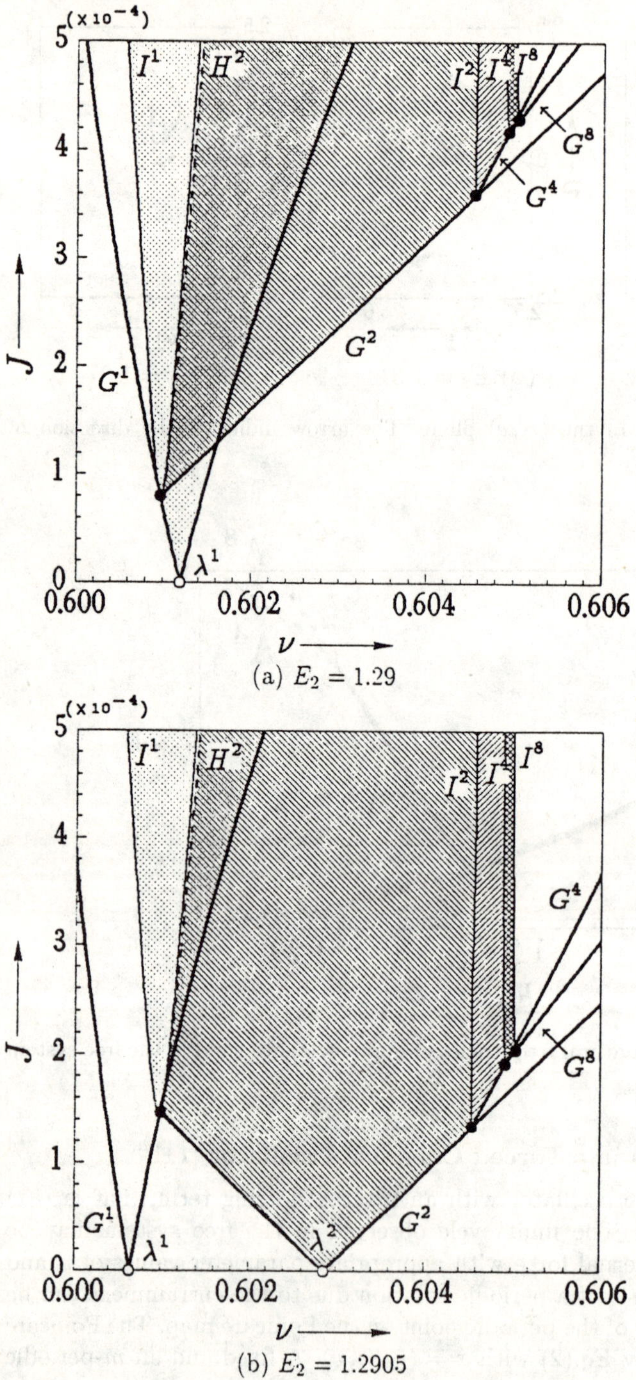

Fig. 13. Bifurcation diagrams of periodic solutions observed in Eqs. (37)

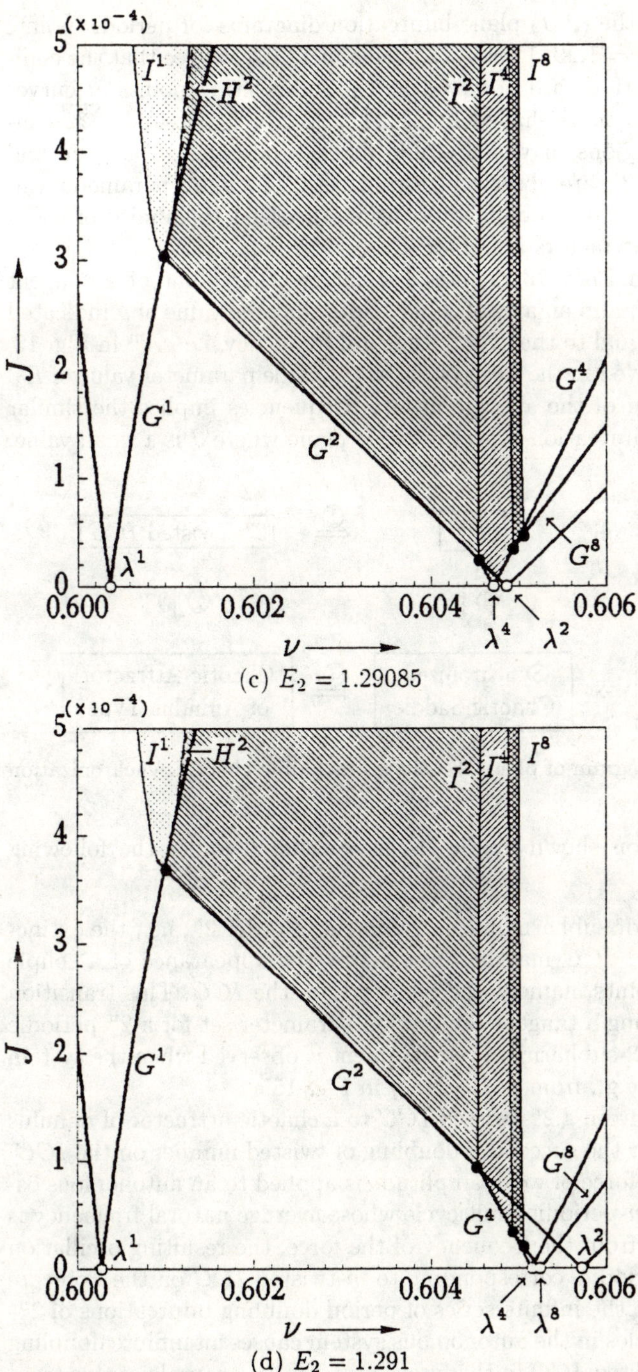

(c) $E_2 = 1.29085$

(d) $E_2 = 1.291$

Fig. 13. Continued

Figures 13 show the (ν, J) plane bifurcation diagrams for periodic points observed in T_λ at $E_2 = 1.29$, 1.2905, 1.29085 and 1.291. We see that the codimension two bifurcations are observed at the intersecting points of curves G^m and I^m. The portions shaded by , , and indicate parameter regions in which stable fixed, 2-periodic, 4-periodic and 8-periodic points of T_λ are observed, respectively. There are parameter values at which the period doubling bifurcations proceed to infinite order of period and chaotic attractors are observed.

The point λ^m in Figs. 13 denotes the intersecting point of a tangent bifurcation curve G^m and an axis of $J = 0$. Note that the value of ν indicated by the point λ^m is equal to the average natural frequency, i.e., Λ^m in Fig. 12, of a limit cycle observed in the free system at the same parameter value of E_2. The tree-like pattern of the average natural frequencies implies the similar pattern of tangent bifurcation curves in (ν, J) plane where J is a small value.

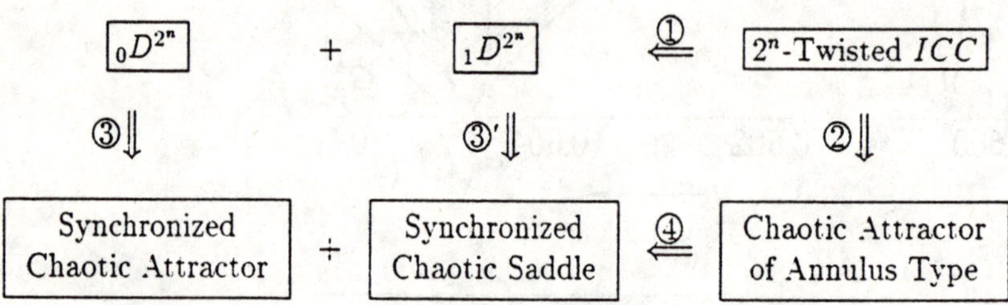

Fig. 14. Schematic diagram of phase transitions including chaotic synchronization

A phase transition shown in Fig. 14 is considered under the following conditions:

① Synchronization or subharmonic entrainment of order 2^n, i.e., the extinction of 2^n-twisted ICC may be observed by the appearance of a couple of 2^n-periodic points, namely $_0D^{2n} + {}_1D^{2n}$, on the ICC. This transition is found by crossing a tangent bifurcation parameter set for a 2^n-periodic point. Indeed, 1/2-subharmonic entrainment is observed when the system parameters vary e.g., from point b to g in Fig. 15.

② Phase transition from a 2^n-twisted ICC to a chaotic attractor of annulus type occurs under the successive doubling of twisted number on the ICC. When a periodic force of weak amplitude is applied to an autonomous oscillator with an m-periodic limit cycle whose average natural frequency is slightly different from the frequency of the force, the resulting oscillation may be quasi-periodic, corresponding to m-twisted ICC on the Poincaré map. Futhermore the infinite series of period doubling bifurcations of 2^n-periodic limit cycles in the autonomous system causes an infinite doubling process of 2^n-twisted ICC in the forced system. An example of this transition is found by continuous variation of E_2 from point a to d in Fig. 15, see Fig. 16 for their phase portraits.

③ Phase transition from a stable periodic point to a chaotic attractor occurs under the successive period doubling bifurcations of 2^n-periodic points. This transition to a chaotic attractor is well known and is observed, e.g., when the system parameters vary from point f to e in Fig. 15.

③' Item ③ for periodic points of saddle type stability. The resulting chaotic state may have both stable and unstable invariant sets. Hence the chaotic state can be called a chaotic saddle.

④ Phase transition between chaotic states may be observed by the appearance of a couple of chaotic attractors and chaotic saddles on a chaotic attractor of annulus type. This transition is considered to be a tangent bifurcation of a 2^n-periodic point, where $n \to \infty$. It is conjectured from the experiment stated above that the tree-like pattern of the tangent bifurcation curves of 2^n-periodic points proceeds to infinite order of period. Therefore the tangent bifurcation curves of 2^n-periodic points may develop in the manner of $n \to \infty$. Indeed, we found that a chaotic attractor of annulus type abruptly changes its shape with continuous variation of the system parameters, e.g., from point d to e in Fig. 15. The phase portraits of the chaotic attractors are shown in Fig. 16 (d) and (e). The transition may be considered to be a synchronization of chaotic state.

4.2 Coupled Oscillator with a Sinusoidal Current Source

4.2.1 A Circuit Model

Figure 17 shows an electrical circuit illustrating the occurrence of all types of codimension two bifurcations. In the circuit, we assume that two voltage controlled conductors $g_1(v)$ and $g_2(v)$ have the same characteristics such that

$$g_1(v) = g_2(v) = g_1 v + g_3 v^3 + g_5 v^5 , \tag{44}$$

and $j(t)$ is a sinusoidal current source

$$j(t) = J\cos(\nu t) . \tag{45}$$

Following the notation in the figure, we have the following circuit equations:

$$\frac{d\phi_1}{dt} = v_1 - R_1 i_1 - E_1 , \tag{46a}$$

$$C_1 \frac{dv_1}{dt} = -i_1 - g_1(v_1) + j(t) , \tag{46b}$$

$$\frac{d\phi_2}{dt} = v_2 - R_2 i_2 - E_2 , \tag{46c}$$

$$C_2 \frac{dv_2}{dt} = -i_2 - g_2(v_2) , \tag{46d}$$

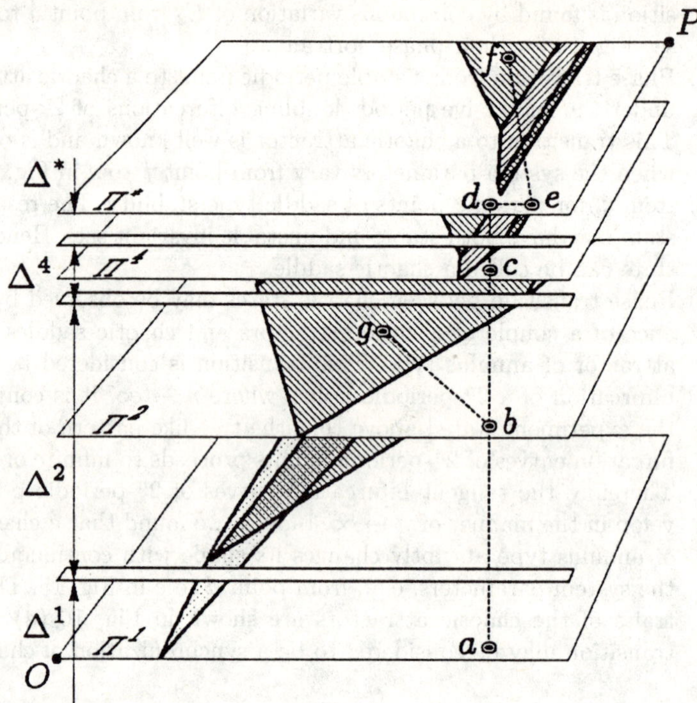

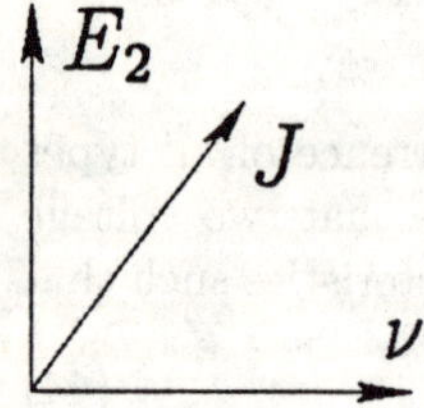

Fig. 15. An example of transitions giving rise to the schematic diagram of Fig. 14. The coordinates of points O and P are the following: O : (0.6, 0, 1.29), P : (0.606, 0.0002, 1.291) in (ν, J, E_2) space. Π^1, Π^2, Π^4 and Π^* are (ν, J) planes satisfying $E_2 = 1.29$, 1.2905, 1.29085 and 1.291, respectively. In the ranges indicated by Δ^1, Δ^2 and Δ^4, stable limit cycles with one, two and four windings are respectively observed in the free system of Eqs. (37). In Δ^*, a chaotic attractor is seen due to the doubling process

where

$$\phi_1 = L_1 i_1 + M i_2 \,, \tag{47a}$$

$$\phi_2 = M i_1 + L_2 i_2 \,. \tag{47b}$$

We fix the several system parameters as

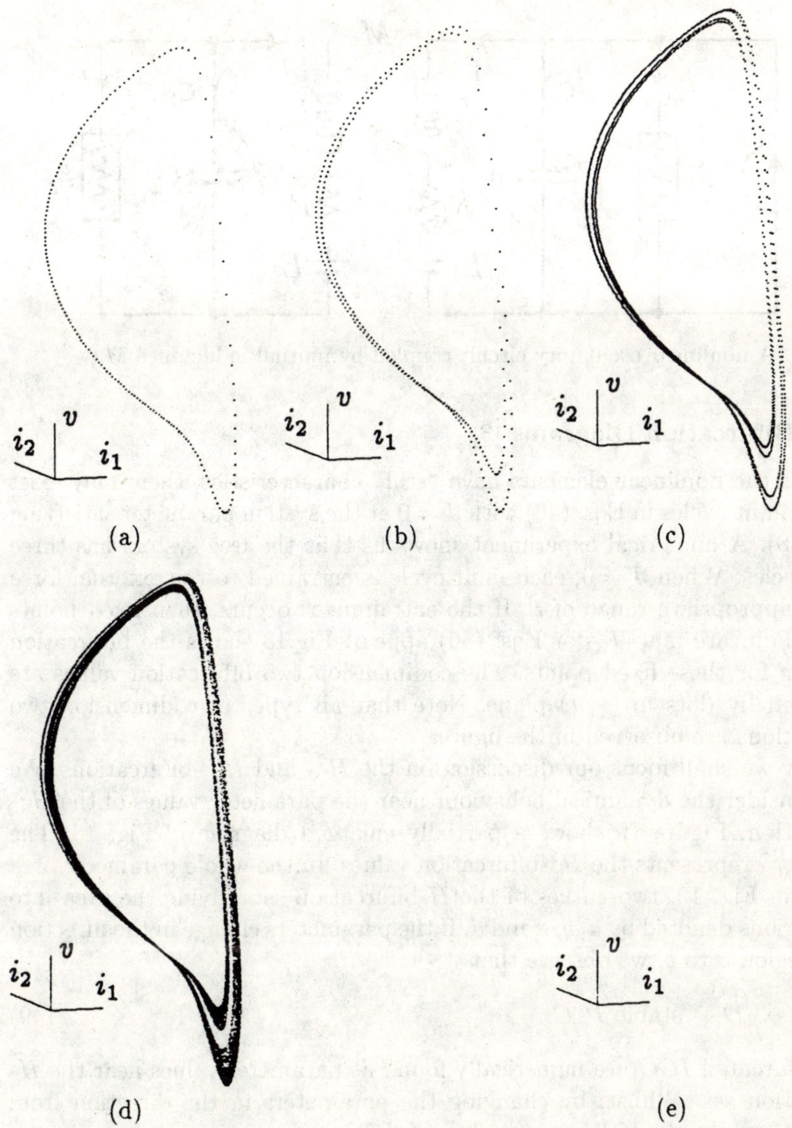

Fig. 16 Phase portraits of T_λ at points labelled as (a) a, (b) b, (c) c, (d) d and (e) e in Fig. 15; (a) simple ICC, (b) 2-twisted ICC, (c) 4-twisted ICC, (d) chaotic attractor of annulus type and (e) synchronized chaotic attractor

$$\begin{aligned} &L_1 = L_2 = 2/3\,, \quad M = -1/3\,, \quad C_1 = C_2 = 1\,, \\ &E_1 = R_2 = 0\,, \quad E_2 = 0.5\,, \quad R_1 = 0.1\,, \\ &g_1 = 0.1\,, \quad g_2 = -0.5\,, \quad g_3 = 0.1\,, \end{aligned} \tag{48}$$

unless we specify otherwise.

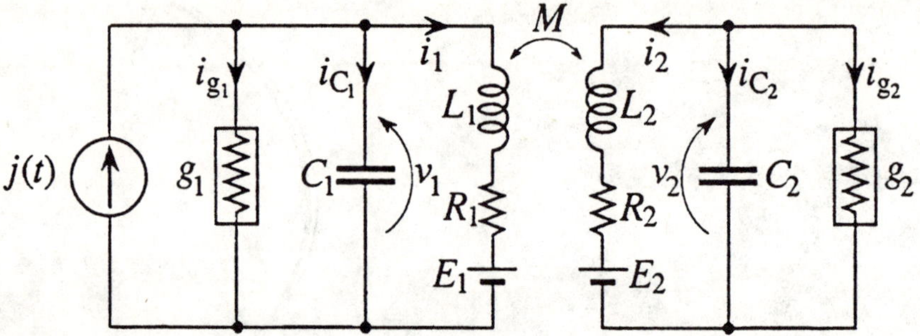

Fig. 17. A nonlinear oscillatory circuit coupled by mutual inductance M

4.2.2 Bifurcation Diagrams [3]

Because the nonlinear elements have "stiff" characteristics, there may exist several limit cycles in Eqs. (46) with $J = 0$ at the system parameter satisfying Eqs. (48). A numerical experiment shows us that the free system has three limit cycles. When $J \neq 0$, each limit cycle is entrained to the external force for an appropriate range of ν. If the entrainment occurs, then fixed points of the Poincaré map T_λ for Eqs. (46) appear. Fig.18 shows the bifurcation diagram for these fixed points. The codimension two bifurcation values are indicated by dots in (ν, J) plane. Note that all types of codimension two bifurcations are observed in the figure.

Now we shall focus our discussion on the H^2- and HP-bifurcations. We first consider the dynamical behaviour near the parameter values of the H^2-bifurcation. Figure 19 shows a partially enlarged diagram of Fig. 18. The point $\lambda_0^{H^2}$ represents the H^2-bifurcation values. In the whole parameter area shown in Fig. 19, two curves of the H-bifurcation sets divide the area into four regions denoted by a, b, c and d. If the parameters change in the direction from region a to b, we observe that

$$ {}_0D \to {}_2D + \text{ stable } ICC \,. \tag{49} $$

The bifurcated ICC are numerically found at parameter values near the H-bifurcation set. Whilst, by changing the parameters in the direction from a to d, we also find the appearance of ICC according to the same type of bifurcation as in Eq.(49). At the parameters in region c near $\lambda_0^{H^2}$, an attracting limit set of points by T_λ may exist, forming an invariant torus in 4-dimensional state space.

Let us discuss the global property of the H-bifurcation sets in three parameter space. When E_2 increases from a value of 0.5, two the H-bifurcation sets are separate as shown in Fig. 20. Namely, in three parameter space, a curve of the H^2-bifurcation set may have an end point without being in contact with any other bifurcation set. At the end point, the location of four

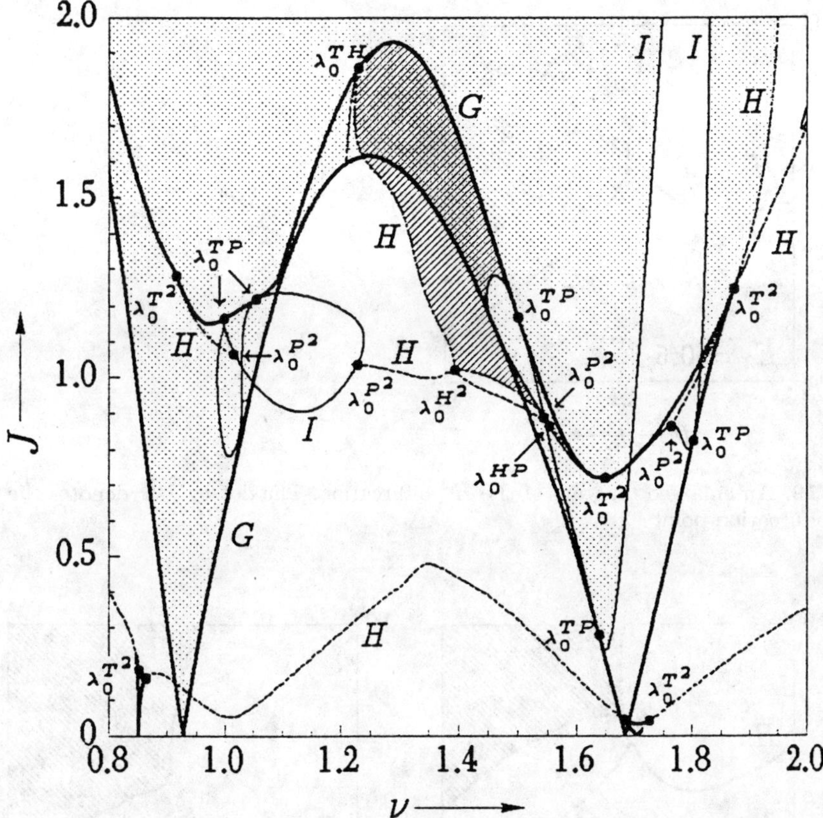

Fig. 18 Bifurcation diagram of fixed points observed in the Poincaré map of Eqs. (46). In the shaded region, we observe a stable fixed point

characteristic multipliers is such that two pairs of complex conjugates exist on the unit circle in the complex plane, see Fig. 21.

Furthermore, we shall consider the HP-bifurcation. Fig. 22 shows the detailed diagram of Fig. 18, the bifurcation sets near the HP-bifurcation point λ_0^{HP}. The point $\lambda_0^{P^2}$ denotes the P^2-bifurcation values. As R_1 varies, the bifurcation diagram in (ν, J) plane topologically changes as shown in Fig. 23. In Fig. 23 (b) the point λ_0^{HP} represents the H^2-bifurcation values. We see that a codimension three bifurcation satisfying both H- and P^2-bifurcation conditions occurs at an intersecting point of the HP-, P^2- and H^2-bifurcation curves in three parameter space.

Fig. 24 shows sets of the bifurcating periodic points, where $R_1 = 0$. In the figure, the points λ_0^{HP}, λ_1^{HP} and λ_2^{HP} exhibit the HP-bifurcation sets of fixed, 2-periodic and 4-periodic points, respectively, and the subscript number of symbol represents a codimension one bifurcation curve indicating the order of the periodic point. Notice that in this figure the HP-bifurcation set for

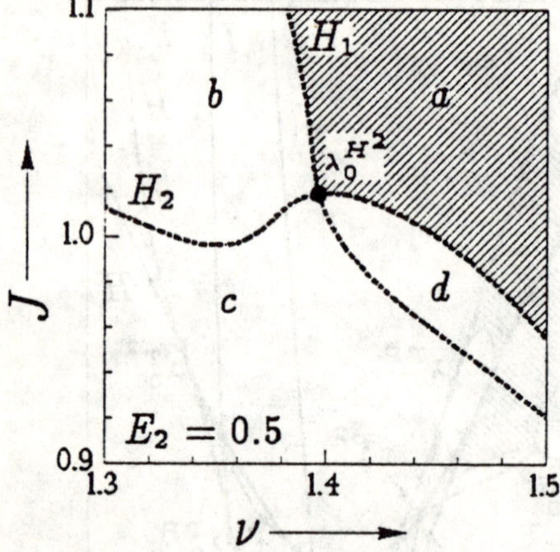

Fig. 19. An enlarged diagram of the H^2-bifurcation. The dot by $\lambda_0^{H^2}$ denotes the H^2-bifurcation point

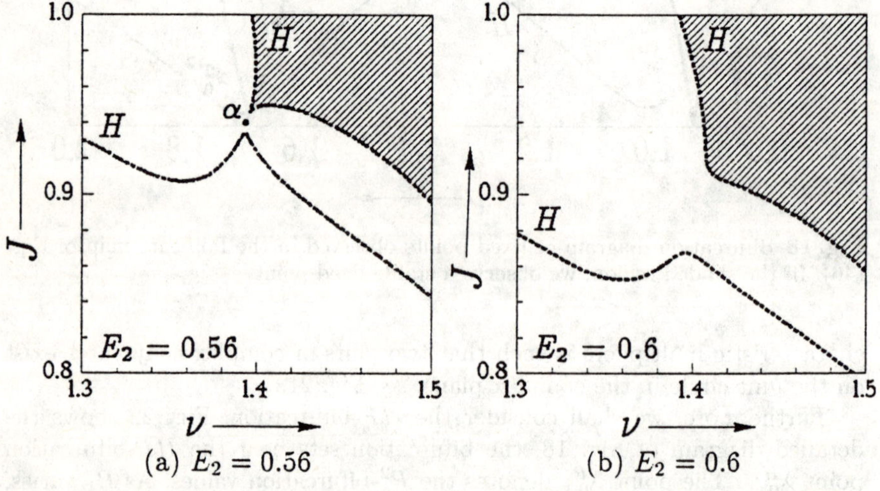

Fig. 20 Extinction of the H^2-bifurcation set

2^m-periodic points are observed successively up to $m = 2$ together with the cascade of the period doubling bifurcations. If the successive HP-bifurcations proceed infinitely, the infinite doubling process of the H-bifurcations necessarily occurs. This phenomena suggests the existence of a chaotic state as ICC bifurcated by the Hopf bifurcation of a 2^m-periodic point, where $m \to \infty$.

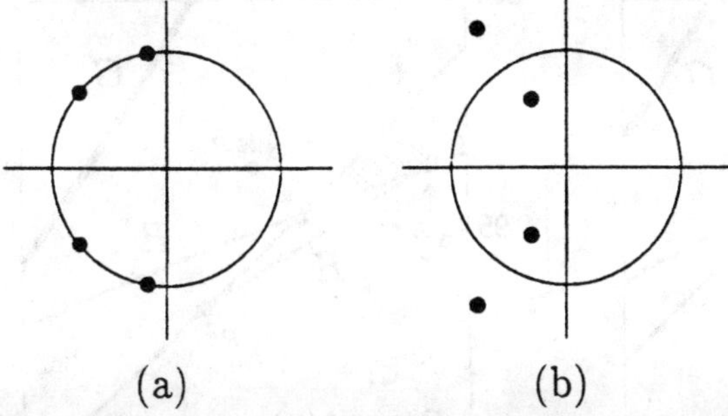

Fig. 21. Schematic diagram of the location of characteristic multipliers at the parameter values denoted by (a) the point $\lambda_0^{H^2}$ in Fig. 19 and (b) the point α in Fig. 20 (a). The circles indicate the unit circle in the complex plane

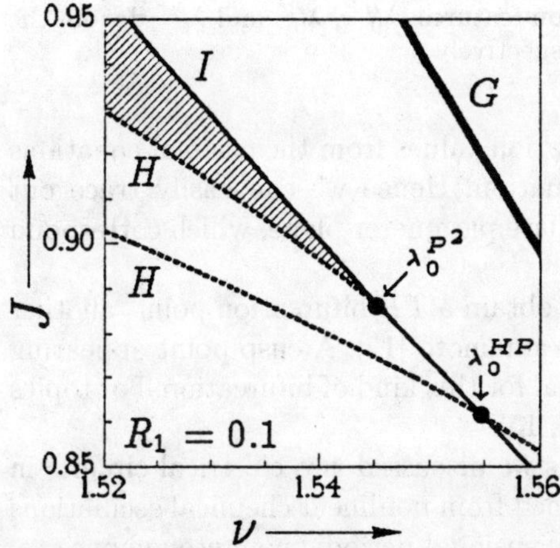

Fig. 22. An Enlarged diagram of Fig. 18 for the HP-bifurcation. The symbols λ_0^{HP} and $\lambda_0^{P^2}$ indicate the HP- and P^2-bifurcation points, respectively

5. Concluding Remarks

In this paper we have discussed a method for calculating the bifurcation value of parameters of periodic solution of nonlinear ordinary differential equations. The three kinds of generic bifurcations stated in Sec. 2.2 are known as codimension one bifurcations. As discussed in Sec. 2.3, if a periodic solution satisfies two bifurcation conditions at a bifurcation value of parameters, then the bifurcation is referred to as a codimension two bifurcation. Possible types of codimension two bifurcations are also classified. A computational algorithm using a shooting method with Newton's iteration is proposed for finding bifurcation values of parameters. Our method enables us to obtain directly

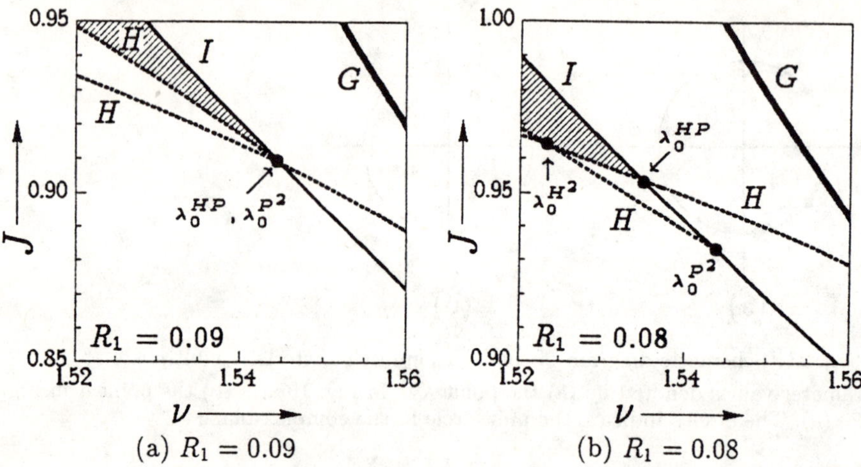

(a) $R_1 = 0.09$ (b) $R_1 = 0.08$

Fig. 23. Structural changes of bifurcation features. λ_0^{HP}, $\lambda_0^{P^2}$ and $\lambda_0^{H^2}$ denote the HP-, P^2- and H^2-bifurcation points, respectively

both codimension one and two bifurcation values from the original equations without special coordinate transformation. Hence we can easily trace out various bifurcation sets in an appropriate parameter plane, which correspond to nonlinear phenomena.

As for the numerical algorithm to obtain a TP bifurcation point, another algorithm was recently proposed by Yamamoto [12]. A cusp point appearing on a tangent bifurcation curve is typical for this kind of bifurcation. For topics related to these problems, see [13, 14, 15].

For purposes of numerical analysis we discussed two electrical circuits in Sec. 4. The first circuit model is obtained from nonlinear chemical oscillations at a water-oil interface. This three dimensional periodic nonautonomous system is however considered as a forced chaotic oscillatory system, that is a chaotic autonomous system driven by a sinusoidal external force. Comparing with a forced second order system, such as a forced van der Pol equation, a forced chaotic system has many interesting properties. Chaotic free oscillation may be entrained with the external force, or with an appropriate subharmonic oscillation, etc. How changes a cascade of period doubling process of a free system when a sinusoidal external force is injected? A typical cascade of this process and a chaotic synchronization are numerically found and illustrated in Sec. 4. We also pointed out that HP- and H^2-bifurcations actually occur in a coupled oscillator circuit with a sinusoidal current source. As this example shows, all types of codimension two bifurcationscan be found in a 4-dimensional dynamical system with a periodic forcing term.

As we have seen in Secs. 2 and 4, some codimension two bifurcations of a fixed point are inevitably correlated with a bifurcation of an invariant closed curve of the Poincaré map. Bifurcation of the invariant curve, i.e., bifurcation

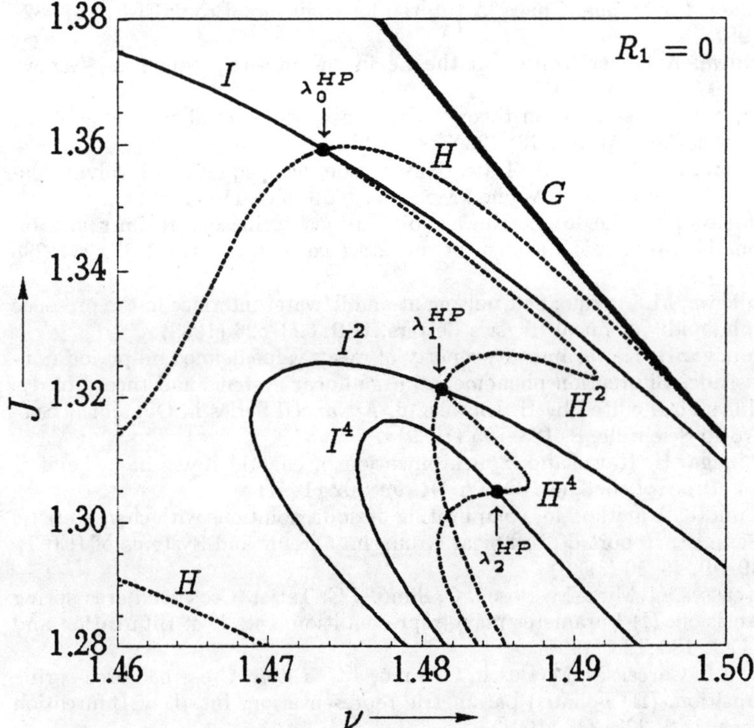

Fig. 24. The successive occurrences of the HP-bifurcation points. λ_0^{HP}, λ_1^{HP} and λ_2^{HP} denote the HP-bifurcation points of fixed point, 2-periodic point and 4-periodic point, respectively

of quasi-periodic solutions, is a global bifurcation and is very difficult for us to find bifurcation conditions. But this kind of bifurcations appear naturally in our circuit models. Hence this is an interesting future problem.

Analytical discussion for codimension two bifurcation, such as normal form representation, is another interesting future problems. Bifurcation problem with higher codimension is also considered as a related topics.

References

1. R. Abraham, J.E. Marsden: Foundation of mechanics. Reading, MA: Benjamin/ Cummings 1974
2. H. Kawakami: Bifurcation of periodic responses in forced dynamic nonlinear circuits. Computation of bifurcation values of the system parameters. IEEE Trans. on Circuits and Systs, **CAS-31**, 248–260 (1984)
3. T. Yoshinaga, H. Kawakami: Codimension two bifurcation in nonlinear circuits with periodically forcing term. Trans. IEICE, **J72-A**, 11, 1821–1828 (1989)

4. T.S. Parker, L.O. Chua: Chaos: A tutorial for engineers. Proc. IEEE, **75**, 982–1008 (1987)
5. K. Shiraiwa: A Generalization of the Levinson-Massera's equalities. Nagoya Math. J., **67**, 121–138 (1977)
6. N. Levinson: Transformation theory of nonlinear differential equations of the second order. Ann. Math., **45**, 723–737 (1944)
7. N. Yamamoto: Bifurcations of solutions of nonlinear equations involving parameters. Theoretical and Applied Mechanics, **33**, 435–444 (1985)
8. K. Yoshikawa, S. Maeda, H. Kawakami: Various oscillatory regimes and bifurcations in a dynamical system at an interface. Ferroelectrics, **86**, 281–298 (1988)
9. K. Yoshikawa, M. Makino: Self-pulsing at an oil/water interface in the presence of phospholipid. Chemical Physics Letters, **160**, 623–626 (1989)
10. T. Yoshinaga, H. Kawakami: A property of mean value defined on period doubling cascade. Bifurcation phenomena in nonlinear systems and theory of dynamical systems. edited by H. Kawakami, Advanced Series in Dynamical Systems, World Scientific, **8**, 183–195 (1989)
11. T. Yoshinaga, H. Kawakami: Synchronization of chaotic states in a chemical oscillator. Proc. of the 2nd NOLTA '91, 99–102 (1991)
12. N. Yamamoto: A method for computating periodic solutions with characteristic multipliers ± 1. Report of Technical Group on Circuit and Systems of IEICE, **CAS 89-99**, 45–50 (1989)
13. J.P. Carcassès, C. Mira, M. Bosch, C. Simó, J.C. Tatjer: Crossroad area-spring area transition. (I) Parameter plane representation. Int. J. of Bifurcation and Chaos, **1**, 1, 183–196 (1991)
14. C. Mira, J.P. Carcassès, M. Bosch, C. Simó, J.C. Tatjer: Crossroad area-spring area transition. (II) Foliated parametric representation. Int. J. of Bifurcation and Chaos, **1**, 2, 339–348 (1991)
15. J.P. Carcassès, H. Kawakami: Condition nécessaire et suffisante d'existence d'un point cuspidal sur une courbe de bifurcation 'fold' dans le plan paramétrique d'une récurrence de dimension n. Preprint of INSA de Toulouse.

Chaos and Its Associated Oscillations in Josephson Circuits

M. Morisue, A. Kanasugi

Department of Electrical and Electronic Engineering,
Saitama University,
Shimo-Ohkubo,
Urawa, 338 Japan

Abstract

This paper summarizes the chaotic phenomena and their associated oscillations produced in Josephson circuits from the view point of nonlinear oscillations. The computer analysis is carried out for the oscillations of two types of a Josephson element, one of which is a tunnel type and the other a bridge type. The Josephson circuits considered here are an rf-driven circuit, an autonomous oscillation circuit and a distributed parameter circuit. The results of analysis show that there are four types of oscillations, i.e. periodic, subharmonic, quasi- periodic and chaotic oscillations in the circuit with the tunnel type Josephson junctions, while the dominant oscillations in the bridge type Josephson junction circuit are the periodic and quasi-periodic oscillations and there are only few chaotic oscillations. In addition to these oscillations, some kinds of relaxation oscillations can be observed. The chaos and its associated oscillations are discussed in relationship with circuit parameters.

1. Introduction

The superconducting devices of Josephson junctions have been monitored with keen interest because of the possibilities of high performance electronic devices such as an ultra fast computer and ultra weak magnetic detector. When a Josephson junction is changed from the superconducting state to the voltage state, it starts to oscillate in various oscillation modes of very high frequency. In some cases these oscillations are utilized for electronic devices such as a high frequency oscillator and high frequency amplifier, and in other cases these oscillations become an unfavorable phenomena such as a significant source of noise. During the past ten years a large number of papers there have been published concerning the chaotic oscillations in Josephson circuits [1, 2]. However, these papers have mainly dealt with the oscillations in rather simple Josephson circuits and have not described the complicated oscillation modes. In this paper we discuss in detail what kinds of oscillation are produced in several typical Josephson circuits in relation to circuit parameters, taking into account the nonlinear current-voltage characteristics of the junctions. There are two types of Josephson junctions, one of which is a tunnel type, the other a bridge type or weak link type. In addition to the Josephson circuit, there are many other kinds of circuit containing these junctions, which are rf-driven forced oscillation circuits, autonomous oscillation circuits, distributed circuits and so on. Therefore, the oscillation mode produced in a Josephson circuit depends on what kind of circuit we choose and what types of junction we consider. In order to investigate these various oscillations, a computer analysis was carried out. The results of analysis show that four oscillation modes exist in the tunnel type Josephson circuit, those being periodic, subharmonic, quasi-periodic and chaotic oscillations. On the other hand, for the bridge type junction circuit, the dominant oscillations are periodic and subharmonic oscillations. The oscillation modes have been analyzed using bifurcation diagram, Poincaré map, return map, time domain diagram, fourier spectrum and so on. However, for the autonomous or free-running periodic oscillations, these procedures can not be applied. This is because the fundamental oscillation frequency required for graphical representation of the solution can not be determined unconditionally. Therefore a novel procedure to determine the fundamental frequency of the oscillation is introduced. The chaos and its associated phenomena produced in Josephson circuits are discussed in relation to the circuit parameter.

2. Model of Josephson Junction

As for Josephson junctions used in practical electronic devices there are two types of junctions, that is, tunnel type junction and the bridge type or weak link junction. The tunnel type junction has a sandwich structure, where a thin insulating sheet is inserted between two superconducting films as shown in Fig. 1 a). The bridge type junction is made so as to form a bridge or a weak link between two parts of a superconducting sheet or a thin film as shown in Fig. 1 b).

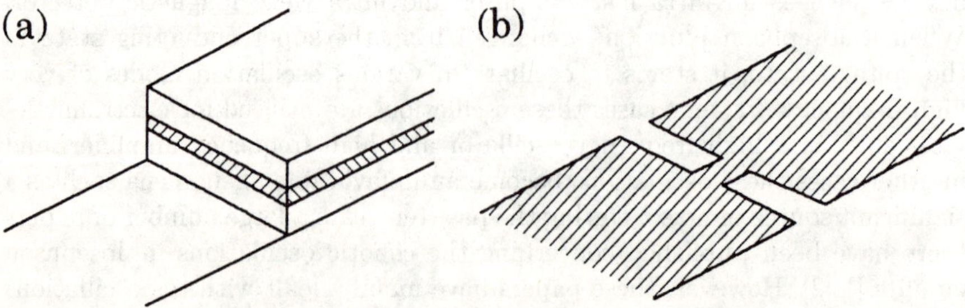

Fig. 1. Structure of Josephson Junction: a) tunnel type; b) bridge type

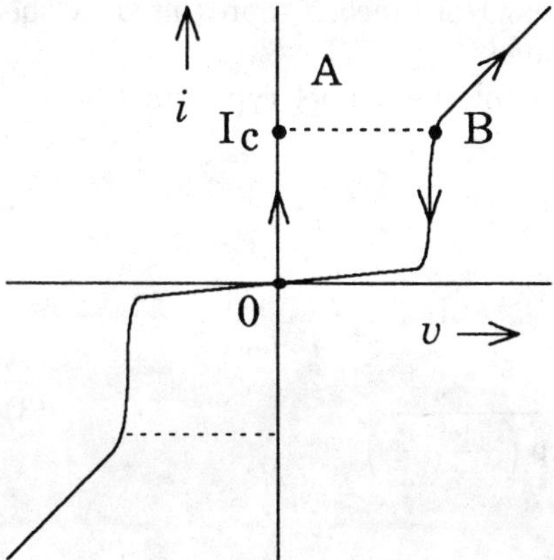

Fig. 2. Current-voltage characteristics of tunnel type Josephson junction

The current-voltage characteristics of a tunnel type junction are shown in Fig. 2. When we apply a current to the junction, the current through the junction increases along the Y axis because the junction remains in the superconducting state. When the current reaches the critical value indicated by the point A in the figure, the junction suddenly switches to the voltage state and produces a voltage across the junction. This causes the operating point jump to the point B. As you increase the current at this point, the voltage across the junction increases. When we decrease the current through the junction, the voltage across the junction decreases along the nonlinear characteristic curve as shown in Fig. 2. When the current decreases to its minimum value, the junction returns to the superconducting state.

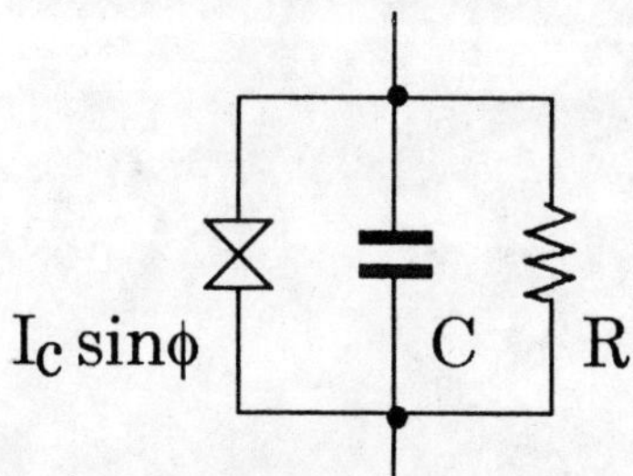

Fig. 3. Equivalent circuit of tunnel type Josephson junction

Consequently, the current-voltage characteristics of a tunnel type junction display a hysteresis curve. The equivalent circuit of a tunnel type junction is shown in Fig. 3, where the current source of $I_C \sin\phi$, the capacitance C for the parallel electrodes and the nonlinear resistance R represented by the characteristic curve are connected in parallel.

The equations describing the behavior of the tunnel type junction are given by:

$$I = I_C \sin\phi \,, \tag{1}$$

$$\frac{d\phi}{dt} = \frac{2e}{\hbar} v \,, \tag{2}$$

$$\frac{1}{R(v)} = \frac{1}{R_{sg}} + \left(\frac{1}{R_{NN}} - \frac{1}{R_{sg}}\right) \frac{1}{1 + \exp\left(-\frac{|v| - V_g}{V_d}\right)} \,, \tag{3}$$

where

ϕ	: phase difference between both superconductors;
v	: voltage across the junction;
C	: capacitance of the junction;
I_C	: critical current of the junction;
$R(v)$	: nonlinear resistance of the junction;
R_{sg}	: subgap resistance of the junction;
R_{NN}	: normal resistance of the junction;
V_g	: gap voltage of the junction;
V_d	: gap depth of the junction.

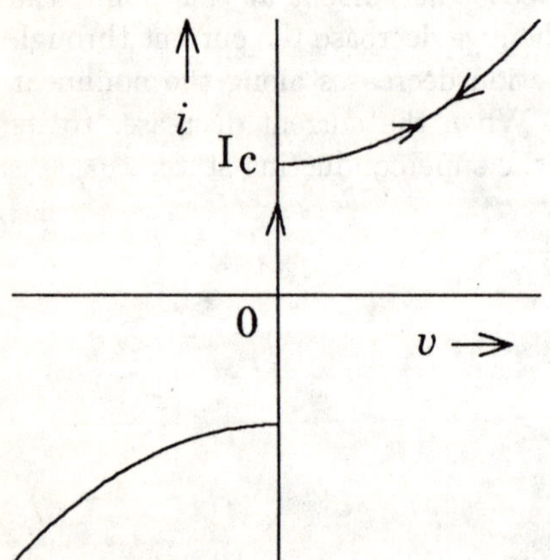

Fig. 4. Current-voltage characteristics of bridge type Josephson junction

The current-voltage characteristics of the bridge type junction are shown in Fig. 4. Although the current through the junction also increases along the Y axis until the current reaches the critical current, the current-voltage characteristics indicate the nonlinear curve shown in the figure after the junction has switched to the voltage state. The equivalent circuits for the bridge type junction are basically represented by the same circuit as shown in Fig. 3, where the capacitance in the parallel connection is ignored. Therefore, the I-V characteristics of bridge type junction are obtained from the equivalent circuit of Fig. 3 by neglecting the capacitance ($C = 0$).

The relation between current and voltage for a bridge type junction is quite different from that of a tunnel type junction after the junction switches to the voltage state.

3. Chaos in a Forced Oscillation Circuit

The chaotic oscillations produced by a Josephson junction have been intensively investigated in order to avoid the chaos in low-noise circuits, because the chaotic noise exceeds the thermal noise by several orders of magnitude. Chaos has been observed in the various circuits shown in Fig. 5 [1].

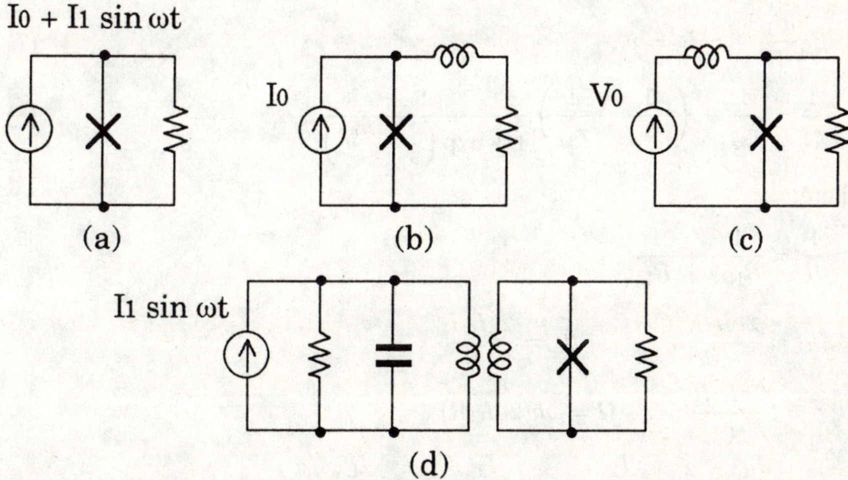

Fig. 5. Josephson circuits for which simulations reveal chaotic behavior

As a most typical circuit we analyzed the chaos and its associated oscillations in a rf-driven Josephson circuit, where a tunnel type Josephson junction with an inductance L_{L} and a load resistance R_{L} is driven by both a rf-current source and a bias dc current source. Fig. 6 shows the equivalent circuit of the model considered here.

The equations describing the circuit are given in dimensionless form as follows:

$$i_{\mathrm{J}} = \sin\phi \,, \tag{4}$$

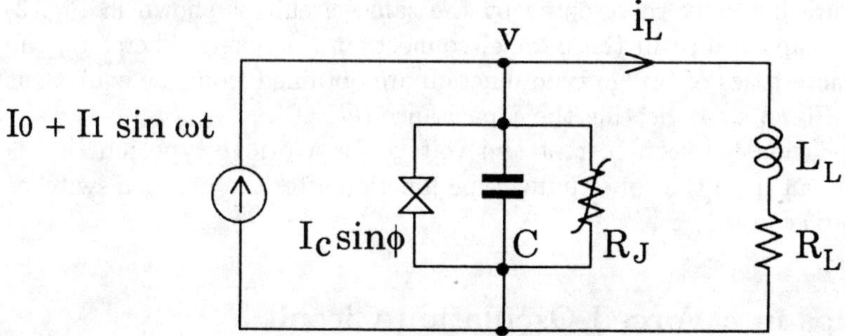

Fig. 6. Equivalent circuit for a rf-driven Josephson circuit with an inductance

$$\frac{d\phi}{d\tau} = v , \tag{5}$$

$$\beta_C \frac{dv}{d\tau} + \frac{1}{r_J} v + i_L + i_J = i_0 + i_1 \sin \Omega\tau , \tag{6}$$

$$\beta_L \frac{di_L}{d\tau} + r_L i_L = v , \tag{7}$$

$$\frac{1}{r_J} = \frac{1}{r_{sg}} + \left(\frac{1}{r_{NN}} - \frac{1}{r_{sg}}\right) \frac{1}{1 + \exp\left(-\frac{|v| - v_g}{v_d}\right)} , \tag{8}$$

where:

$$\frac{1}{R} = \frac{1}{R_{NN}} + \frac{1}{R_L} ,$$

$$\beta_C = \frac{2eI_C R^2 C}{\hbar} , \qquad \beta_L = \frac{2eI_C I_L}{\hbar} ,$$

$$\tau = t \frac{2eI_C R}{\hbar} , \qquad \Omega = \omega\hbar(2eI_C R) ,$$

$$i_0 = \frac{I_0}{I_C} , \quad i_1 = \frac{I_1}{I_C} , \quad i_J = \frac{I_J}{I_C} , \quad i_L = \frac{I_L}{I_C} ,$$

$$v = \frac{V}{I_C R} , \quad v_g = \frac{V_g}{I_C R} , \quad v_d = \frac{V_d}{I_C R} ,$$

$$r_{NN} = \frac{R_{NN}}{R} , \quad r_{sg} = \frac{R_{sg}}{R} , \quad r_L = \frac{R_L}{R} .$$

The notation of the symbols in the above equations are as follows: i_J is the current through the Josephson junction, I_C and ϕ are the critical current and the phase difference of the junction, I_0, I_1 and ω are the dc current, ac current and angular frequency of the driving current source, respectively, R_{NN} and R_{sg} are the normal resistance and subgap resistance of the Josephson junction, V_g and V_d are the gap-voltage of the junction and the voltage associated with v, respectively.

Since the above equations are nonlinear differential equations, it is rather difficult to solve these equations by an analytical method. Therefore, we have made simulations using the Runge-Kutta method.

The type of oscillation produced in a Josephson circuit depends on the circuit parameters. We have investigated the oscillation modes by changing one of the circuit parameters, while the other circuit parameters remain fixed. At first we made simulations by changing the dc component i_0 of the external force in intervals of 0.002 from 0 to 3.0. Other fixed circuit parameters are $\beta_C = 23.1$, $\beta_L = 30.4$, $r_L = 7.25$, $i_1 = 10$, and $\Omega = 0.2$.

Fig. 7 shows the results of the simulations, where a) reveals the voltage across the junction as a function of the dc component i_0 of the driving force, b) the average voltage across the junction vs. i_0. At any point of i_0 in the figure, sixty solutions for v are plotted at the discrete time of $t = n(2\pi/\Omega)$ where n is the integer from 61 to 120. Therefore a point or a single line in the plane is evidence of a periodic oscillation where the junctions are phase-locked with the rf-external force.

It is seen from these results that the following four types of oscillations exist.

(1) Periodic oscillation: This type of oscillation is indicated by the straight lines of A in Fig. 7.
(2) Subharmonic oscillation: This type of oscillation is represented by the bifurcation lines or higher even-fold lines indicated as B.
(3) Chaotic oscillation: The regions of chaos are shown by C in the figure. In this oscillation the voltage across the junction does not return to the same position at the given phase angle of each rf cycle.
(4) Quasi-periodic oscillation: This is indicated by D. The oscillation is a long term periodic oscillation although it seems like a chaos oscillation. This type of oscillation is shown in Fig. 7 b), where the $v - i_0$ curve becomes an increasing straight line as i_0 increases.

These four types of oscillation are more clearly distinguished by plotting the terminal voltages across the junction against the phase ϕ in the phase plane as shown in Fig. 8.

It is noted that the stripe-pattern known as the strange attractor is observed in the phase plane for the chaotic oscillation. Furthermore, we show the oscillation waveforms in time domain and their frequency spectrum. Figs. 9 (a), (b), (c) and (d) indicate the frequency spectrum corresponding to the periodic, double periodic, chaotic and quasi-periodic oscillations as shown in Fig. 8 (a), (b), (c) and (d), respectively.

The effect of load inductance on oscillation modes has been analyzed. Even if the load inductance is neglected in the Josephson circuit of Fig. 5, the oscillation mode does not change so as to produce an oscillation similar to that in the circuit with load inductance. A quite similar pattern could be observed for the bifurcation diagram and Poincaré map. The most we can say is that the range for chaotic oscillation widens a little by increasing the

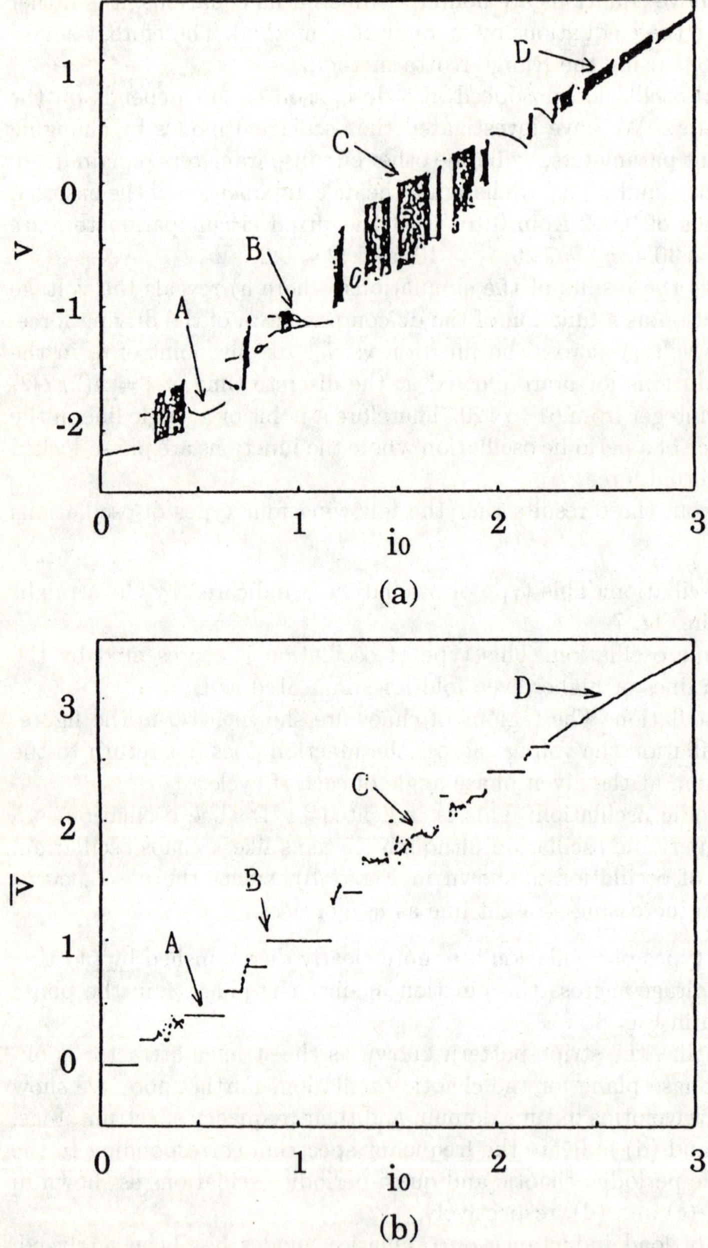

Fig. 7. A transition between chaotic and periodic oscillation, where $\Omega = 0.2$, $i_1 = 10$: (a) relation between v and i_0; (b) relation between $\bar{v}$ and i_0

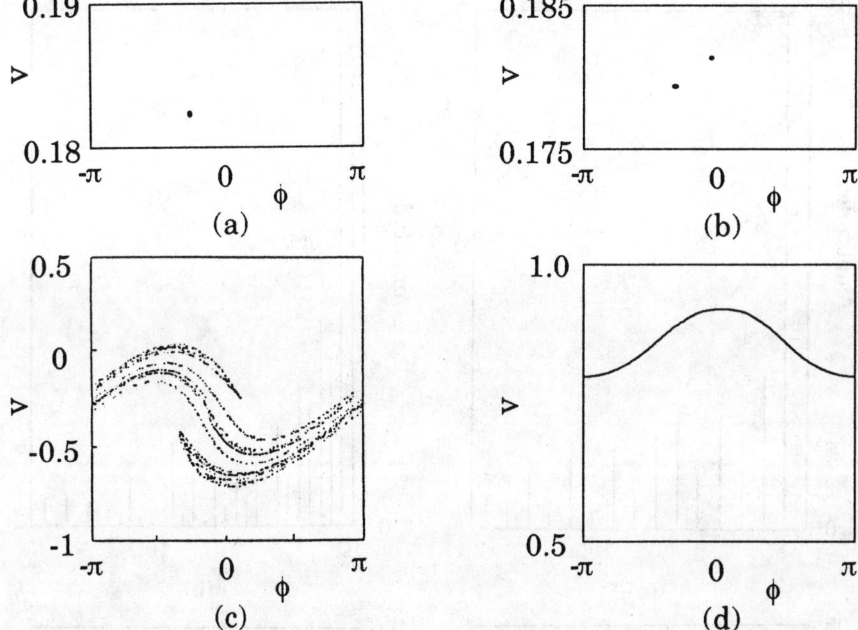

Fig. 8. Attractors produced in the circuit of Fig. 6: (a) periodic oscillation; (b) double-periodic oscillation; (c) chaotic oscillation; (d) quasi-periodic oscillation

load inductance. Fig. 10 shows how the bifurcation diagram is changed by increasing the load inductance. Fig. 10(a) is the diagram for the Josephson circuit without the load inductance. As the value of inductance increases, the bias current needed to start chaotic oscillation decreases.

4. Autonomous Josephson Circuit

4.1 Introduction

In this section we consider an autonomous Josephson circuit, where the rf-driven force is eliminated as shown in Fig. 11.

For the analysis of oscillation modes in autonomous circuits, bifurcation diagram and Poincaré map are not easily obtainable because the oscillation frequency required for graphical representation of solutions for oscillations can not be unconditionally determined. Therefore, to begin with, we describe the procedure to determine the period of free-running periodic oscillations. We then characterize what types of oscillations occur in autonomous circuits as clearly as in a forced oscillation circuit. Furthermore, we describe the effect of load resistance on the oscillation modes, one of which is a novel

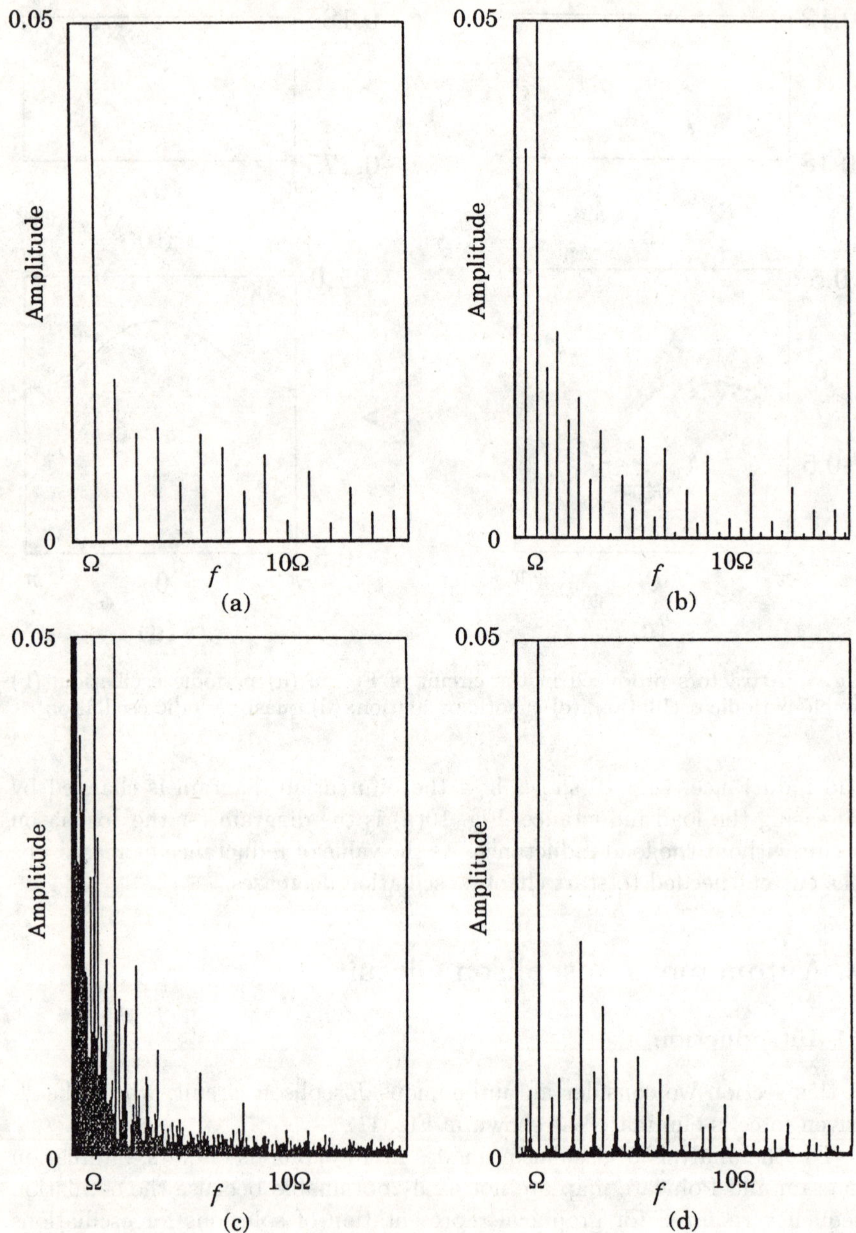

Fig. 9. Frequency spectrum corresponding to oscillations in Fig. 8: (a) periodic oscillation; (b) double-periodic oscillation; (c) chaotic oscillation; (d) quasi-periodic oscillation

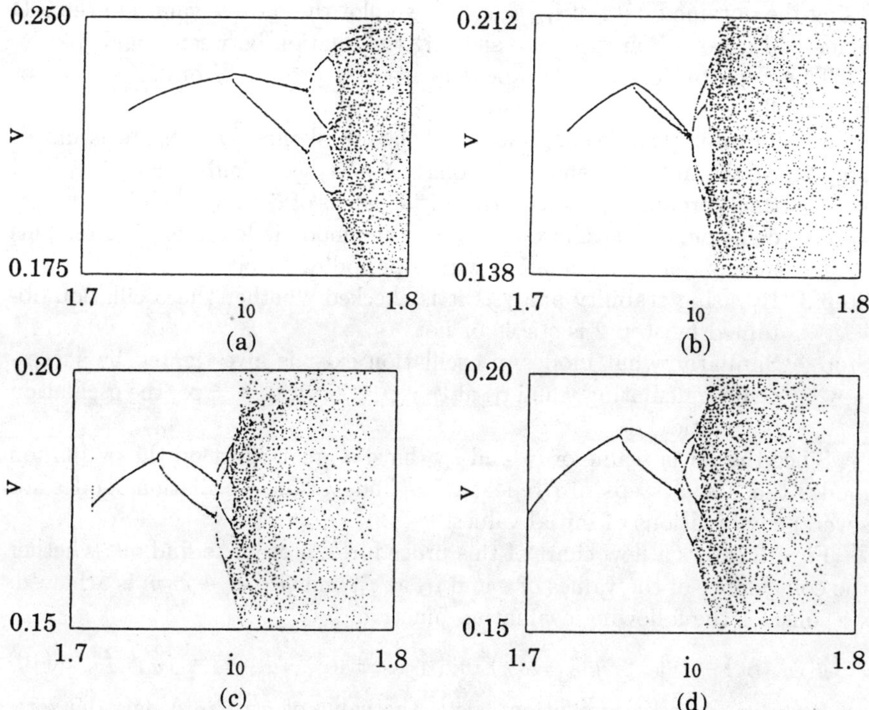

Fig. 10. Effects of load inductance on oscillation mode: (a) $\beta_L = 0$; (b) $\beta_L = 30.4$; (c) $\beta_L = 91.2$; (d) $\beta_L = 152$

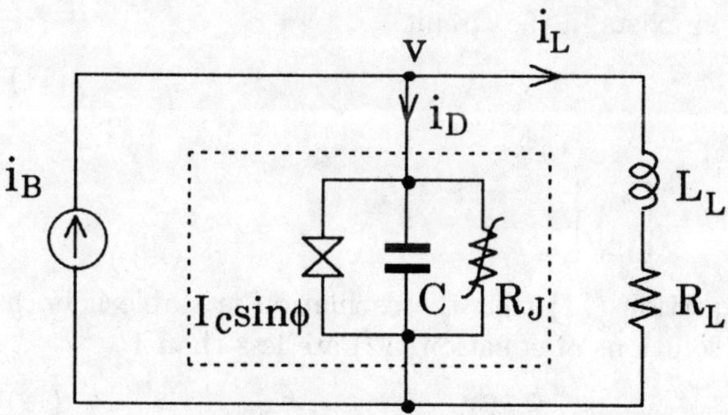

Fig. 11. Equivalent circuit of a Josephson autonomous circuit

oscillation. For the analysis, we adapt the novel approach to plot the solutions in a 3-dimensional plane instead of a bifurcation diagram. For graphical representation we introduce the following current i_D instead i_L,

$$i_D(\phi) = i_0 - i_L(\phi) . \tag{9}$$

Using the obtained data of i_D, v and ϕ, we plot these three values in the 3-D plane and draw a Poincaré map showing the relation between v and i_D.

The steps to describe the above-mentioned procedure on the computer are as follows.

Step 1. The initial value of v_0 and i_{D0} at the initial phase $\phi = \phi_0$ are assumed.

Step 2. At the moment when ϕ becomes $\phi + 2\pi$, we calculate v and i_D, and then compare them with the initial values of v_0 and i_{D0}. If both values coincide, the oscillation for which the mode index m is 1 exists. This means that an oscillation with a period of 2π occurs.

Step 3. By using stability analysis it is checked whether the oscillation obtained by step 2 is stable or not.

Step 4. Similarly, what modes of oscillation exist is investigated by successively calculating v and i_D at $\phi = \phi_0 + 2\pi m$. If $m = \infty$, the oscillation is chaotic.

Since the initial value of v_0 and i_{D_0} have a great effect on the oscillation mode, the above steps are repeated to find various oscillation modes for several combinations of initial values.

Fig. 12 shows a flow chart of this procedure. In order to find out whether the coincidence of the values of v and i_D at ϕ_0 and $\phi = \phi_0 + 2\pi m$ is achieved, we consider the following evaluating function

$$g(v_0, i_{D_0}) = (v|_{\phi_0} - v|_{\phi_0+2\pi m})^2 + (i_D|_{\phi_0} - i_D|_{\phi_0+2\pi m})^2 = f_V^2 + f_{i_D}^2 \,. \quad (10)$$

If the m-periodic oscillation exists, the value of $g(v_0, i_{D_0})$ must be zero. Then, after assuming the initial value of v_0 and i_{D_0} at the beginning, the modified values are calculated using the Powell method, so that the condition of $g(v_0, i_{D_0}) = 0$ is satisfied. For investigation of the stability of the oscillation, the stability analysis is carried out. When the following equation holds, we can say a periodic oscillation exists in the circuit

$$T_\phi X = Y = X \,, \quad (11)$$

where,

$$X = \begin{bmatrix} v_0 \\ i_{D_0} \end{bmatrix}_{\phi=\phi_0} , \qquad Y = \begin{bmatrix} v_0 \\ i_{D_0} \end{bmatrix}_{\phi=\phi_0+2\pi m} .$$

It is concluded from equation (11) that the oscillation is stable if both absolute values of the two solutions of equation (12) are less than 1.

$$DT_\phi X - \mu I = 0 \,, \quad (12)$$

where

$$DT_\phi = \begin{bmatrix} \frac{\partial f_V}{\partial V} & \frac{\partial f_V}{\partial i_D} \\ \frac{\partial f_i}{\partial V} & \frac{\partial f_i}{\partial i_D} \end{bmatrix} , \qquad I = \begin{bmatrix} 1 & 0 \\ 0 & 1 \end{bmatrix} .$$

We have developed the program to calculate these steps indicated in Fig. 12 on a computer.

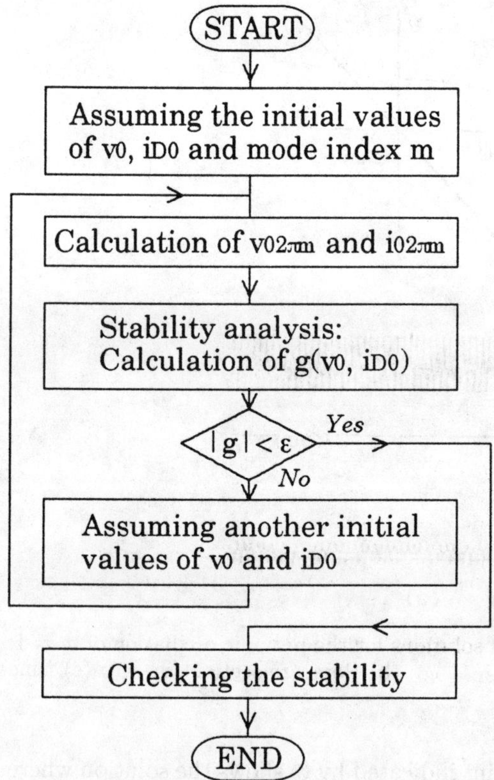

Fig. 12. Flow chart of analysis for an autonomous periodic oscillation circuit

4.2 Results of Calculation

In order to investigate what type of oscillation is produced in an Josephson autonomous circuit, we apply our procedure to the model shown in Fig. 11. For simplicity we have fixed the circuit parameters as follows, $\beta_c = 10.0$, $r_L = 0.2$ and changed the bias current i_B from 1.1 to 1.4 in intervals of 0.001. The results of these calculations are shown in Figs. 13 to 15. We can clearly deduce what type of oscillation is produced from these results. In Fig. 13 a periodic oscillation of $m = 1$ at $i_B = 1.49$ is illustrated in a 3-D plane (a) and a Poincaré section (b) and a time domain (c). In a 3-D plane where the values of i, v, and ϕ are plotted, the trajectory of the state variables becomes a single line, and in a Poincaré map where the values of v and ϕ are plotted at discrete time of $t = n\frac{2\pi}{\Omega}$, a single point is observed. Here Ω is an angular frequency. The results of Fig. 14 correspond to a bias point of $i_B = 1.28$, where the periodic oscillation of $m = 4$ is observed. In a 3-D plane of Fig. 14(a), the four trajectories which revealed the existence of the periodic oscillation of $m = 4$ are observed.

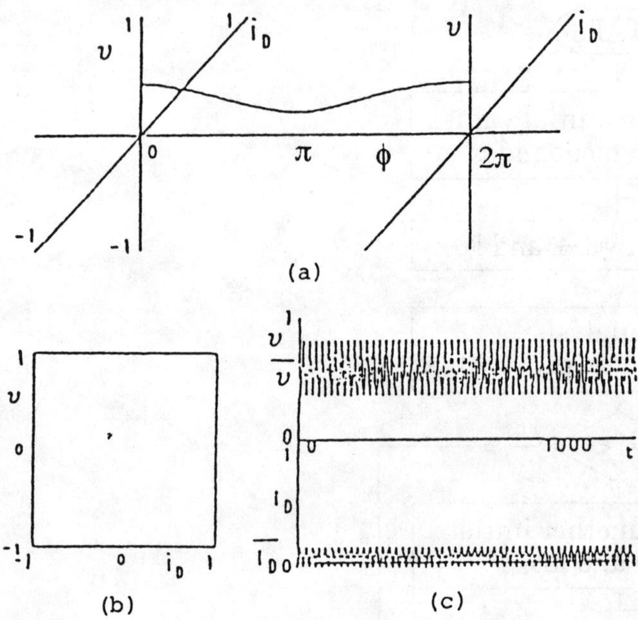

Fig. 13. Graphical representation of solutions for the periodic oscillation of $m = 1$: (a) 3-dimensional phase plane ($v - i_D - \phi$); (b) Poincaré-section ($v - \phi$); (c) time domain

In the Poincaré section the point indicated by ⊗ shows the solution where the value of ϕ becomes negative, and therefore does not contribute to the oscillation mode m, since the point ⊗ cancels out the single point in the positive direction of ϕ. Fig. 15 shows the results for $i_B = 1.18$ where a chaotic oscillation is observed. In this case the voltage does not return to the same position at the given phase of $\phi_0 + 2\pi m$. The chaotic oscillation is especially characterized by the strip-pattern in a Poincaré section.

It is seen from these results that three types of oscillation exist in an autonomous Josephson circuit: (1) periodic oscillation, (2) subharmonic oscillation and (3) chaotic oscillation (exception: the quasi-periodic oscillation observed in rf-driven circuit).

It is well known, that in a Josephson circuit with load resistance in parallel connection some kinds of oscillation modes exist [5], especially relaxation oscillation mode. We analyzed these oscillation modes in detail and we observed a new oscillation which had not been found before. For the oscillation modes of the circuit, Van Duzer [4] showed three types of modes: a latching periodic oscillation, a non-latching periodic oscillation, and a relaxation oscillation, which are produced by decreasing the load resistance in a circuit while other parameters are kept constant. We observed a kind of periodic oscillation followed by a relaxation oscillation as shown in Fig. 16 instead of a non-latching periodic oscillation.

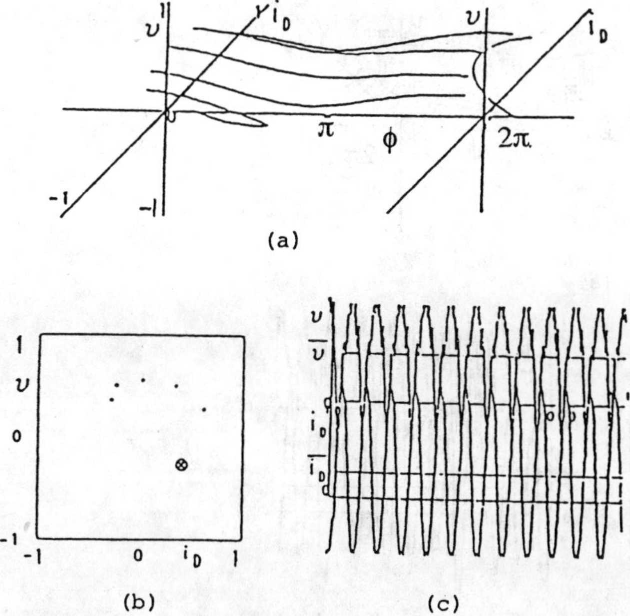

Fig. 14. Graphical representation of solutions for the oscillation of $m = 4$: (a) 3-dimensional phase plane ($v - i_D - \phi$); (b) Poincaré-section ($v - \phi$); (c) time domain

Fig. 17 and Fig. 18 show the region where the novel oscillation mode can occur. In these figures, the region of periodic oscillation is shown by A, the region of novel oscillation is shown by B, and the region of relaxation oscillation is shown by C.

We obtained two eigenvalues, $|\mu_1| = |\mu_1| = 0.96$ for the case shown in Fig. 16. As these eigenvalues approach the value of 1, the stability of the oscillation becomes worse. So it is to be expected that the stability of the novel oscillation mode is worse although the oscillation can still exist.

5. Distributed Parameter Circuit

The model to be considered here is a Josephson junction circuit containing a transmission line as shown in Fig. 19, where R_L is the load resistance, Z_0 is the characteristic impedance of the transmission line, i_0 and i_1 are the dc and rf currents of the source, Γ is the reflection coefficient of the line at the receiving end, I_C and ϕ are the critical current and phase angle of the Josephson junction, C and R are the capacitance and nonlinear resistance of the Josephson junction.

Assuming $F_1(t)$ and $F_2(t)$ are the transmitting wave and the reflected wave on the transmission line, respectively, we introduce the equations de-

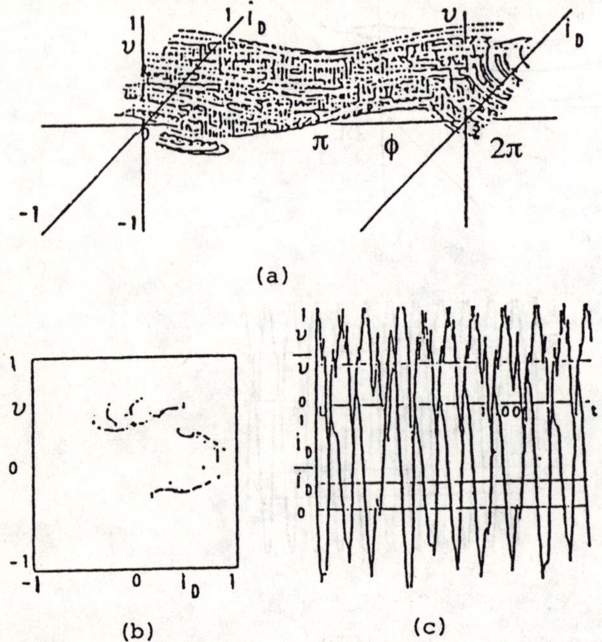

Fig. 15. Graphical representation of solutions for the chaotic oscillation: (a) 3-dimensional phase plane $(v - i_{\mathrm{D}} - \phi)$; (b) Poincaré -section $(v - \phi)$; (c) time domain

scribing the model as follows:

$$C\frac{dv}{dt} + \frac{v}{R} + \sin\phi + i_{\mathrm{d}} = i_0 + i_1 \sin \Omega t, \tag{13}$$

$$\frac{d\phi}{dt} = v\ , \tag{14}$$

$$\frac{1}{R(v)} = \frac{1}{R_{\mathrm{sg}}} + \left(\frac{1}{R_{\mathrm{NN}}} - \frac{1}{R_{\mathrm{sg}}}\right)\frac{1}{1 + \exp\left(-\frac{|v| - V_{\mathrm{g}}}{V_{\mathrm{d}}}\right)}, \tag{15}$$

$$i_{\mathrm{d}} = \frac{F_1(t) - F_2(t)}{Z_0}\ , \tag{16}$$

$$c = F_1(t) + F_2(t)\ , \tag{17}$$

$$F_2(t) = \Gamma F_1(t)\ , \tag{18}$$

$$\Gamma = \frac{R_{\mathrm{L}} - Z_0}{R_{\mathrm{L}} + Z_0}\ , \tag{19}$$

where i_{d} is the current to the transmission line. For simulation we have solved the above simultaneous nonlinear differential equations by the Runge-Kutta method using a computer.

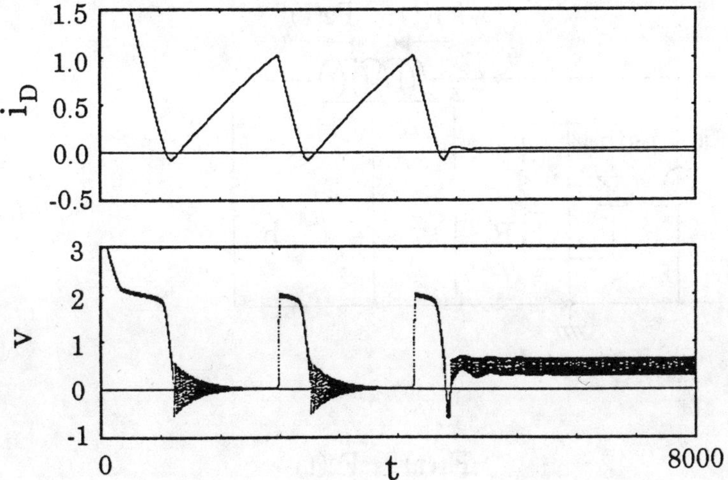

Fig. 16. Calculated waveforms

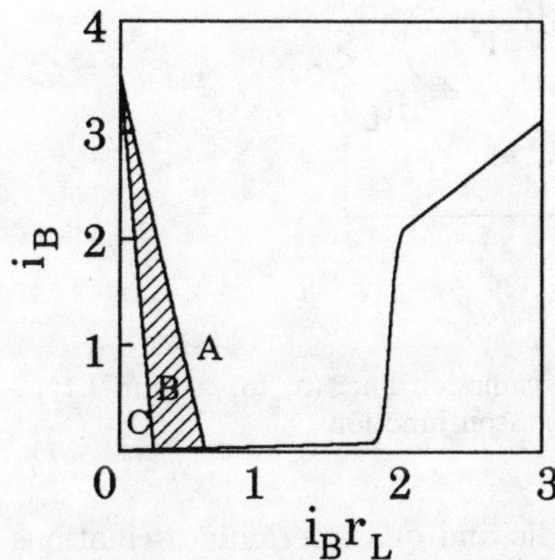

Fig. 17. Relation between load resistance and oscillation mode

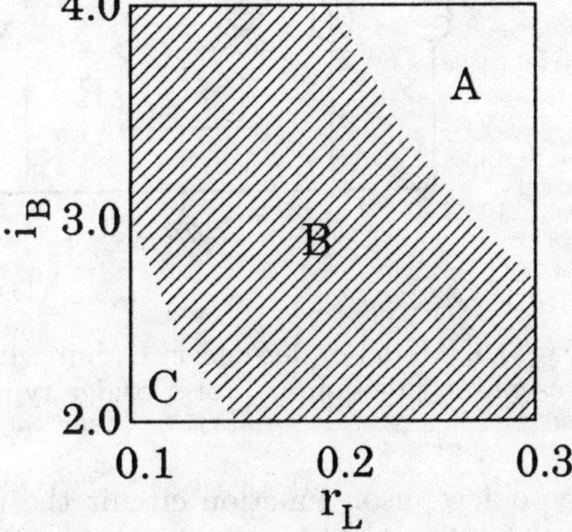

Fig. 18. Relation between circuit parameters and oscillation mode

In order to investigate what type of oscillation occurs in a Josephson circuit with a transmission line, we have carried out simulations by changing the circuit parameters. For simplicity, normalized parameters were used. Fig. 20 shows a comparison of the bifurcation diagrams of the tunnel and bridge type junctions. For any value of i_0 in the figure, thirty points of instantaneous voltage v are plotted as discrete time of $t = n\frac{2\pi}{\Omega}$. It is clearly seen from the results that in the tunnel type Josephson junction circuit four types of oscillation mode exist, those being periodic oscillation, subharmonic oscillation, chaotic oscillation and quasi-periodic oscillation, while in the bridge

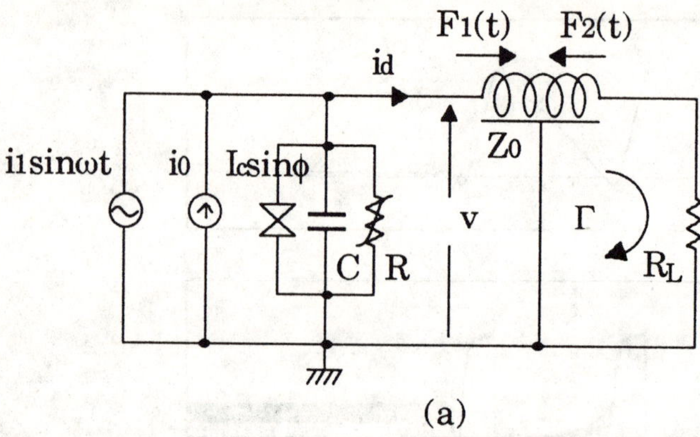

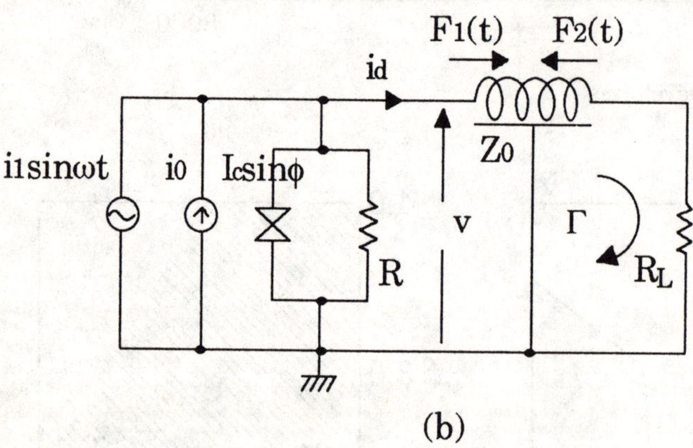

Fig. 19. Model of Josephson circuit with transmission line: (a) for a tunnel type Josephson junction; (b) for a bridge type Josephson junction

type Josephson junction circuit the periodic and quasi-periodic oscillations are dominant. Although chaotic oscillations only rarely occur in circuits with a bridge type Josephson junction, we can observe the chaotic oscillation in a distributed parameter circuit in a special case, that is when $\Gamma \approx 1$. Fig. 21 shows the bifurcation diagram and average voltage diagram to indicate the chaotic oscillation in a bridge type Josephson junction circuit with transmission line for which the range of Γ is from 0.9 to 1. The route to chaotic oscillation from the periodic oscillation is via two, four, eight and sixteen subharmonic oscillations. The classification of the oscillation mode can be made clear by using a Poincaré map.

It is concluded from the simulation results as follows:

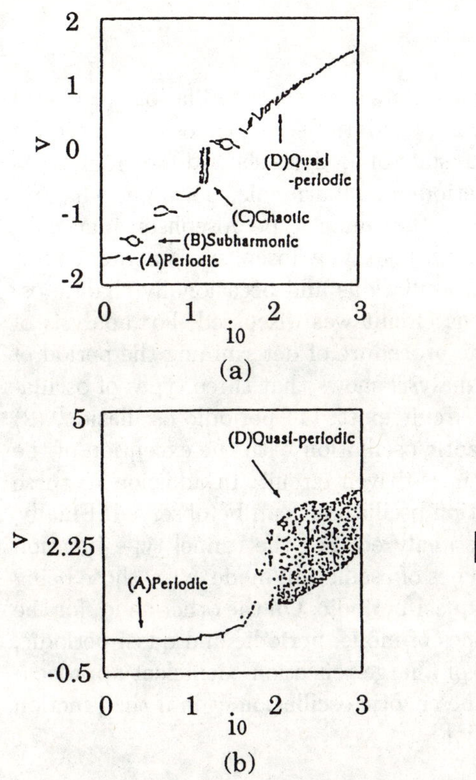

Fig. 20. Comparison of bifurcation diagram between a tunnel type junction and a bridge type junction circuit: (a) for a tunnel type junction circuit; (b) for a bridge type junction circuit

Fig. 21. A bifurcation diagram and an average voltage diagram showing a chaotic oscillation in a bridge type junction circuit: (a) bifurcation diagram; (b) average voltage diagram

(1) Chaotic oscillation occurs when the amplitude of the oscillation across the Josephson junction just crosses the zero-voltage axis into the negative region.

(2) Quasi-periodic oscillation occurs when the amplitude of oscillation waveform across the Josephson junction never crosses into the negative region, and the relation between the average voltage and the bias current of i_0 monotonously increases.

(3) For the tunnel type junction circuit with transmission line four types of oscillation mode exist, those being periodic, subharmonic, chaotic and quasi- periodic. On the other hand, for the bridge type junction circuit, two types of mode, periodic and quasi-periodic, are dominant. However, in the case of a large reflection coefficient approaching the value of 1, we can observe the chaotic oscillation even if the junction in the circuit is a bridge type junction.

6. Conclusion

In this paper, the chaotic phenomena and its associated oscillations produced in a Josephson circuit have been summarized. In Sect. 3, chaos in a forced oscillation circuit was discussed. The results of analysis showed that four types of oscillations exist, those being periodic, subharmonic, quasi-periodic and chaotic oscillations in the circuit with the tunnel type Josephson junctions, while the dominant oscillations in the bridge type Josephson junction circuit are the periodic and quasi-periodic oscillations and occasionally chaotic oscillations. In Sect. 4, an autonomous circuit was discussed. For analysis of oscillation modes, we introduced the procedure of determining the period of free-running periodic oscillations. Analysis shows that three types of oscillation in an autonomous Josephson circuit exist: (1) periodic oscillation, (2) subharmonic oscillation and (3) chaotic oscillation with the exception of the quasi-periodic oscillation observed in rf-driven circuit. In addition to these oscillations, various kinds of relaxation oscillations can be observed. Finally, a distributed parameter circuit was analyzed. For the tunnel type junction circuit with transmission line four types of oscillation mode exist, those being periodic, subharmonic, chaotic and quasi-periodic. On the other hand, for the bridge type junction circuit, two types of mode, periodic and quasi-periodic, are dominant. However, in the case of a large reflection coefficient approaching the value of 1, we can observe the chaotic oscillation even if the junction in the circuit is a bridge type junction.

Acknowledgements. The authors would like to thank the staff in our laboratories for their distinguished contributions to this work.

References

1. R.L. Kautz: Chaos in Josephson circuits. IEEE Tans. Magn. **MAG–19**, 3 (1983)
2. M. Morisue, H. Fukuzawa, Y. Marushima: Chaos in superconducting element circuit. Proc. ISCAS '85, 855–858 (1985)
3. K. Araki, M. Morisue, H. Kasahara, S. Yamamoto: Oscillation modes in the tunnel type and the weak-link type Josephson junction. IEEE Trans. Magn., **MAG-23**, 2 (1987)
4. M. Morisue, H. Kasahara, K. Araki: An approach for analysis of chaos in Josephson autonomous circuit. Proc. of ISCAS '87 (1987)
5. H.W. Chan, T. Van Duzer: Josephson nonlatching logic circuits. IEEE J. Solid-State Circuits, **SC-12**, 1, 73–79 (1977)
6. M. Morisue, J. Gu, T. Morimae, M. Otake: Nonlinear oscillations produced in complex Josephson circuits. Proc. ISCAS '91, 2979–2982 (1991)
7. K. Araki, M. Morisue, Y. Takahashi, K. Katori: Oscillation mode analysis for Josephson junction circuits containing transmission lines. Proc. of ISEC'87, **SP-33** (1987)

Chaos in Systems with Magnetic Force

J. Tani

Institute of Fluid Science,
Tohoku University,
2–1–1 Katahira,
Aoba-ku,
Sendai, 980 Japan

Abstract

The paper includes the nonlinear phenomena investigation of a system with two parallel wires carrying currents, of a multi-well-potential magnetic system and of a system with magnetic levitation. Some special bifurcation diagrams exhibiting a variety of nonlinear phenomena as well as numerical evidence of chaos are given.

1. Introduction

Chaotic phenomena have been a subject of interest for researchers all over the world for the last twenty years [1–5]. In systems with magnetic forces, chaotic vibrations occur due to the strong nonlinearity of magnetic force [6]. The magnetic attractive and repulsive forces exhibit nonlinearity with respect to the distance from the magnetic pole. The magnetic body couple of ferromagnetic materials—ferromagnetic due to magnetization—also has the nonlinearity with respect to the deformation of the ferromagnetic materials. While the governing equation admits no random parameters or inputs, the resulting vibration of these systems appears to be chaotic and unpredictable.

In this chapter, the chaos in three systems is explained; (1) a system with two parallel wires carrying currents, (2) a multi-well-potential magnetoelastic system, (3) a system with magnetic levitation. Through the numerical simulation chaotic time histories, phase portraits, Poincaré maps, bifurcation diagrams, Lyapunov exponents and fractal dimensions can be displayed. Some experimental results also are shown and compared to the theoretical prediction.

2. System of Two Conducting Wires

In magnetoelastic systems, chaotic phenomena easily occurs because the magnetic force implicitly contains a nonlinear dependence on distance between source and observation points. Much work has been devoted to giving experimental or theoretical evidence of chaos in such systems [6]. A new system with magnetic forces as the origin of chaotic behavior is analyzed numerically and results of regular and nonregular types are presented in this section [7].

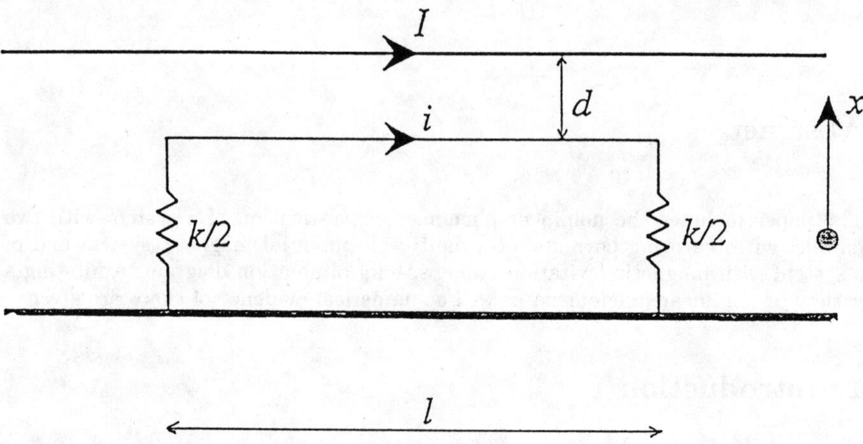

Fig. 1. Wire-spring system

The system under consideration consists of two nonflexible wires carrying currents in the same direction separated by a distance d as shown in Fig. 1. The lower wire is attached to the support by the springs with a total spring constant k. The two wires are of the same length l and the currents in the top and lower wires are I and i, respectively. The transverse vibration in the system is analyzed with respect to the coordinate x, the origin of which is put on the lower wire.

2.1 Formulation of Dynamical Equations

Let us consider the static conditions. There are two forces interacting in the system. That is the magnetic force between the two wires carrying currents, which can be written in the form

$$f_m = -\frac{b}{d-x} \quad \text{with} \quad b = \frac{\mu_0 I i l}{2\pi} , \tag{1}$$

and the following spring restoring force:

$$f_s = kx . \tag{2}$$

The total force is $f = f_m + f_s$ and it is easy to find the corresponding potential energy of the system as follows:

$$U(x) = \frac{1}{2}kx^2 - b\ln(d-x)\,. \tag{3}$$

The balance of above forces, $f = 0$, describes the equilibrium position in the system, which can also be expressed as

$$x_{\mathrm{e}}^{\pm} = \frac{d}{2} \pm \sqrt{\frac{d^2}{4} - \frac{b}{k}}\,. \tag{4}$$

Three different cases are possible depending on the parameters. Choosing the distance between wires d as a variable, the bordering value is the limit point $d_l = 2\sqrt{b/k}$. If $d > d_l$, there are two equilibria in the system (one stable x_{e}^-, and one unstable x_{e}^+), whereas if $d < d_l$ there is no equilibrium. The spring restoring force together with the magnetic forces for three values of the distance d are shown in Fig. 2. The corresponding potential energy is presented in Fig. 3. The three cases described above can be clearly seen on these pictures.

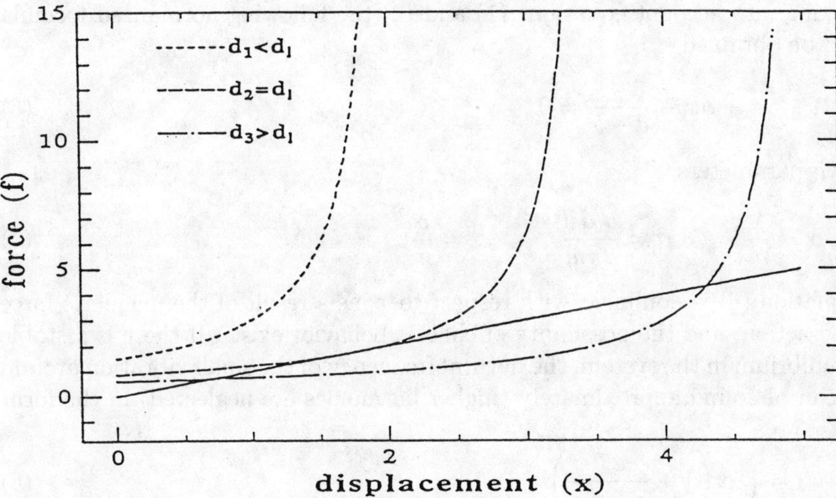

Fig. 2. Spring restoring force (solid line) and magnetic forces for three cases versus displacement

Let us now consider the dynamic behavior of the system. The equation of motion is

$$m\ddot{x} + \delta\dot{x} + f = 0\,, \tag{5}$$

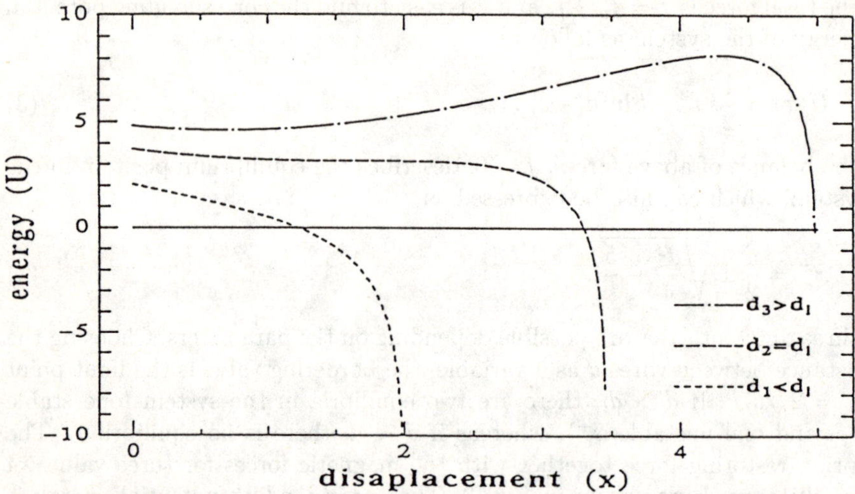

Fig. 3. Potential energy of the system versus displacement

where m is the mass of the lower wire, and δ is the damping coefficient. Taking into account equations (1) and (2) the following normalized formula can be obtained

$$\ddot{x} + \gamma\dot{x} + \alpha x - \frac{\beta}{d - x} = 0\,, \tag{6}$$

with parameters

$$\alpha = \frac{k}{m}\,, \qquad \beta = \frac{\mu_0 I i l}{2\pi m}\,, \qquad \gamma = \frac{\delta}{m}\,. \tag{7}$$

Equation (6) is nonlinear with respect to x, as a result of the magnetic force interaction, and the possibility of chaotic behavior exists. If there is a stable equilibrium in the system, the natural frequency of the small vibration around it can be found approximately (higher harmonics are neglected) in the form

$$\omega_0 = \sqrt{\alpha\left(1 + \frac{x_{\mathrm{e}}^-}{x_{\mathrm{e}}^- - d}\right)}\,. \tag{8}$$

The excitation term is still missing in equation (6). Two types of excitation are considered for system. For the alternating current in the top wire (I) one should put $I + A\sin\Omega t$ instead of I, and for the mechanical transverse excitation of the support (II) one should put $x + A\sin\Omega t$ instead of x. Then, the final set of equations suitable for numerical simulation can be obtained.

2.2 Analytical Procedure

Equation (6) can be rewritten as a system of first order differential equations in the form

$$\dot{\boldsymbol{x}} = \boldsymbol{f}(\boldsymbol{x}, \boldsymbol{\mu}) , \tag{9}$$

where $\boldsymbol{x} = (x, y, z)^T \in \mathbb{R}^3$ is a vector of variables (displacement, velocity, and phase), $\boldsymbol{f} = [f_1, f_2, f_3]^T : \mathbb{R}^3 \to \mathbb{R}^3$ is a vector function given by force components, and $\boldsymbol{\mu} = [\alpha, \beta, \gamma, \eta]^T \in \mathbb{R}^4$ is a vector of parameters of the system. Functions $f_1 = y$ and $f_3 = \Omega$ are the same for both kinds of excitation, whereas for the case I

$$f_2 = -\gamma y - \alpha x + \frac{\beta}{d - x}(1 + \eta \sin z) ; \qquad \eta = \frac{A}{I} , \tag{10}$$

and for the case II

$$f_2 = -\gamma y - \alpha x + \frac{\beta}{d - (x + \eta \sin z)} - \eta[(\alpha - \Omega^2) \sin z + \gamma \cos z] ; \quad \eta = A. \tag{11}$$

Equation (9) will be studied now using the center manifold theory [1, 8, 9, 10]. The linearized equation around equilibrium position $\overline{\boldsymbol{x}}_e$ can be written as

$$\dot{\boldsymbol{x}} = \mathring{\boldsymbol{A}} \boldsymbol{x} , \tag{12}$$

where $\mathring{\boldsymbol{A}} = D\boldsymbol{f}|_{x=\overline{x}_e, y, z=0}$. Hence the matrix $\mathring{\boldsymbol{A}}$ has the form

$$\mathring{\boldsymbol{A}} = \begin{bmatrix} 0 & 1 & 0 \\ -\alpha + s_1 & -\gamma & s_2 \\ 0 & 0 & 0 \end{bmatrix} , \text{with} \quad s_1 = -\frac{\beta}{(d - \overline{x}_e)^2} , \tag{13}$$

and for the case I

$$s_2 = -\frac{\beta \eta}{d - \overline{x}_e} , \tag{14}$$

and for the case II

$$s_2 = -\frac{\beta \eta}{(d - \overline{x}_e)^2} - \eta(\alpha - \Omega^2) . \tag{15}$$

The eigenvalue problem of $\mathring{\boldsymbol{A}}$,

$$(\mathring{\boldsymbol{A}} - \xi \boldsymbol{I}) \boldsymbol{v} = 0 , \tag{16}$$

yields a characteristic equation of the form

$$\xi(\xi^2 + \xi\gamma + \alpha - s_1) = 0 , \tag{17}$$

hence the set of eigenvalues is

$$\xi_1 = 0\,, \qquad \xi_2 = \frac{-\gamma - \sqrt{\Delta}}{2}\,, \qquad \xi_3 = \frac{-\gamma + \sqrt{\Delta}}{2}\,, \tag{18}$$

where $\Delta = \gamma^2 - \tilde{\delta}$, and $\tilde{\delta} = 4(\alpha - s_1)$. Respective eigenvectors are

$$\boldsymbol{v_1} = \left[\frac{s_2}{s_1 - \alpha}p, 0, p\right]^T\,, \quad \boldsymbol{v_{2,3}} = [q, q\xi_{2,3}, 0]^T\,, \quad p, q \in \mathbb{R}\,. \tag{19}$$

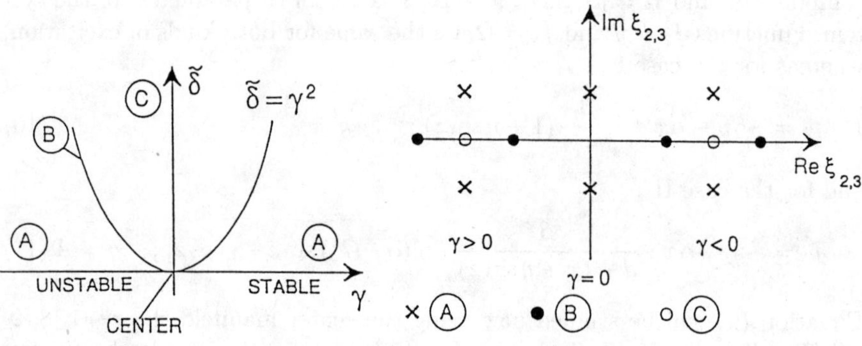

Fig. 4. Behavior of the solution and the loci of eigenvalues

The center manifold W^{C} for equation (9) is always nonempty, as it is transient to the center eigensubspace E^{C} of the linearized problem (12), spanned by at least one eigenvector $\boldsymbol{v_1}$, and the form of other manifolds (stable and unstable) depends on the relations between the parameters. If γ is positive, the stable manifold W^{S} exists, and is transient to the stable eigensubspace E^{S} spanned by vectors $\boldsymbol{v_2}$ and $\boldsymbol{v_3}$. Conversely, the unstable manifold W^{U} exists for negative damping, and is transient to E^{U} spanned by the same vectors. If $\gamma = 0$, the centre eigensubspace is spanned by the whole eigenvector triplet. The locations of the eigenvalues in different ranges of parameters are shown in Fig. 4. It can be seen from the above that the main values of parameters when bifurcation occurs are $\tilde{\delta} = \gamma^2$ and $\gamma = 0$. For $\gamma = 0$ the dimension of the centre manifold is one greater than for $\gamma \neq 0$, hence even the structure of the motion is different, which was to be expected.

2.3 Numerical Simulation of Chaos

On the basis of the mathematical formulation given in equations (9), (10) and (11), the numerical simulation of chaotic motion for the two cases of excitation was carried out using a CRAY supercomputer. The differential system of equations was changed into a discrete system and the mathematical feedback iteration process was used to solve the problem. The following parameters were fixed: $\alpha = 1.0$, $\beta = 3.0$, $\gamma = 0.1$, $d = 5.0$, resulting in a limit value of d as

$d_1 = 3.46$, so the system has two equilibria. The stable equilibrium position is $\overline{x}_{\mathrm{e}} = 0.697$ and natural frequency of the small vibration around it is $\omega_0/2\pi = 0.146$. The excitation amplitude η was chosen as 2.0 for support excitation and 20.0 for excitation by the current, while the excitation frequency $\Omega/2\pi$ was varied from 0.1 up to 10.0.

The simulation time was taken as 50 periods of excitation (phase from 0 to 314) with four different frequencies $\Omega/2\pi = 0.4, 0.7, 1.0, 3.0$. It was found that chaos is more or less transient for this kind of excitation. This means that after some time (20, 30 periods) the motion switches from chaotic to regular. The Lyapunov exponents [10] calculated are negative (approx. -0.002).

A Poincaré map is a special set of points in the picture obtained from a long time simulation (15000 periods of excitation) by taking samples of x and y in every coincident point of excitation periods, and plotting it as a point on the map. Such maps are presented in Fig. 5 for case I. The structure of the strange attractor is many-fold spiral, and it was observed that after some time the motion goes towards the center point due to the transient behavior in time. The fractal dimensions obtained are 0.22, 0.19, 0.17, 0.10 for the corresponding frequencies of 3.0, 1.0, 0.7, 0.4.

Similar results were obtained for the transverse support excitation (II). For the frequency 0.7 the chaotic response is not transient. The Lyapunov exponent is 0.044. Poincaré maps are shown in Fig. 6. The structure of the strange attractor looks similar to the one in case I, but now points appear in the center of picture and outside it during the entire period of simulation for the frequencies 0.7 and 1.0. This proves the long time stationary chaotic behavior. Chaos is transient for a frequency of 0.4, whereas subharmonic vibration occurs at a frequency of 3.0. The fractal dimensions obtained are 0.45, 1.83, 2.45, 0.28 for the corresponding frequencies of 3.0, 1.0, 0.7, 0.4.

The global dependence of chaos on the excitation frequency as a variable parameter can be obtained by preparing bifurcation diagrams. Two of them are presented in Fig. 7a for the case I, and Fig. 7b for the case II. The motion was sampled during 1000 periods of excitation for 100 different values of excitation frequency (in the range from 0.5 to 6), and one point of the period was put in the picture, so each one represents 100,000 points.

In conclusion, the numerical evidence of the chaos in the wire-spring system shows that for the excitation by an alternating current the chaotic behavior is always transient, whereas for the transverse support excitation non-transient chaos occurs for a certain range of excitation frequency.

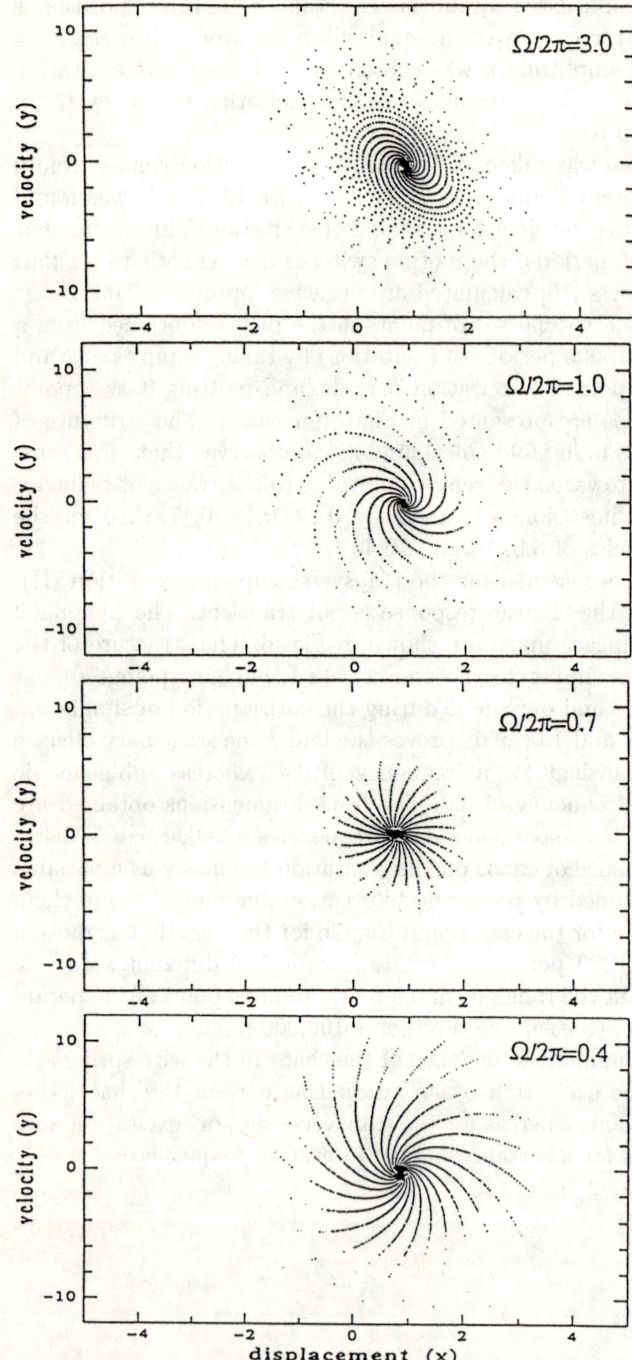

Fig. 5. Poincaré maps for the case I

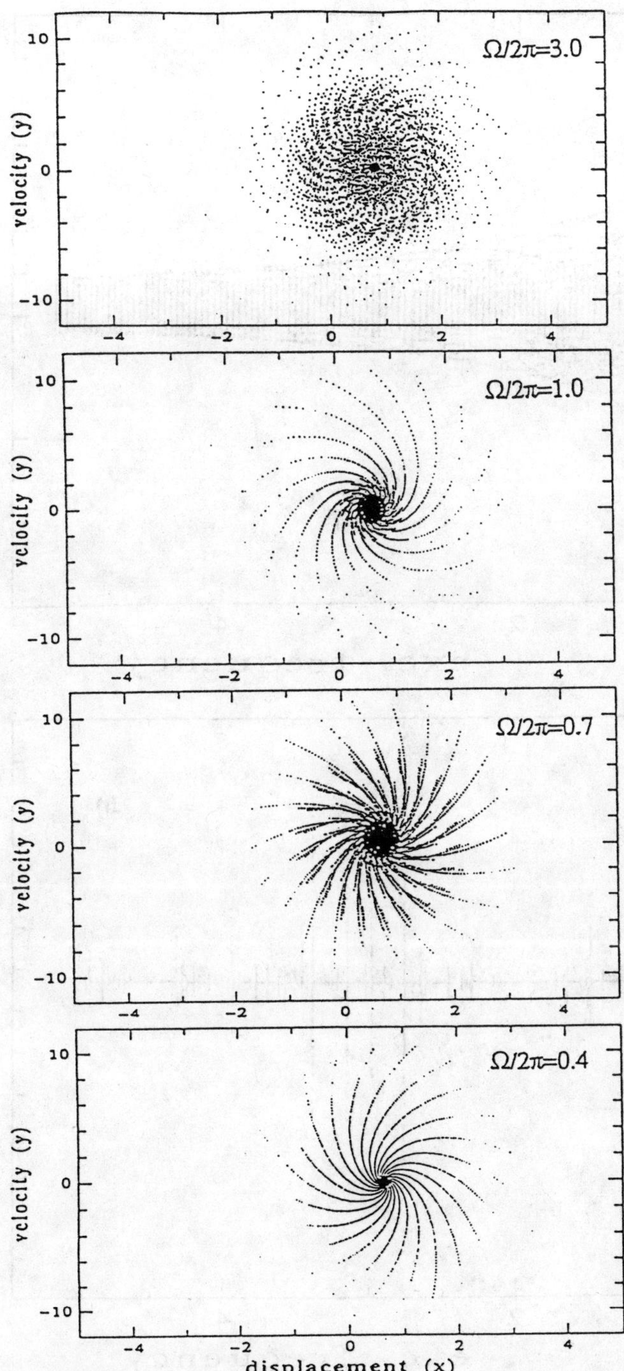

Fig. 6. Poincaré maps for the case II

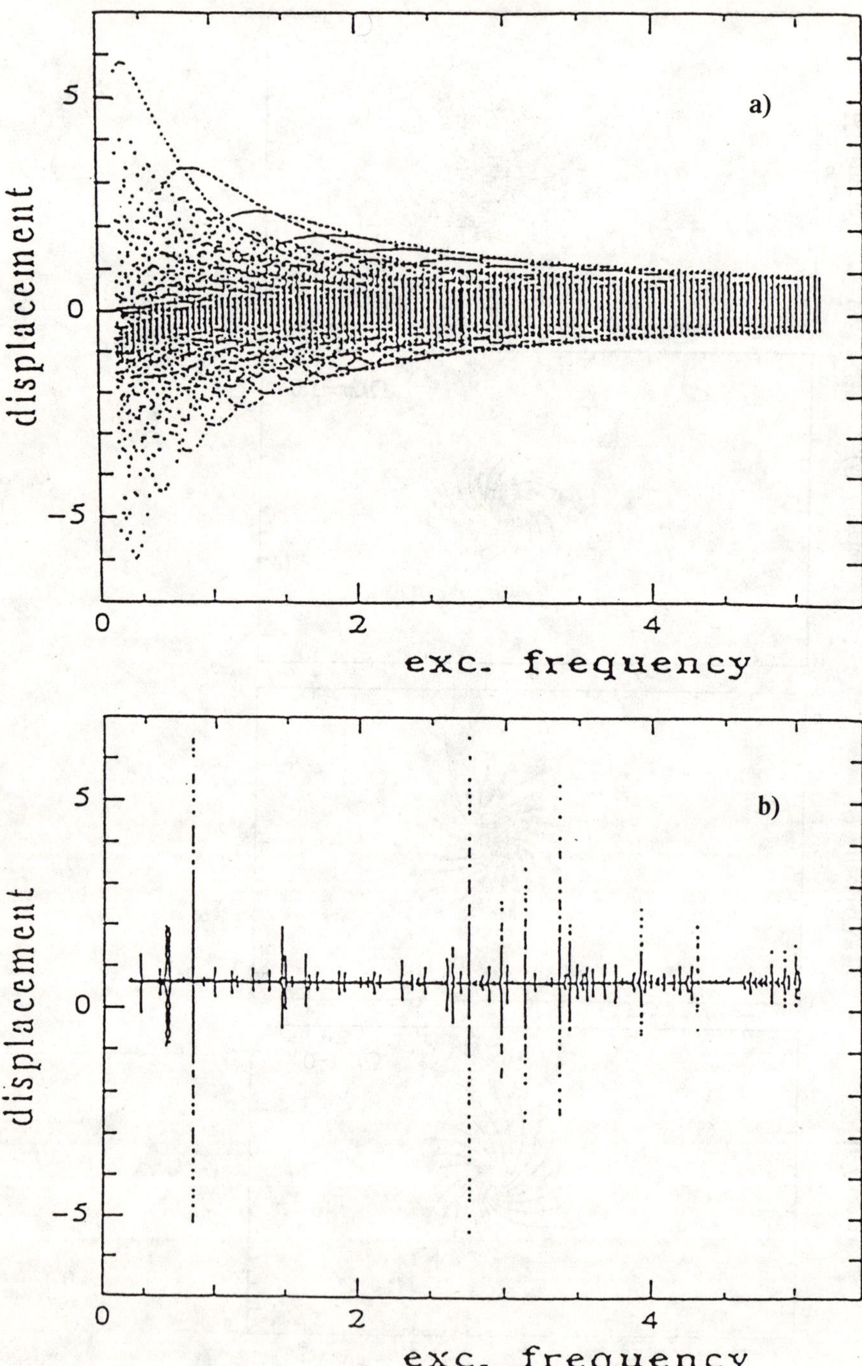

Fig. 7. Bifurcation diagram for cases I (a) and II (b)

3. Multi-Equilibrium Magnetoelastic Systems

The study of the interaction between an elastic beam-plate and magnetic forces usually leads to nonlinear differential equations [6]. The motion described by these equations can be chaotic in certain parameter ranges. An selected set of systems described by multi-well potential equations has been analyzed and numerical evidence of chaos in these systems is given by such descriptors as the Poincaré map, time history, phase portrait and bifurcation diagram. Some Lyapunov exponents and fractal dimensions of Poincaré maps are also mentioned for completeness.

The system under consideration is a vertical elastic beam-plate (paramagnetic) with additional mass and permanent magnet at the lower tip of the beam-plate. The beam-plate is clamped at the upper tip and harmonically excited there. The magnetic force acting upon the magnet at the beam-plate tip comes from one of the three magnetic systems:

I. The magnet of the same pole under the free position of the beam-plate tip;

II. Two same pole magnets separated by a distance *d* located centrally under the free position of the beam-plate tip;

III. Three same pole magnets located under the beam-plate.

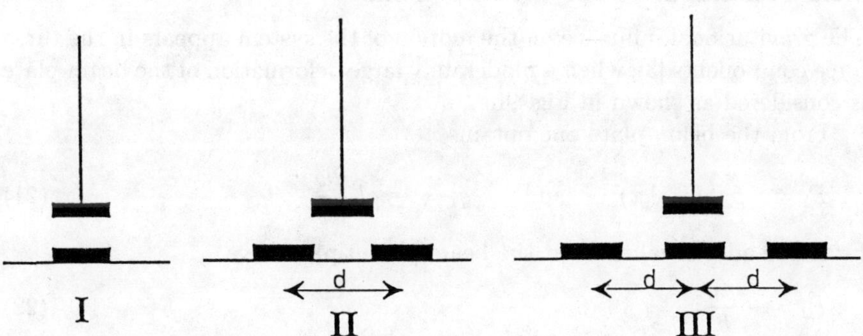

Fig. 8. Simple view of magnetoelastic systems

Each structure is depicted in Fig. 8.

It can easily be predicted that the system will have 3 equilibria (2 stable, 1 unstable) in the first case, 5 equilibria (3 stable, 2 unstable) in the second case, and finally 7 equilibria (4 stable, 3 unstable) in the third case. For each system the magnetoelastic origin of nonlinear governing equations is given. The set of first order differential equations is built, which is convenient for achieving numerical simulation. Computed results of response motion for appropriate parameters of the chaos region are also shown. Results were obtained using a CRAY computer. Furthermore, the experimental results for case I are also shown and these correspond well with the numerical simulation.

3.1 Theoretical Models

The elastic beam-plate under axial and lateral loads can be described by the following differential equation [11]

$$D\frac{\partial^4 w}{\partial x^4} + P\frac{\partial^2 w}{\partial x^2} = q\,, \tag{20}$$

which can be derived from the balance of moments equation. A respective coordinate system is shown in Fig 9a, and w is the deflection of the beam-plate (z direction). The parameters are as follows: b – width, L – longitude, h – thickness of the beam-plate, ρ – mass density of the beam-plate media, and $D = \frac{Eh^3}{12(1-\nu^2)}$, where E is the Young modulus and ν is the Poisson ratio. The force components which appeared in equation (20) are: q – lateral load intensity (acting in z direction) in the elemental strip (of y direction), P – compressive forces (acting in x direction) in the elemental strip.

All force components of q or P type existing in the system will be collected and put in the basic equation (1) to construct the magnetoelastic differential equations of motion for each of the three systems under consideration.

3.1.1 Nonmagnetic Force Components

The gravitational influence on the motion of the system appears in the three force components [3], when a moderately large deformation of the beam-plate is considered as shown in Fig. 9b

From the beam-plate one obtains

$$P_1 = -\rho g h(L - x) \qquad \text{and} \qquad q_1 = -\rho g h\frac{\partial w}{\partial x}\,. \tag{21}$$

From the additional mass at the beam-plate tip we have

$$P_2 = -\frac{mg}{b}\,. \tag{22}$$

The next two components come from the inertia. The inertia force of the beam-plate is

$$q_2 = -\rho h\frac{\partial^2 w}{\partial t^2}\,, \tag{23}$$

and the inertia force of the additional mass is

$$q_3 = -\frac{m}{bL}\frac{\partial^2 w}{\partial t^2}\delta(x - L)\,, \tag{24}$$

where δ is the Dirac delta-function.

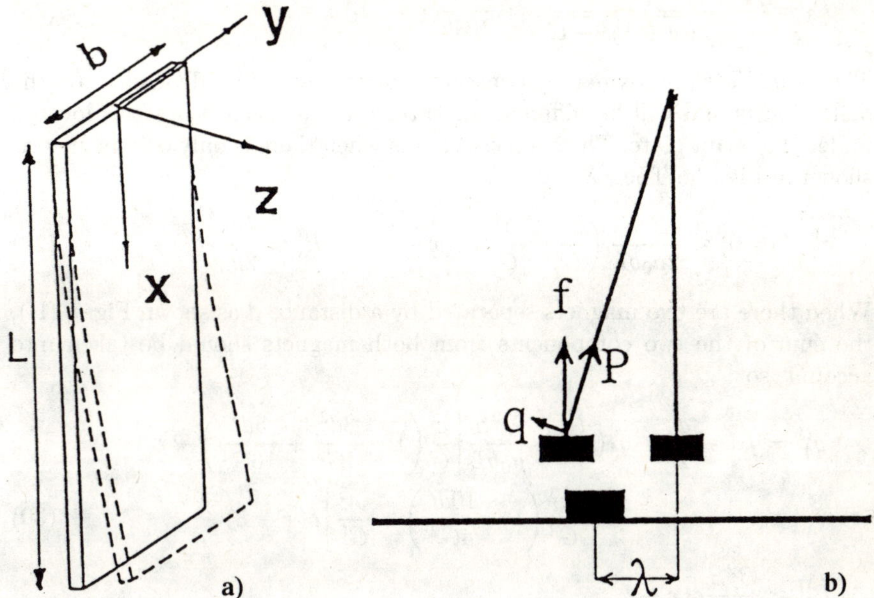

Fig. 9. Parameters and coordinates of the beam-plate (a) and magnetic force between two magnets (b)

3.1.2 Magnetic Forces Between Magnets

At first, the interaction between two magnets is only considered in order to obtain a suitable description. If the magnet under the beam-plate tip is misplaced by a distance λ with respect to the reference state as shown in Fig. 9b, the magnetic pressure between the like poles of cross-section S is:

$$f = \frac{B^2}{2\mu_0} \frac{G^2}{(w+\lambda)^2 + G^2} , \tag{25}$$

where B is the magnetic induction measured when the poles are separated by the gap G. μ_0 is the magnetic permeability of the vacuum. When $|w+\lambda| < G$, f can be expressed as

$$f = \frac{B^2}{2\mu_0} \left(1 - \frac{(w+\lambda)^2}{G^2} + \frac{(w+\lambda)^4}{G^4} - \frac{(w+\lambda)^6}{G^6} + \dots \right) , \tag{26}$$

using the formula of the sum of a geometrical series.

Two force components appear as a result of magnetic interaction:

$$q_4(\lambda) \approx \frac{fS}{bL} \frac{w+\lambda}{G} \delta(x-L) \tag{27}$$

and

$$P_3 = \frac{fS}{b}\frac{G\delta(x-L)}{\sqrt{(w+\lambda)^2+G^2}} \approx \frac{B^2S}{2\mu_0 b}\delta(x-L)\,. \tag{28}$$

The term P_3 is approximately constant, but the term q_4 is dependent on the deflection w and will be different for three cases of magnet systems located under the beam-plate. The simplest case is when there is only one magnet as shown in Fig. 7a. Then $\lambda = 0$ and

$$q_4^{\rm I} = q_4(0) \approx \frac{B^2S}{2\mu_0 bL}\left(\frac{w}{G}-\frac{w^3}{G}\right)\delta(x-L)\,,\quad P_3^{\rm I} \approx \frac{B^2S}{2\mu_0 b}\delta(x-L)\,. \tag{29}$$

When there are two magnets separated by a distance d as shown Fig. 8 (II), the sum of the two components from both magnets should be taken into account, so

$$\begin{aligned} q_4^{\rm II} = q_4\Big(-\frac{d}{2}\Big) + q_4\Big(\frac{d}{2}\Big) \approx \frac{B^2S}{\mu_0 bL}\bigg[\frac{w}{G}\left(1-\frac{3d^2}{4G^2}+\frac{5d^4}{16G^4}\right)- \\ -\frac{w^3}{G^3}\left(1-\frac{10d^2}{4G^2}\right)+\frac{w^5}{G^5}\bigg]\delta(x-L)\,, \end{aligned} \tag{30}$$

$$P_3^{\rm II} \approx \frac{B^2S}{\mu_0 b}\delta(x-L)\,.$$

For the three magnets under the beam-plate as shown in Fig. 8 (III), the sum of the three components from each magnet should be considered, so

$$\begin{aligned} q_4^{\rm III} &= q_4(-d) + q_4(0) + q_4(d) \approx \\ &\approx \frac{B^2S}{2\mu_0 bL}\bigg[\frac{w}{G}\left(3-\frac{6d^2}{G^2}+\frac{10d^4}{G^4}-\frac{14d^6}{G^6}\right)- \\ &-\frac{w^3}{G^3}\left(3-\frac{20d^2}{G^2}+\frac{70d^4}{G^4}\right)+\frac{w^5}{G^5}\left(3-\frac{42d^2}{G^2}\right)-3\frac{w^7}{G^7}\bigg]\delta(x-L)\,, \end{aligned} \tag{31}$$

$$P_3^{\rm III} \approx \frac{3B^2S}{2\mu_0 b}\delta(x-L)\,.$$

The appropriate number of terms are only kept in equations (29), (30) and (31) to obtain a real number of equilibria for each system.

3.1.3 Motion of a Two-Well Potential System

Collecting nonmagnetic force components together and putting them into equation (1) one can obtain the equation of motion for each system. The first case (I) will be considered here and the equation of motion is given as follows:

$$\begin{aligned} D\frac{\partial^4 w}{\partial x^4} + \left[\frac{B^2S}{2\mu_0 b}\delta(x-L) - \frac{mg}{b} - \rho g h(L-x)\right]\frac{\partial^2 w}{\partial x^2} + \rho g h\frac{\partial w}{\partial x} + \\ -\frac{B^2S}{2\mu_0 bL}\left(\frac{w}{G}-\frac{w^3}{G^3}\right)\delta(x-L) + \left[\rho h + \frac{m}{bL}\delta(x-L)\right]\frac{\partial^2 w}{\partial t^2} = 0\,. \end{aligned} \tag{32}$$

After incorporating the boundary conditions for the beam-plate ends

$$w(0,t) = \frac{\partial w}{\partial x}(0,t) = 0\,, \quad \frac{\partial^2 w}{\partial x^2}(L,t) = \frac{\partial^3 w}{\partial x^3}(L,t) = 0\,,$$

and using only the first mode of the Galerkin representation of the form

$$w(x,t) = u(t)\Phi(x)\,, \tag{33}$$

with normalization

$$\int_0^L \Phi^2(x)dx = 1\,, \qquad \Phi(L) = 1\,, \tag{34}$$

the following equation can be obtained (based on equation (13) and the Galerkin procedure):

$$\ddot{u} - \alpha u + \beta u^3 = 0\,, \tag{35}$$

where parameters are

$$\alpha = \frac{B^2 S - 2\mu_0 bLaG}{2\mu_0(\rho bLh + m)G}\,, \quad \beta = \frac{B^2 S}{2\mu_0(\rho bLh + m)G^3}\,,$$
$$a = D\int_0^L (\Phi'')^2 dx + \rho gh \int_0^L (L-x)(\Phi')^2 dx\,. \tag{36}$$

The system has three equilibria described by the equation $u(\beta u^2 - \alpha) = 0$, so an unstable equilibrium is obtained when $u = 0$ and two stable equilibria are obtained when $u = \pm\sqrt{\alpha/\beta}$. To find the natural frequency around one of the stable equilibria one should substitute $u = u_0 + A\sin\omega t$ into equation (35) and will deduce on calculation that it is $\omega_0 = \sqrt{2\alpha}$.

Damping and excitation terms should be added to equation (32) to obtain a full description of the system, so that it can be written as

$$\ddot{u} + \delta\dot{u} - \alpha u + \beta u^3 = A\Omega^2 \sin\Omega t\,, \tag{37}$$

where δ is the damping coefficient, A the displacement amplitude, and Ω the frequency of harmonic excitation. Equation (37) can be normalized into a dimensionless one after substituting $\tilde{u} = u/u_0$, $\tau = \omega_0 t$, $\omega = \Omega/\omega_0$

$$\frac{\partial^2 \tilde{u}}{\partial \tau^2} + \gamma\frac{\partial \tilde{u}}{\partial \tau} - \frac{1}{2}(1 - \tilde{u}^2)\tilde{u} = \tilde{A}\sin\omega\tau\,, \tag{38}$$

where $\tilde{A} = \Omega^2 A/\omega_0^2 u_0$, which is the well-known formula of the Duffing type equation.

Equation (37) can also be rearranged into a system of first order differential equations, suitable for numerical simulation. After substitution of $x_1 = u$, $x_3 = \Omega t$, one can see that

$$\begin{aligned}\dot{x}_1 &= x_2\,,\\ \dot{x}_2 &= -\delta x_2 + \alpha x_1 - \beta x_1^3 + A\Omega^2 \sin x_3\,,\\ \dot{x}_3 &= \Omega\,.\end{aligned} \tag{39}$$

3.1.4 Equations of Motion for a Three-Well Potential System

Using q^{II}, P^{II} and the same procedure as in the above subsection, the analogous equations of motion in the five equilibrium system can be derived as follows:

$$\begin{aligned}\dot{x}_1 &= x_2\,,\\ \dot{x}_2 &= -\delta x_2 + \alpha' x_1 - \beta' x_1^3 + \gamma' x_1^5 + A\Omega^2 \sin x_3\,,\\ \dot{x}_3 &= \Omega\,,\end{aligned} \tag{40}$$

where

$$\begin{aligned}\alpha' &= \frac{B^2 S(4G^4 - 3d^2G^2 + 20d^4) - 2\mu_0 bLaG^5}{4G^5(\rho hbL + m)\mu_0}\,,\\ \beta' &= \frac{B^2 S(2G^2 - 5d^2)}{4G^5(\rho hbL + m)\mu_0} = \frac{\gamma'}{2}(2G^2 - 5d^2)\,,\\ \gamma' &= \frac{b^2 S}{2G^5(\rho hbL + m)\mu_0}\,.\end{aligned} \tag{41}$$

This system has two unstable equilibria when $x_u = \pm\sqrt{(\beta' - \sqrt{\beta'^2 - 4\gamma'\alpha'})/2\gamma'}$ and three stable equilibria when $x_0 = 0$ and $x_0 = \pm\sqrt{(\beta' + \sqrt{\beta'^2 - 4\gamma'\alpha'})/2\gamma'}$.

3.1.5 Motion of a Three-Well Potential System

Taking into account q^{III} and P^{III}, the following equations of motion can be obtained for the system with seven equilibria:

$$\begin{aligned}\dot{x}_1 &= x_2\,,\\ \dot{x}_2 &= -\delta x_2 + \alpha'' x_1 - \beta'' x_1^3 + \gamma'' x_1^5 - \delta'' x_1^7 + A\Omega^2 \sin x_3\,,\\ \dot{x}_3 &= \Omega\,,\end{aligned} \tag{42}$$

where

$$\begin{aligned}\delta'' &= \frac{3B^2 S}{2\mu_0 G^7(\rho bLh + m)}\,, \quad \gamma'' = \frac{\delta''}{3}(3G^2 - 42d^2)\,,\\ \beta'' &= \frac{\delta''}{3}(3G^4 - 20d^2G^2 + 70d^4)\,,\\ \alpha'' &= \frac{\delta''}{3}(3G^6 - 6d^2G^4 + 10d^4G^2 - 14d^6) - \frac{2\delta''}{3B^2 S}\mu_0 bLaG^7\,.\end{aligned} \tag{43}$$

The system has three unstable equilibria at $-x_u$, 0, x_u and four stable at $-x_{o1}$, $-x_{o2}$, x_{o2}, x_{o1}, but it is impossible to give analytical solutions for them, because the order of equation is too high.

3.2 Numerical Simulation

On the basis of equations (39), (40) and (42), the numerical simulation of motion for each system was carried out. At first the differential system of equations was changed into a discrete system and the mathematical feedback iteration process was used to solve the equations. The accuracy was about 300 samples per period, which is about 250–400 μs in these cases. The appropriate number of stable positions for each system, and a small vibration in a short time around these positions can be easily observed as well as the change of position. One can also find small orbit cycles of vibration around each stable equilibrium and also a large orbit cycle, where the change of position occurs.

The Poincaré map is a special set of points in the picture obtained from a long time simulation by taking a sample of $x_1 = w$ and x_2 at every coincident moment of exciting periods, and plotting this as a point in the map. Such maps are presented in Fig. 10 (I–III) for each system. One can recognize the typical shape of the strange attractor for the Duffing equation in the Fig. 10 (I) [3, 5, 6], as in that case when the system has two stable positions. Other shapes occur for the systems with three and four stable equilibria, which can be clearly seen in Fig. 10 (II) and Fig. 10 (III). The motion was sampled at the beginning of the excitation period (phase equal to 0°). When the sampling point is changed (phase: 45°, 90°, 135°, 180°) the picture will also be changed. The evolution with the phase of the Poincaré map for each system is presented in Fig. 11 (I–III).

The Lyapunov exponents calculated from time histories corresponding to each Poincaré map in Fig. 10 (I–III) [7] are: $\lambda_1 = 0.127$ for case I, $\lambda_1 = 0.489$ for case II, and $\lambda_1 = 0.439$ for case III. Two bifurcation diagrams are presented in the next two figures, Figs. 12(a),(b) for the first system also. A bifurcation of a normalized excitation frequency ω in Fig. 12a and bifurcation of a normalized excitation amplitude $\tilde{A}$ in Fig. 12b are presented. The horizontal width of the Poincaré map, corresponding to the depicted parameter value, is plotted as a vertical bar in the diagrams.

To complete the chaos description in a considered multi-equilibrium system, the fractal dimensions of a set of Poincaré maps were computed, using the correlation definition [3]. They are shown in Table 1. As is evident the fractal dimension is not very sensitive to the phase of excitation.

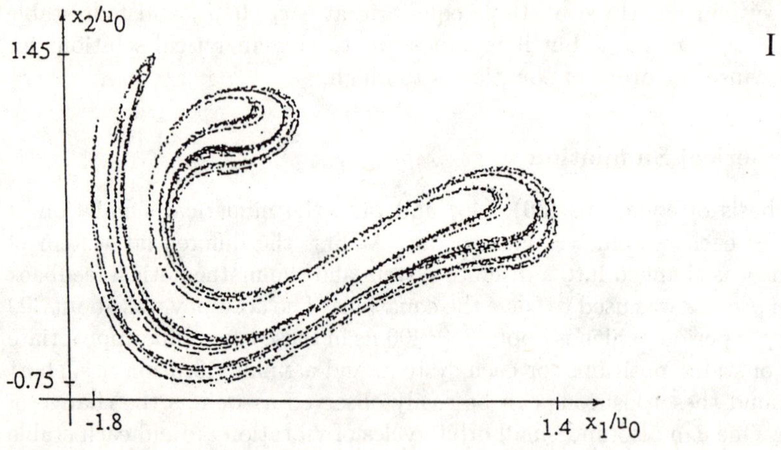

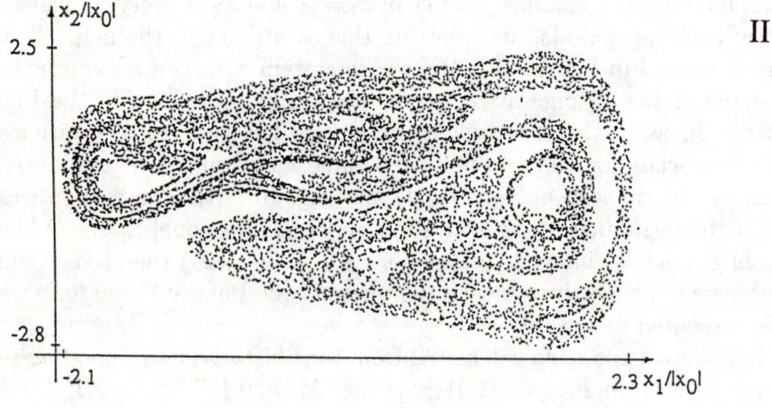

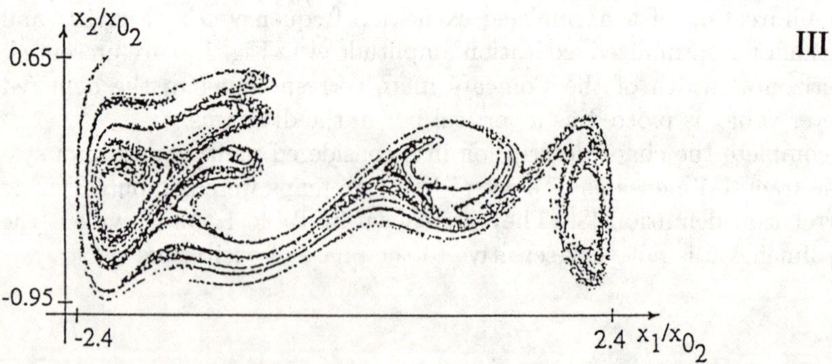

Fig. 10. Poincaré map for systems (I): $\gamma = 0.14$, $\tilde{A} = 0.5$, $\omega = 1.05$; (II): $\gamma = 0.08$, $\tilde{A} = 0.25$, $\omega = 0.65$; (III): $\gamma = 0.09$, $\tilde{A} = 0.155$, $\omega = 0.7$

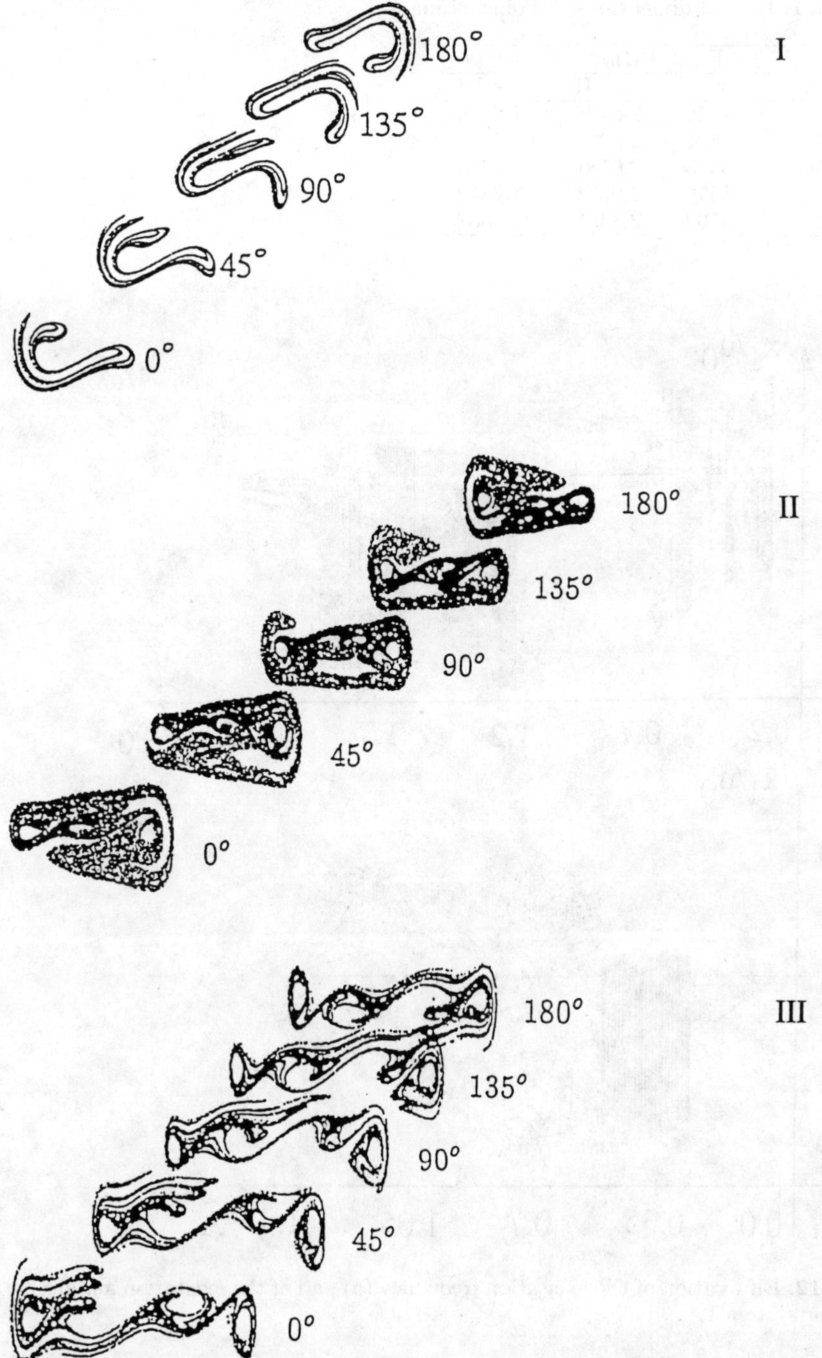

Fig. 11. Evolution of the Poincaré map with phase for systems I, II and III

Table 1. Fractal dimensions of Poincaré maps

	Fractal Dimension for		
phase	case I	case II	case III
0°	1.8020	2.0199	1.7206
45°	1.8606	2.0267	1.7206
90°	1.8623	2.0581	1.7706
135°	1.8575	2.0071	1.7312
180°	1.9103	2.1408	1.7587

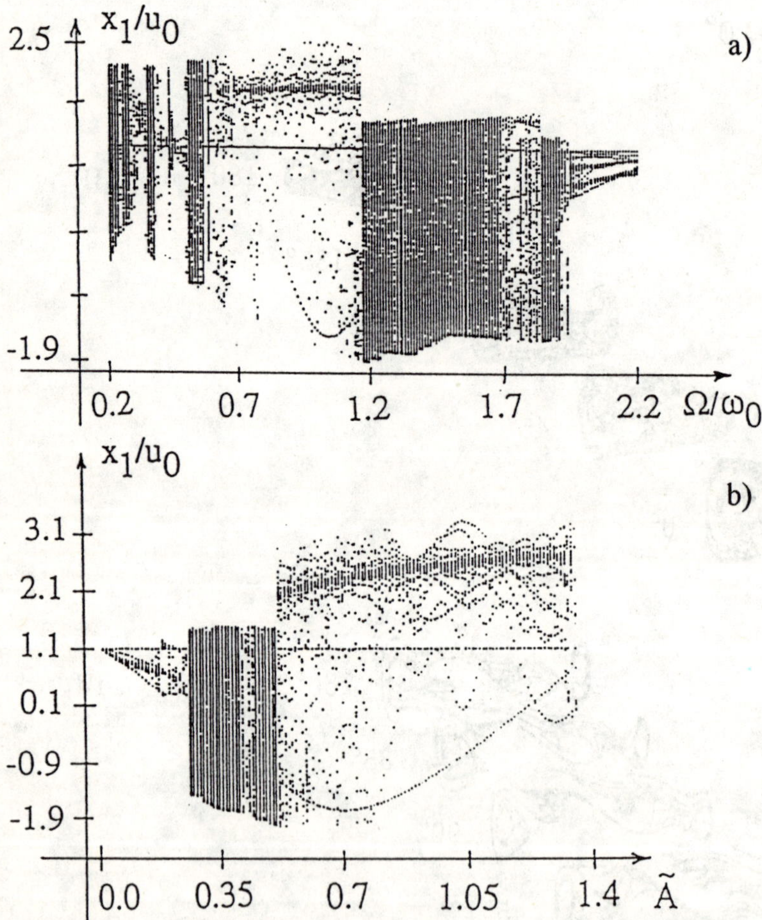

Fig. 12. Bifurcation of the excitation frequency (a) and of the excitation amplitude (b) for system I

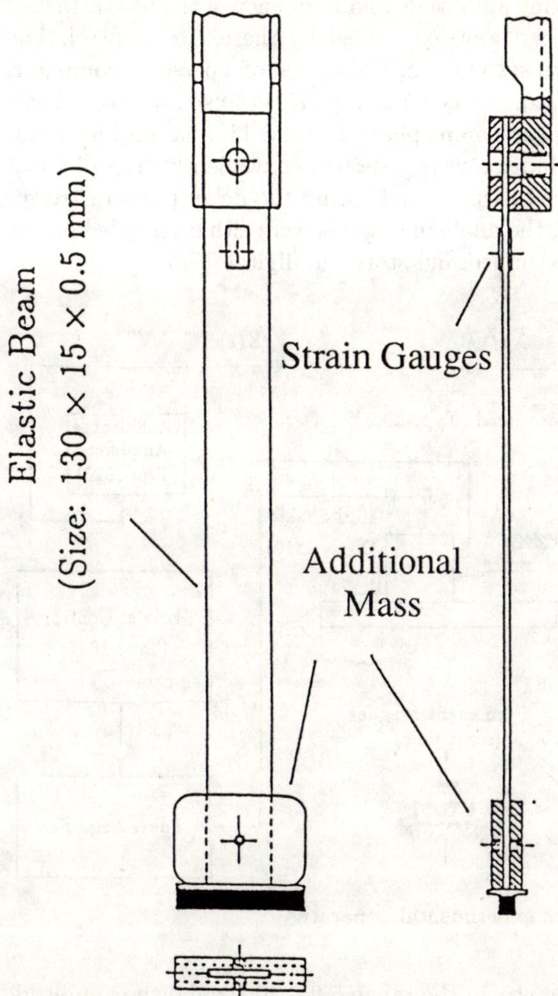

Fig. 13. Cantilever beam

3.3 Experiment

3.3.1 Method of Experiment

To confirm the validity of the theoretical analysis the experiment for case (I) was performed by using a cantilever beam with a permanent magnet and an additional mass at its free tip. The beam of length 130.0 mm, width 15.0 mm and thickness 0.5 mm was made of aluminium and installed vertically as shown in Figs. 13 and 14. Another permanent magnet was fixed and the free tip of the beam. The response of the beam was measured by a stain gage bonded to the beam near its clamped end. The magnetic field was measured by a Gauss Meter. The exciting amplitude and frequency of the base attached to the beam at the upper end were controlled by the exciter control. The computer-aided experimental system which consisted of a personal computer, A/D board and a differentiator as shown in Fig. 15 was used to trace chaos. With this system, the Poincaré map, phase history, bifurcation and other characteristics of chaotic vibration were obtained conveniently. The phase of the Poincaré map was chosen easily by adjusting the delay point of trigger in the software. In addition, the analogue signals were either recorded on the tape for the further use or sampled and stored in digital form.

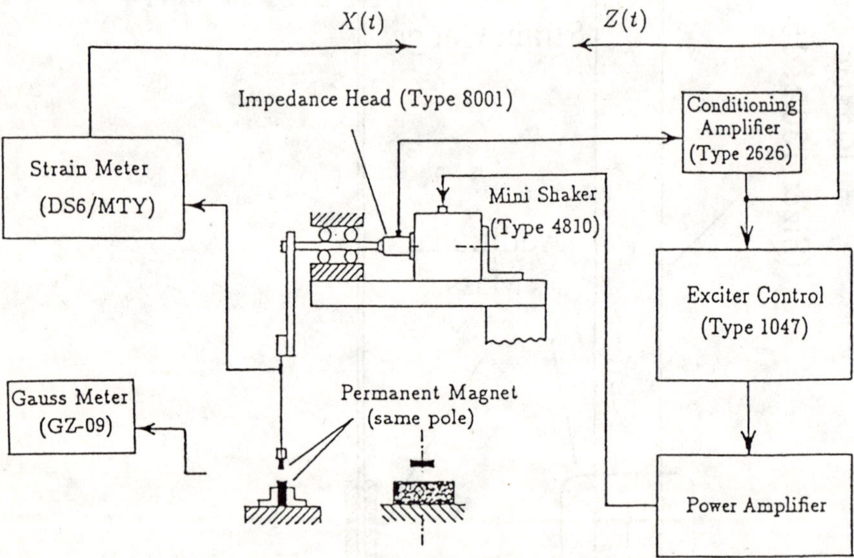

Fig. 14. Schematic diagram of experimental apparatus

The excitation frequency was in the range 5.0 – 40 Hz and the amplitude 1.0 – 4.0 mm. For a fixed separation between the two magnets, the exciting amplitude was fixed and the frequency was then increased little by little. The response of the beam was observed at all times. Secondly, the exciting amplitude was increased slightly and the frequency was again increased. In this way, the chaotic motion of the beam was investigated using an FFT analyzer and a computer.

3.3.2 Experimental Results

The time history and power spectra of the bending strain of the beam under the condition of the magnet gap $G = 5$ mm, the exciting amplitude $A = 1.5$ mm were shown in Fig. 16. In this figure (a) is the resonance state and (b) the chaotic state. Chaotic motion of the beam under the periodic excitation

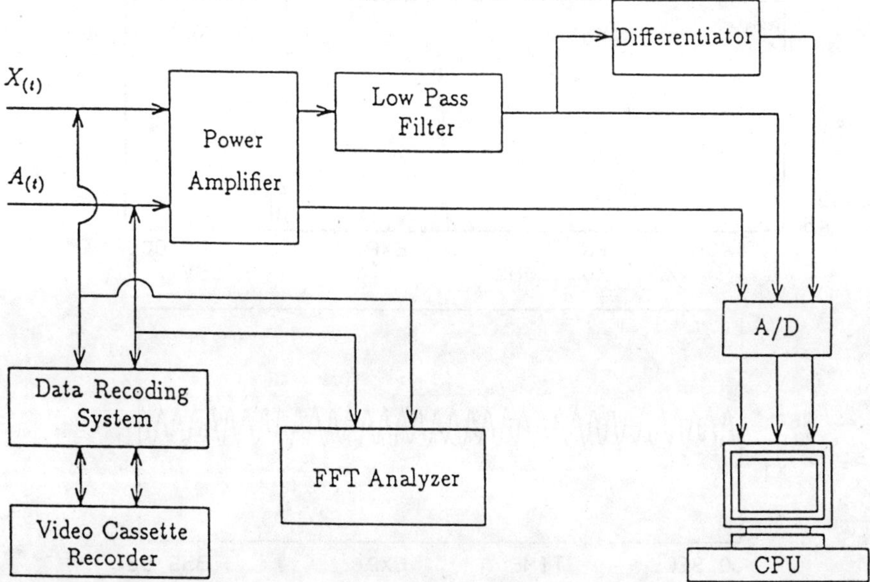

Fig. 15. Diagram of computer-aided chaos observation system

is characterized by a broadband power spectrum. The chaotic response is larger than that of the resonance. This is because the resonance is the regular vibration around one equilibrium point and the chaos is the random-like vibration around two equilibrium points with jump phenomena. The Poincaré map in the case with $G = 5$ mm, $A = 2.0$ and $\Omega = 10.6$ Hz is shown in Fig. 17. In comparison with Fig. 11 (I), the experimental Poincaré map agrees almost with the theoretical one in quality. Using the Poincaré map with phase zero for the various exciting amplitudes, a bifurcation diagram was obtained as shown in Fig. 18. When the exciting amplitude increases, the chaos occurs suddenly in this case. The exponent was obtained by sampling the response between the amplitude 1.0 and 4.0 mm and using the method of Wolf, Swift, Swinney and Vastano [10] as shown in Fig. 19. The positive value of the Lyapunov exponent indicates the occurrence of the chaos. This result of the Lyapunov exponent corresponds to the bifurcation diagram. The regions in which chaos occurs are shown in Fig. 20. The first natural frequencies are 15.37 and 10.45 Hz for the gaps $G = 3$ and 5 mm, respectively. The chaotic region is near the first natural frequency.

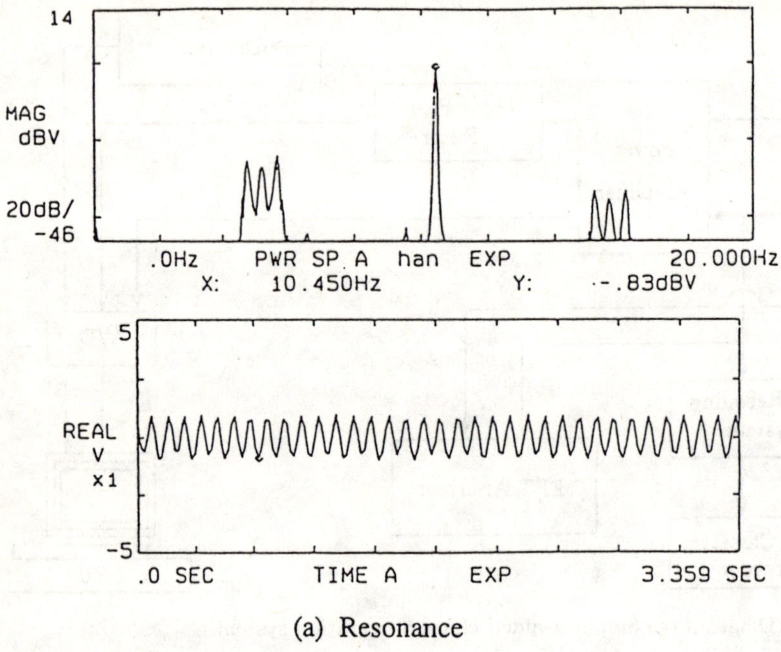

(a) Resonance

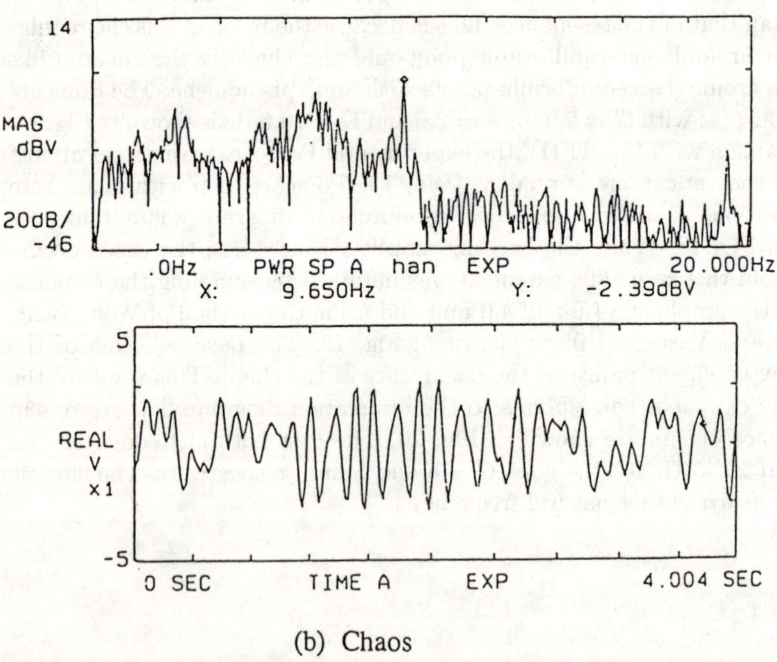

(b) Chaos

Fig. 16. Time history and power spectra

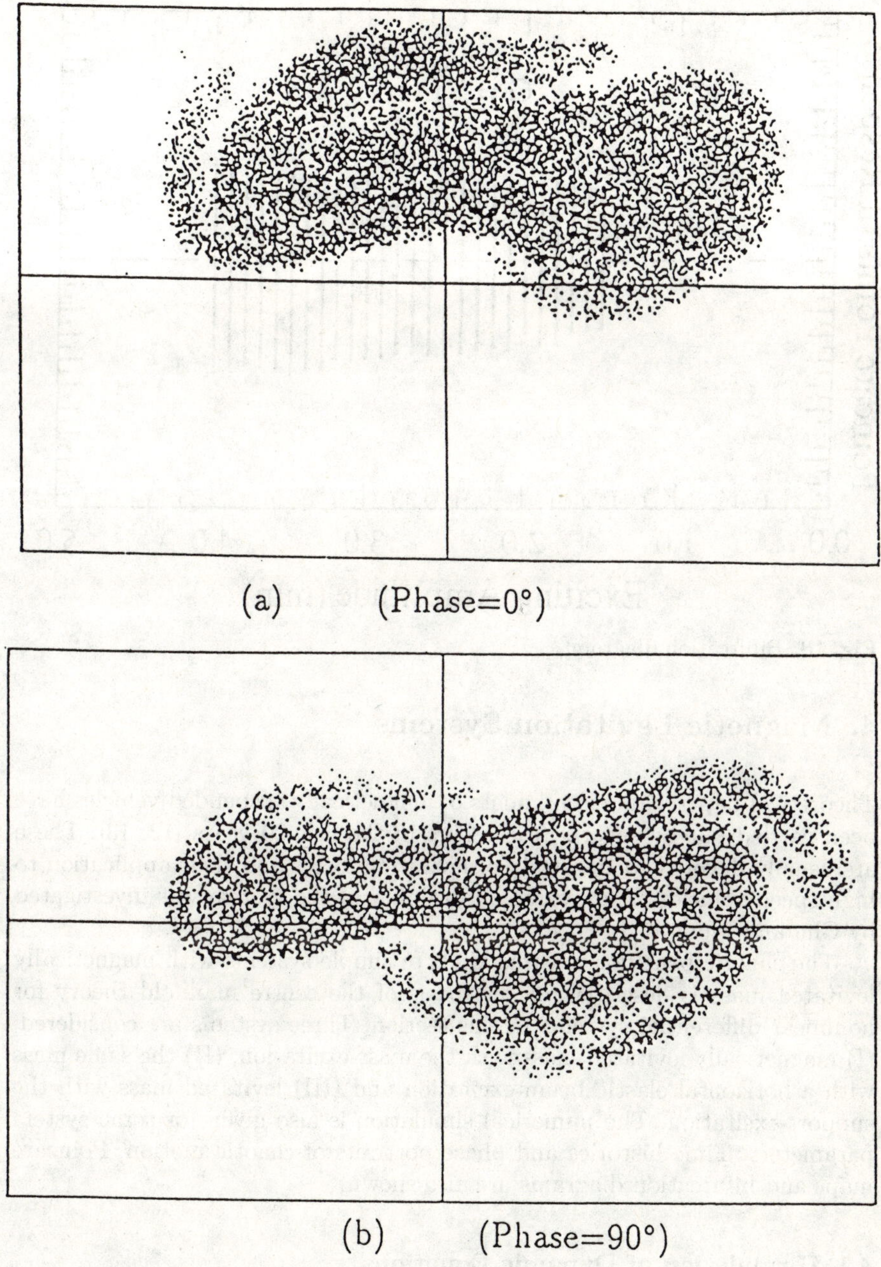

Fig. 17. Poincaré map: (a) phase = 0°, (b) phase = 90°

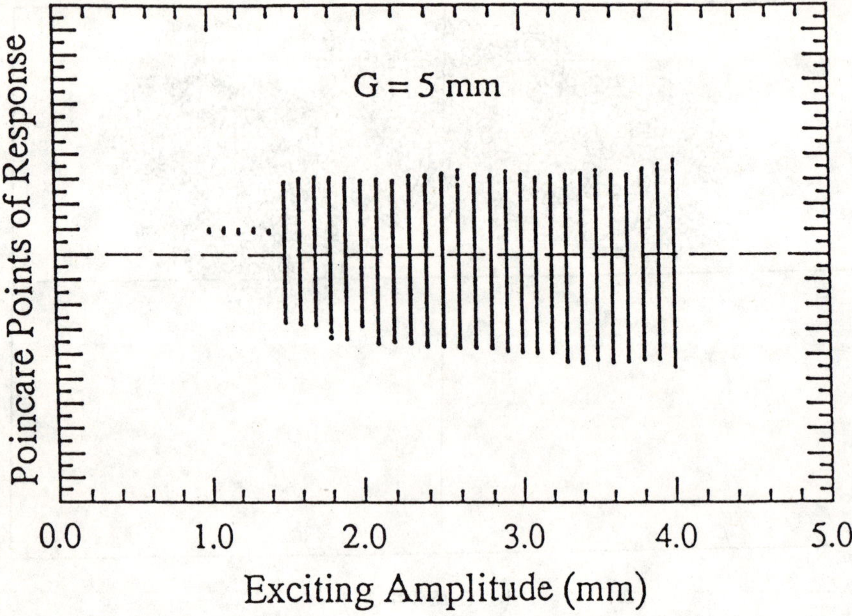

Fig. 18. Bifurcation diagram

4. Magnetic Levitation Systems

Theoretical papers and experiments of magnetically suspended vehicles have been carried out by several researchers in the last 20 years [12, 13]. These are usually devoted to the study of levitation forces and their application to high speed vehicles. The dynamic instability of the system was investigated by Chu and Moon [14].

The chaotic behavior of three kinds of simple systems with magnetically levitated mass is described on the basis of the centre manifold theory for nonlinear differential systems in this section. Three systems are considered: (I) magnetically levitated mass with the mass-excitation, (II) the same mass with a horizontal elastic beam-excitation and (III) levitated mass with the support-excitation. The numerical simulation is also given for some system parameters. Time histories and phase portraits of chaotic motion, Poincaré maps and bifurcation diagrams are also shown.

4.1 Formulation of Dynamic Equations

The system II shown in Fig. 21 is considered to obtain the formulation of dynamic equations, and after equating some parameters to zero, formulations for systems I and III are obtained. The system is composed of the mass m with the permanent magnet, connected to the elastic beam described by: bending

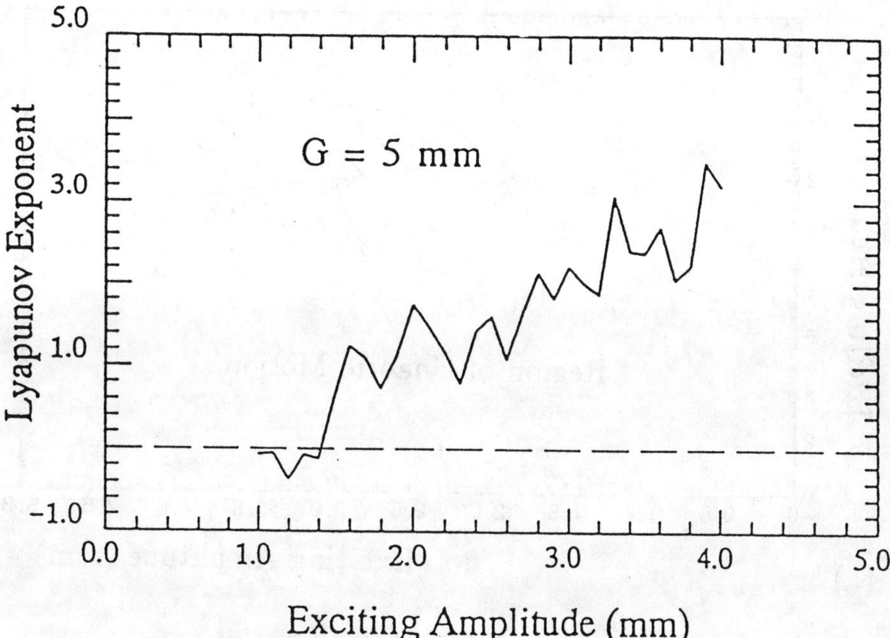

Fig. 19. Lyapunov exponent

rigidity $D = Eh^3/12(1-\nu^2)$, where E is Young modulus, ν Poisson ratio and h thickness of the beam. Additional parameters are: L length, b width, and M mass of the beam. The mass with the permanent magnet is subjected to the magnetic force from the second magnet of the same pole situated on the support under the first one (S is the cross-section of the pole). The magnetic induction is taken as B, when the gap between the magnets is G (this is the free initial position, the beam is underformed and then horizontal). The coordinate system is also depicted in Fig. 21. The beam in case (II) of Fig. 21 is excited at its clamped end.

The motion of an elastic beam can be described by the following differential equation:

$$D\frac{\partial^4 w}{\partial y^4} = q\,, \tag{44}$$

which can be derived from the balance of moments equation. Denotations are as follows: $w = w(y,t)$ is the deflection of the beam in the x direction, q is the lateral load intensity (acting in x direction) per unit length (of z direction) in $[\mathrm{N/m^2}]$. To obtain a proper description every force component existing in the considered system should be taken into account and entered into equation (44). There are four components in the system:

- gravitational force of the mass

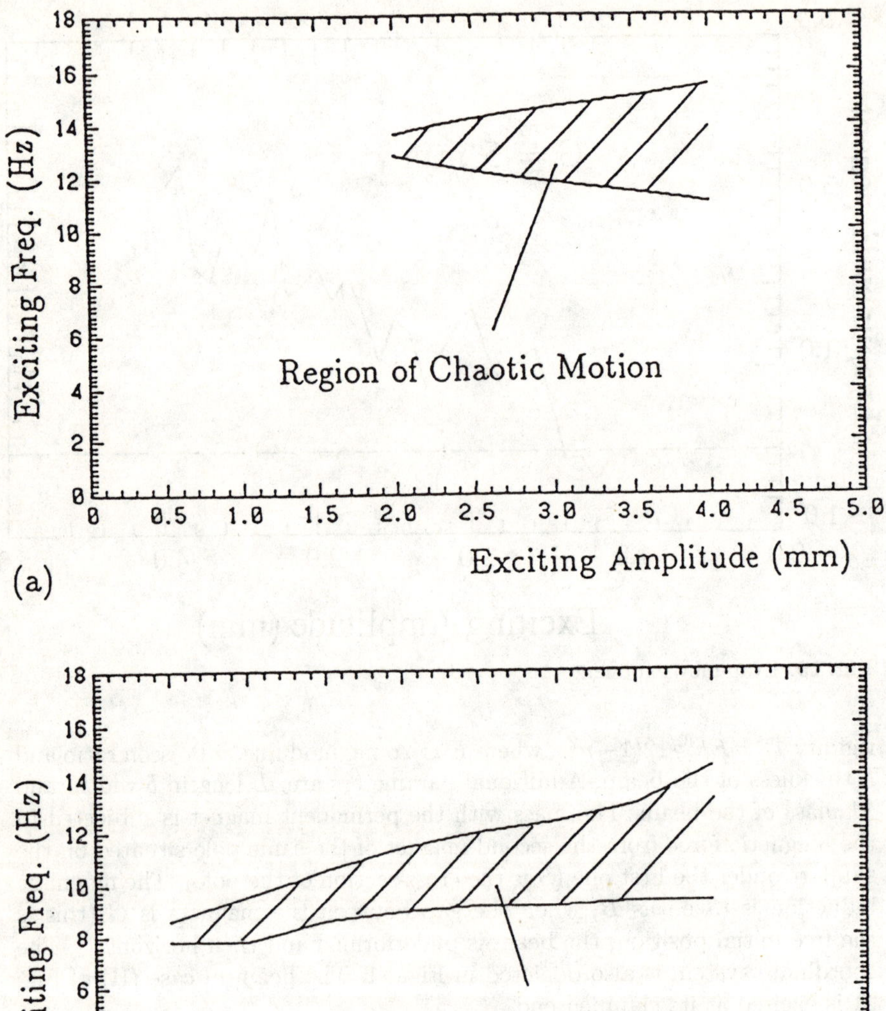

Fig. 20. Chaos regions: (a) $G = 3$ mm, (b) $G = 5$ mm

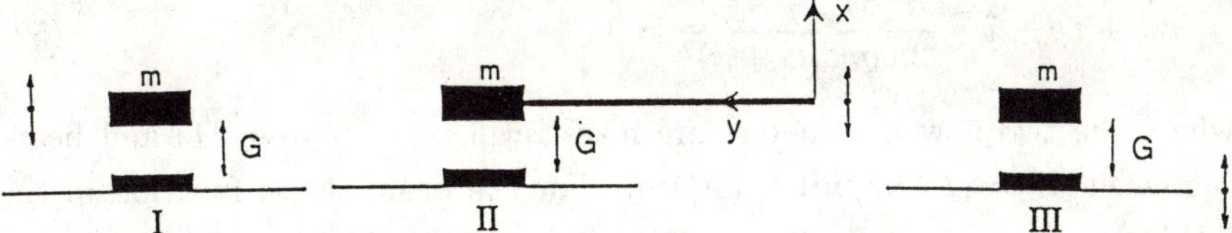

Fig. 21. Simple view of system considered (detailed description is given in the text)

$$q_1 = -\frac{mg}{bL}\delta(y-L)\,; \tag{45}$$

- inertial force of the mass

$$q_2 = -\frac{m}{bL}\frac{\partial^2 w(y,t)}{\partial t^2}\delta(y-L)\,; \tag{46}$$

- magnetic force between magnets [15]

$$q_3 = \frac{B^2 S}{2\mu_0 bL}\frac{G^2\delta(y-L)}{(G+w)^2}\,; \tag{47}$$

- inertial force of the beam

$$q_4 = -\frac{M}{bL}\frac{\partial^2 w(y,t)}{\partial t^2}\,. \tag{48}$$

Substitution of the sum $q = q_1 + q_2 + q_3 + q_4$ into equation (44) gives

$$DbL\frac{\partial^4 w}{\partial y^4} + mg\delta(y-L) + [m\delta(y-L)+M]\frac{\partial^2 w}{\partial t^2} - \frac{B^2 S}{2\mu_0}\frac{G^2\delta(y-L)}{(G+w)^2} = 0\,. \tag{49}$$

Incorporating the boundary conditions of the beam ends

$$w(0,y) = \frac{\partial w}{\partial y}(0,t) = 0 \quad \text{and} \quad \frac{\partial^2 w}{\partial y^2}(L,t) = \frac{\partial^3 w}{\partial y^3}(L,t) = 0\,, \tag{50}$$

and only one (first) mode of the Galerkin representation

$$\begin{gathered} w(y,t) = \Phi(y)x(t)\,, \\ \int_0^L \Phi^2(y)dy = 1\,, \\ \Phi(L) = 1\,, \end{gathered} \tag{51}$$

you can see after integration of the left hand side of equation (49) multiplied by $\Phi(y)$ (Galerkin procedure), that:

$$\alpha x + \tau g + \ddot{x} - \frac{B^2 S \tau}{2\mu_0 m} \frac{G^2}{(G+x)^2} = 0 , \tag{52}$$

where the two new parameters are mass ratio $\tau = m/(m+M)$ and beam coefficient $\alpha = (DbL\tau/m) \int_0^L (\Phi'')^2 dy$. For the beam shape function in the form of

$$\Phi(y) = \frac{1}{2}\left[\cosh\frac{\lambda}{L}y - \cos\frac{\lambda}{L}y - a\left(\sinh\frac{\lambda}{L}y - \sin\frac{\lambda}{L}y\right)\right] , \quad \lambda = 1.875104 , \tag{53}$$

α can be calculated as $\alpha = Db\tau\lambda^4/mL^3$.

The equation describing equilibria (static solution of equation (52)) is of the third order, so it is hard to obtain analytical solutions, but if the beam is assumed to be weak (D small, or $L \gg b$, and $M \ll m$), then $\alpha \approx 0$, $\tau \approx 1$, and one stable equilibrium position can be found easily (note, that this is the same case as for the systems I and III in Fig. 21 without beam), so

$$x_{\mathrm{e}} = GB\sqrt{\frac{S}{2\mu_0 m g}} - G . \tag{54}$$

Natural frequency around this equilibrium can also be found to be

$$\omega_n = \sqrt{2\frac{g}{x_{\mathrm{E}}}} , \tag{55}$$

where $x_{\mathrm{E}} = G + x_{\mathrm{e}}$ (always positive).

Knowing one equilibrium approximately, it is possible to find the remaining two (for system II in Fig. 21 only) by reduction of the order of the equilibrium equation by one and taking advantage of the assumptions $0 \neq \alpha \ll 1$ and $\tau \approx 1$. They are

$$x'_{\mathrm{e}} \approx x_{\mathrm{e}} , \qquad x''_{\mathrm{e}} \approx \frac{\tau}{\alpha} g . \tag{56}$$

The locations of these equilibria can be found on the Poincaré map for the system.

To complete the description of the system, the damping term should be involved into equation (52) and excitation of the harmonic form should be assumed, hence $(x + A\sin\Omega t)$ is substituted in place of x. After regrouping terms equation (52) can be expressed as

$$\ddot{x} + \gamma\dot{x} + \alpha x + \frac{\tau x_{\mathrm{E}}\omega_n^2}{2}\left[1 - \frac{x_{\mathrm{E}}^2}{(G + x + A\sin\Omega t)^2}\right] = A(\beta\Omega^2 - \alpha)\sin\Omega t , \tag{57}$$

where γ is the damping coefficient (can be measured), and a new parameter β was involved, which will be useful later to obtain solutions for the system

(III) with support excitation (Fig. 21), where $\beta = 0$. For systems in Fig. 21 (I) and Fig. 21 (II) $\beta = 1$.

Equation (57) can be rewritten as a system of first order differential equations, which will be studied using centre manifold theory in the next section. Substituting $x_1 = x$, $x_2 = \dot{x}$, and $x_3 = \Omega t$, one can obtain

$$\dot{\boldsymbol{x}} = \boldsymbol{f}(\boldsymbol{x}, \boldsymbol{\mu}) \tag{58}$$

where $\boldsymbol{x} = [x_1, x_2, x_3]^T \in \mathbb{R}^3$ and the vector of independent parameters is $\boldsymbol{\mu} = [G, \omega_n, \gamma, A, \Omega, \alpha, \tau] \in \mathbb{R}^7$. The vector function $\boldsymbol{f} = [f_1, f_2, f_3]^T : \mathbb{R}^3 \to \mathbb{R}^3$ is

$$\begin{aligned} f_1 &= x_2 \\ f_2 &= -\gamma x_2 - \alpha x_1 + \frac{\tau x_{\mathrm{E}} \omega_n^2}{2} \left[\frac{x_{\mathrm{E}}^2}{(G + x + A \sin x_3)^2} - 1 \right] + \\ &\quad + A(\beta \Omega^2 - \alpha) \sin x_3 \\ f_3 &= \Omega \,. \end{aligned} \tag{59}$$

It should be noted that the equation (59) describes the forced oscillations of the mass in the system in Fig. 21 (II) if $\beta = 1$, in Fig. 21 (I) if $\beta = 1$, $\tau = 1$, $\alpha = 0$, and in Fig. 21 (III) if $\beta = 0$, $\tau = 1$, $\alpha = 0$. The last two systems only have one equilibrium position (54).

4.2 Linearization in Terms of Manifolds

The linearized equation of (58) around equilibrium position x_{e} can be written as [1, 8]

$$\dot{\boldsymbol{X}} = \mathring{\boldsymbol{A}} \boldsymbol{X} \,, \tag{60}$$

where $\mathring{\boldsymbol{A}} = D\boldsymbol{f}|_{x_1 = x_{\mathrm{e}}, x_{2,3} = 0}$, hence the matrix $\mathring{\boldsymbol{A}}$ has the form

$$\mathring{\boldsymbol{A}} = - \begin{bmatrix} 0 & -1 & 0 \\ \alpha + \tau \omega_n^2 & \gamma & A(\alpha + \tau \omega_n^2) \\ 0 & 0 & 0 \end{bmatrix} \,. \tag{61}$$

The eigenvalue problem of $\mathring{\boldsymbol{A}}$,

$$(\mathring{\boldsymbol{A}} - \xi \boldsymbol{I}) \boldsymbol{v} = \boldsymbol{0} \tag{62}$$

yields a characteristic equation of the form

$$\xi(\xi^2 + \xi\gamma + \alpha + \tau\omega_n^2) = 0 \,, \tag{63}$$

hence the set of eigenvalues is

$$\xi_1 = 0 \,, \qquad \xi_2 = \frac{-\gamma - \sqrt{\Delta}}{2} \,, \qquad \xi_3 = \frac{-\gamma + \sqrt{\Delta}}{2} \,, \tag{64}$$

where $\Delta = \gamma^2 - \delta$ and $\delta = 4(\alpha + \tau\omega_n^2)$. Respective eigenvectors are

$$\boldsymbol{v_1} = [-Ap, 0, p]^T , \quad \boldsymbol{v_{2,3}} = [q, q\xi_{2,3}, 0]^T , \ p, q \in \mathbb{R} . \tag{65}$$

The centre manifold W^{C} for the equation (58) is always nonempty, as it is transient to the centre eigensubspace E^{C} of the linearized problem (60), spanned by at least one eigenvector $\boldsymbol{v_1}$. If γ is positive the stable manifold W^{S} exists, and is transient to the stable eigensubspace E^{S} spanned by vectors $\boldsymbol{v_2}$ and $\boldsymbol{v_3}$. Conversely the unstable manifold W^{U} exists for negative damping, and is transient to E^{U} spanned by the same vectors. If $\gamma = 0$, the centre eigensubspace is spanned by the whole eigenvector triplet. The locations of the eigenvalues in different ranges of parameters are similar to those presented in Fig. 4. It can be seen from the above that the main values of parameters when bifurcation occurs are $\delta = \gamma^2$ and $\gamma = 0$. For $\gamma = 0$ the dimension of the centre manifold is one greater than for $\gamma \neq 0$, hence even the structure of the motion is different, which was to be expected.

A similar derivation could be carried out around the third equilibrium position x_{e}'' for the system (II) with the beam (Fig. 21), but the formula is much more complicated, hence it is omitted here, because the behavior is generally the same. The following relations are valid for systems I and III in Fig. 21.

$$\mathring{\boldsymbol{A}}^{\mathrm{ac}} = -\begin{bmatrix} 0 & -1 & 0 \\ \omega_n^2 & \gamma & A\omega_n^2 \\ 0 & 0 & 0 \end{bmatrix} , \qquad \delta^{\mathrm{ac}} = 4\omega_n^2 , \tag{66}$$

hence the behavior of the system is implicated in general by the relations between damping and natural frequency with a small beam parameter modification. Linearized matrices for the two systems I and III with different excitation (Fig. 21) are the same, but nonlinear equations (59) are of a significantly different form (because of β), hence the motions described by these equations are also different. This will become clear in the next section.

4.3 Numerical Simulation

Based upon the mathematical formulation in equations (58) and (59), numerical simulation of chaotic motion for each system considered is carried out. The CRAY computer was used with a Silicon Graphics IRIS workstation, hence graphics of figures are obtained from Multi-Purpose Graphics System. Results are presented simultaneously for all three systems to enable a comparison. The phase portraits (long time simulation, about 500 cycles) are shown in Fig. 22 (I, II, III). There is now a displacement x_1 on the horizontal axis and a velocity of the displacement x_2 on the vertical one. The parameters are chosen so that a chaos occurs (see captions), and one can see that, the chaos descriptors for every system are different. For system III with the support excitation the influence of the magnet on the support appears in form of strong deformation of circular orbits on the side of the magnetic force. Also a transient chaos was found, which means the motion switches from chaotic to harmonic or subharmonic after some time of simulation. Such a response obtained for appropriate parameter values. The simplest way is to assume a high damping coefficient ($\gamma = 3 - 4$). The harmonic motion appears around the equilibrium position $(x_e, x_2(x_e))$.

Let us now introduce more complicated descriptors of chaos, such as Poincaré maps and bifurcation diagrams. A Poincaré map is a special set of points in the picture obtained from a long time simulation by taking samples of x_1 and x_2 whenever the moments of exciting periods ($x_3 = 2n\pi + \varphi$) coincide, and plotting this as a point on the map. Such maps are presented in Fig. 23 (I–III) for each system. In the middle of the graph in Fig. 23 (I) one can see a three-fold spiral around the equilibrium position for this system, whereas in Fig. 23 (II) one can see three one-fold spirals, as this system has three equilibria (x_e, x_e', x_e''). The simulation approves the theoretical derivation that two of them (x_e, x_e') are very close. Another interesting fact is that the velocities associated with each equilibrium are different $x_2(x_e) \approx -x_2(x_e')$, $x_2(x_e'') \approx 0$. The motion was sampled at the beginning of the excitation period ($\varphi = 0°$). When the sampling point is changed ($\varphi = 45°, 90°, 135°, 180°$) the graph will also be changed. Evolution with phase of Poincaré map for system I is shown in Fig. 24. The largest structure (with a contour) of the map is for $\varphi = 0°$. A similar one can be seen for $\varphi = 180°$, but it has not been included. For $\varphi = 90°$ only the spiral is left, however the number of points in each graph (number of periods of excitation) is the same (10,000). For system II, if certain parameters of the system are changed the response in the form of a Poincaré map can be slightly different, especially if it is a frontier of the chaotic motion region. Figure 25 presents a Poincaré map for system III for a higher excitation frequency ($\Omega/\omega_n = 2.1$). Chaos region in that case is $\Omega/\omega_n = 1.6 - 2.1$. Tamura [16] obtained similar results.

The global dependence and sensitivity of motion on the system parameters can be obtained by preparing bifurcation diagrams. A parameter is varied on the horizontal axis and the range of sampled ($x_3 = 2n\pi$) displacement x_1 is varied on the vertical one. The graph can be obtained by taking the Poincaré map step by step slowly changing one of the system parameters. Some examples of such diagrams are shown below. Bifurcation diagrams for system III are presented in Fig. 26 with the normalized frequency of excitation Ω/ω_n taken as a parameter.

There are two digital descriptors of chaos, which can be calculated based on the long time history (Lyapunov exponent) or Poincaré map (fractal dimension). If the Lyapunov exponent is positive it means the motion is unpredictable far in the future. The highest Lyapunov exponents are calculated for the simulated time histories of every system. They are respectively: I – $\lambda_1 = 0.36$, II – $\lambda_1 = 0.30$, and III – $\lambda_1 = 0.126$. To complete the chaos description of considered systems, the fractal dimensions of a set of Poincaré

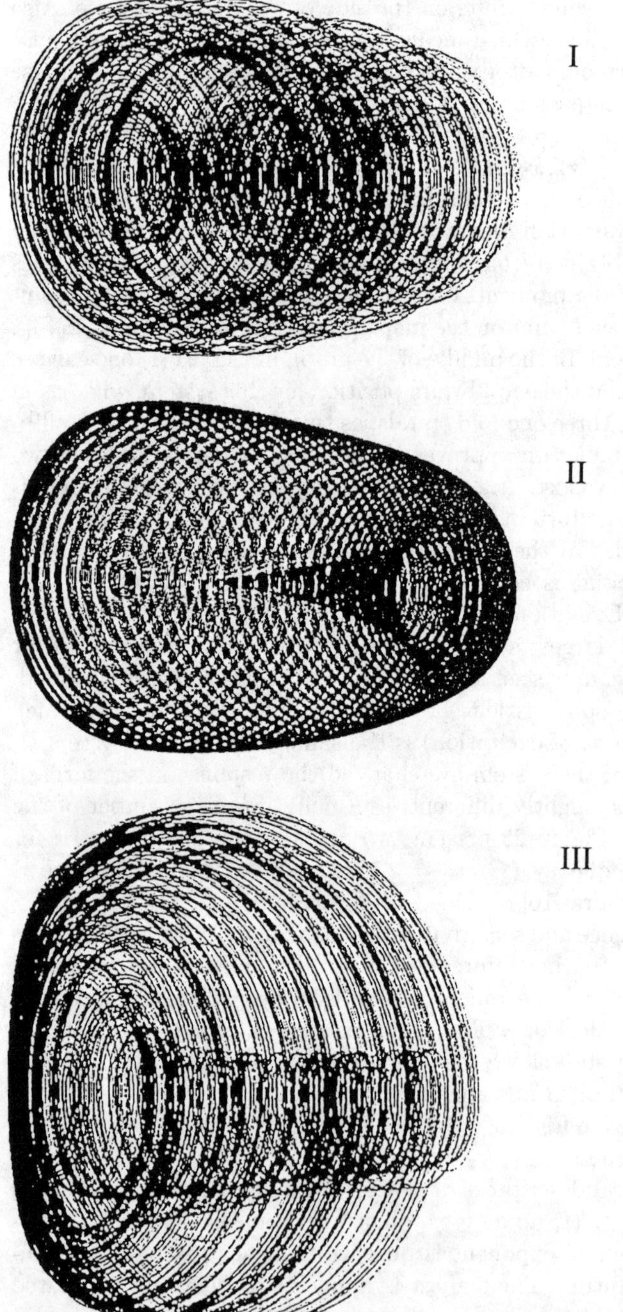

Fig. 22. Long time phase portraits of the systems I, II and III

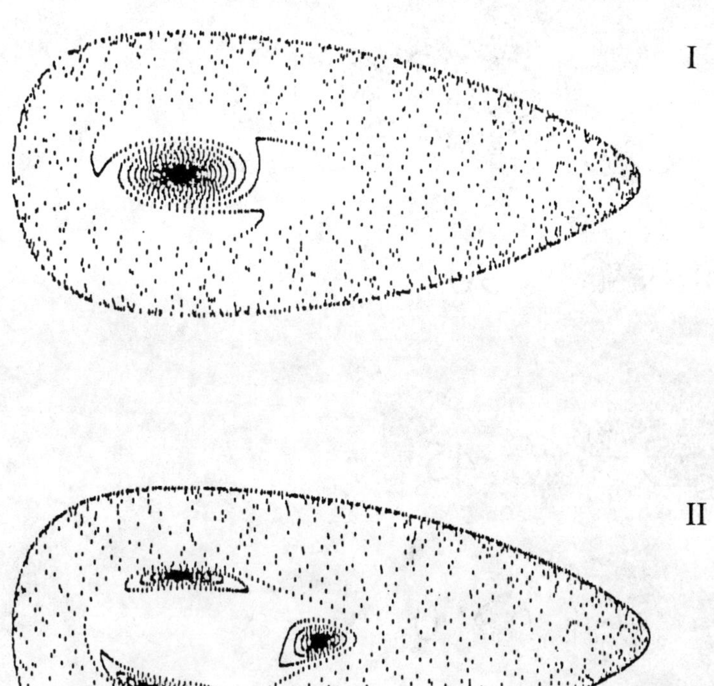

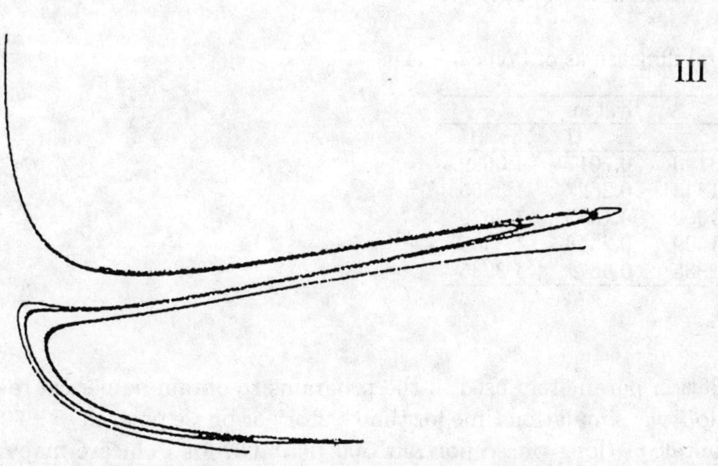

Fig. 23. Poincaré map for the system (I): $\gamma = 0.53$, $\Omega/2\pi = 23$, $A = 2.7$; (II): $\gamma = 0.53$, $\Omega/2\pi = 23$, $A = 2.5$, $\alpha = 0.9$; (III): $\gamma = 0.2$, $\Omega/\omega_n = 1.9$, $A/x_e = 0.4$

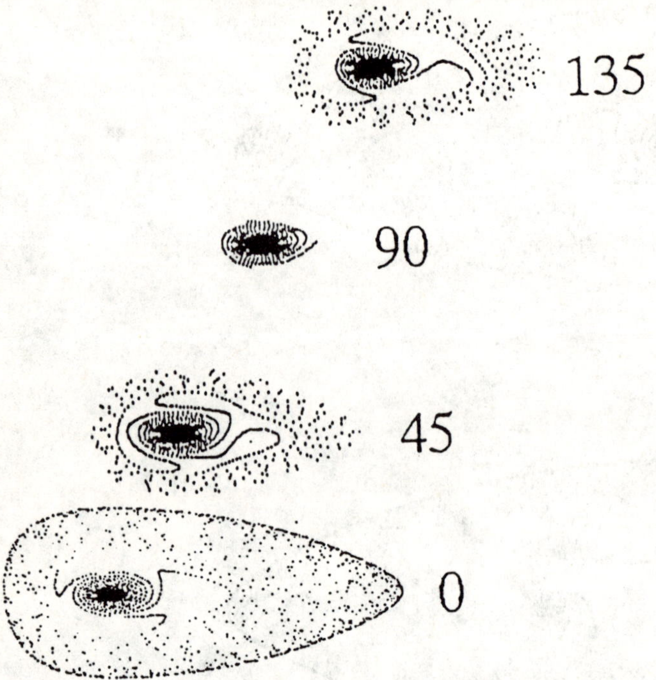

Fig. 24. Evolution of Poincaré map with phase for system I

maps are computed using correlation definition [6]. They are reported in the Table 2. It can be seen that the fractal dimension for the system III is not very sensitive to the phase of excitation.

Table 2. Fractal dimensions of Poincaré maps

phase	System		
φ	I	II	III
0°	0.3140	0.7017	1.3689
45°	0.1343	0.2627	1.3565
90°	0.1000	0.1532	1.3629
135°	0.1509	0.2358	1.3355
180°	0.2935	0.5672	1.3265

The simulation parameters used in the programs to obtain numerical results were as follows: simulation time for time history or phase portrait, 50–70 periods of excitation (long phase portrait 500 periods), for Poincaré maps, 10,000–15,000 periods (points on the graph), and for bifurcation diagrams, 500–1000 periods for each of the 100–200 values of the parameter (about

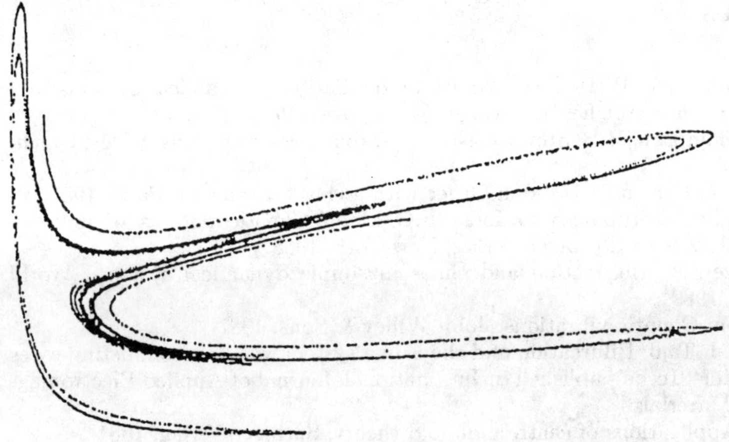

Fig. 25. Poincaré map for the system III, $\gamma = 0.2$, $\Omega/\omega_n = 2.1$, $A/x_e = 0.4$

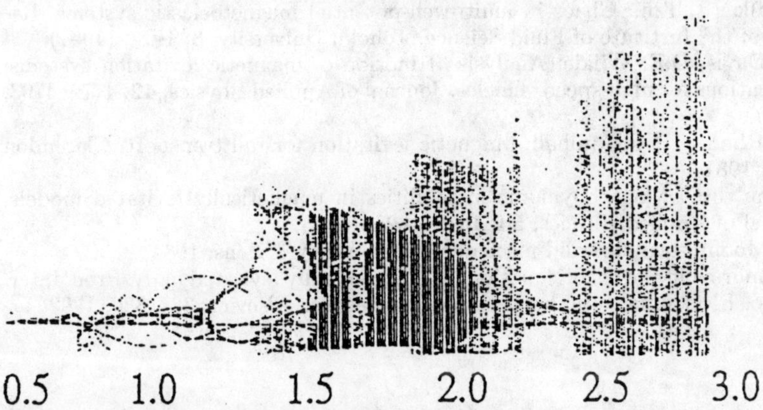

Fig. 26. Bifurcation of normalized excitation frequency (Ω/ω_n) for the system III, $\gamma = 0.2$, $A/x_e = 0.4$ (176,400 points)

50,000–180,000 points on the graph). The sampling was 250–300 samples per excitation period.

4.4 Conclusion

Numerical evidence of chaos in three different systems with magnetic levitation is given. It should be noted that the range of chaos for first two systems (I and II) is very narrow, so it is very sensitive to the parameters, and the structure of strange attractor (Poincaré map) is very different.

References

1. J. Guckenheimer, P. Holmes: Nonlinear oscillations, dynamical systems and bifurcations of vector fields. Springer-Verlag, New York, 1983
2. J.M.T. Thompson, H.B. Stewart: Nonlinear dynamics and chaos. Wiley, Chichester, 1986
3. P. Berge, Y. Pomeau, C. Vidal: Order within chaos. Herinann, Paris, 1985
4. W. Szemplinska-Stupnicka, G. Iooss, F.C. Moon: Chaotic motions in nonlinear dynamical systems. Springer-Verlag, New York, 1988
5. J. Awrejcewicz: Bifurcation and chaos in simple dynamical systems. World Scientific, 1989
6. F.C. Moon: Chaotic vibrations. John Willey & Sons, 1987
7. D. Gafka, J. Tani: Bifurcation and chaos in a system with two conducting wires and a spring. To be published in International Journal of Applied Electromagnetics in Materials
8. J. Carr: Applications of centre manifold theory. Springer-Verlag, 1981
9. M. Berger, B. Gostiaux: Differential geometry: manifolds, curves and surfaces. Springer-Verlag, 1988
10. A. Wolf, J.B. Swift, H.I. Swinney, J.A. Vastano: Determining Lyapunov exponents from a time series. Physica D, **16**, 285–317, (1985)
11. D. Gafka, J. Tani: Chaos in multi-well-potential magnetoelastic systems. Reports of the Institute of Fluid Science. Tohoku University, **3**, 1–17, (1991)
12. L.C. Davis, D.F. Wilkie: Analysis of motion of magnetic levitation systems: implications for high-speed vehicles. Journal of Applied Physics, **42**, 4779–4793, (1971)
13. R.G. Rhodes, R.G. Mulhall: Magnetic levitation for rail transport. Clarendon Press, 1981
14. D. Chu, F.C. Moon: Dynamic instabilities in magnetically levitated models. Journal of Applied Physics, **54**, 1619–1625, (1983)
15. F.C. Moon: Magneto-solid mechanics. John Wiley & Sons, 1984
16. H. Tamura: On the chaotic motion of a magnetically levitated body. Proc. Inter. Sympo. Electromagnetic Forces and Applications, Elsevier, 327–330, 1992

Bifurcation and Chaos in the Helmholtz-Duffing Oscillator

G. Rega

Dipartimento di Ingegneria delle Strutture, delle Acque e del Terreno
Università dell'Aquila, L'Aquila, Italy

Abstract

An oscillator with asymmetric (Helmholtz) and symmetric (Duffing) nonlinearities, representative of an interesting problem in nonlinear structural dynamics, is discussed as a simple but prototypical model to describe bifurcation from regular to chaotic behaviour. It is considered: (i) to check the reliability and computational efficiency of numerical procedures for obtaining the system response and of quantitative measures for identifying chaos; (ii) to examine the bifurcation predictive capability of the stability analysis of periodic solutions; (iii) to interpret response unpredictability and chaos in the light of a geometrical description of the system's global dynamics.

1. Mechanical System and Mathematical Model

Interest in chaotic phenomena in the nonlinear dynamics of continuous mechanical systems has developed in recent years. Restricting our attention to elastic one-dimensional structures undergoing problems of geometric nonlinearity and to the description of the relevant dynamic phenomena of single-degree-of-freedom oscillators, systems with exclusively symmetric nonlinearities and systems with both asymmetric and symmetric nonlinearities can be distinguished.

Duffing-type oscillators are symmetric and have been widely studied for chaos under both harmonic and parametric excitation [1–12]. The unstable Duffing oscillator exhibits two stable static equilibria corresponding to a two well potential. It has negative linear stiffness and hardening cubic stiffness and can be referred to in structural dynamics when describing the single-mode finite vibrations of a periodically forced buckled beam. Oscillators with symmetric nonlinearities of higher order as used in solid mechanics applications have been considered, too. Stable Duffing oscillators exhibit only one equilibrium state. If they exhibit positive linear stiffness, they can represent the single-mode finite vibrations of a taut string or of a straight beam.

Oscillators with both asymmetric (Helmholtz) and symmetric (Duffing) nonlinearities have received attention for chaos only in recent years [13–17], notwithstanding their special interest just due to the presence of both kinds of nonlinear terms. In structural applications, an even term arises in addition to the odd one due to the initial curvature of the elastic system (curved beams, shallow arches, suspended cables). Heavy shallow arches have been considered in some applications. They undergo both tensile and compressive internal forces and can thus exhibit three physically admissible positions of static equilibrium, two stable and one unstable.

Here, following previous works [16–18], a harmonically forced asymmetric oscillator with quadratic and cubic nonlinearities described by the dimensionless differential equation

$$\ddot{q} + \mu\dot{q} + c_2 q^2 + c_3 q^3 = P \cos \Omega t \tag{1}$$

is considered. The coefficients of nonlinear terms are given quite high values, $c_2 = 35.952$ and $c_3 = 534.53$, which characterize the system as a strongly nonlinear one.

They correspond to the geometrical and mechanical properties of a heavy parabolic cable suspended between two fixed supports at the same level, with a sag-to-span ratio d/l of about 1/45 and axial rigidity-initial tension ratio EA/H of about 500, vibrating in its plane with the first symmetric mode under uniform vertical forcing $p(x,t)$ (Fig. 1). A more realistic description of the cable's finite dynamics would require the use of at least a two-degree-of-freedom model for studying coupling between in-plane and out-plane oscillations. However, restricting our investigation to moderately large planar motions, the cable dynamics can be described by the only partial equation in the vertical displacement $v(x,t)$ [19]

$$\left(Hv' + \frac{EA}{l} \cdot (y' + v') \int_0^l \left(y'v' + \frac{v'^2}{2} \right) dx \right)' + p - \mu\dot{v} = m\ddot{v}, \tag{2}$$

which is a first level approximation to the analysis of cables used in overhead transmission lines. Using nondimensional quantities, representing the displacement through one eigenfunction and considering an excitation with given spatial distribution, the Galerkin method leads to equation (1) which retains both the quadratic and the cubic term occurring in the original continuous problem - the former associated with initial curvature of system, the latter with the stretching of the cable's axis.

It is worth observing that the Hamiltonian system associated with equation (1) has three equilibrium points, $q = 0$ and $q = (-c_2 \pm (c_2^2 - 4c_3)^{1/2})/2c_3$, corresponding to an asymmetric two-well potential $V(q)$ (Fig. 2). The situation is thus mathematically similar to that occurring for a shallow arch or a two pin-ended bars system [14] but mechanically different, since the cable can only resist tensile forces and is not capable of maintaining its geometrical

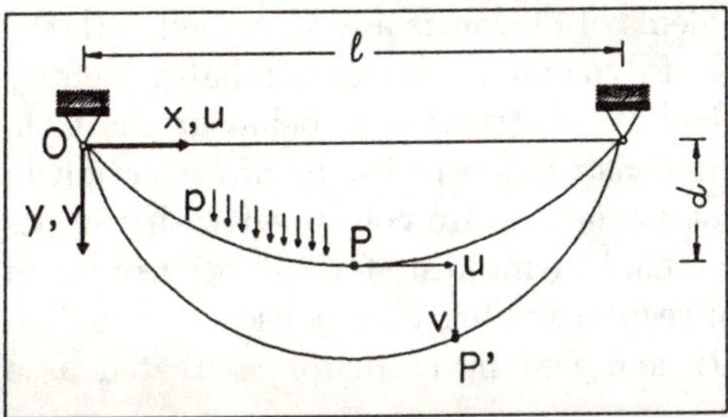

Fig. 1. Cable configurations

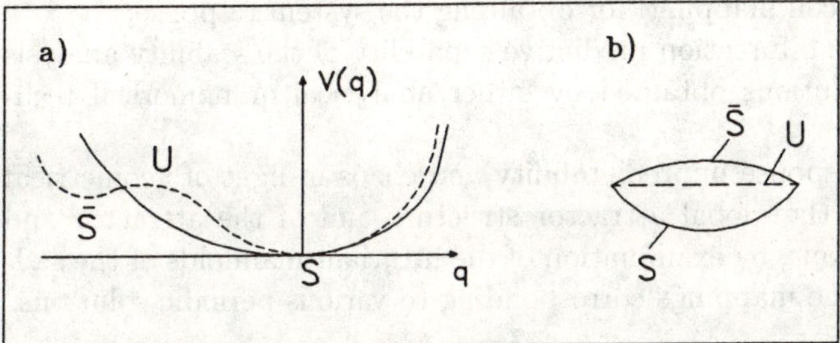

Fig. 2. Cable potential (a) and static equilibrium configurations (b)

shape in the horizontal and upward configurations, which are thus not actually admissible. For the numerical case considered here, the quantity $(c_2^2 - 4c_3)$ is negative, so the oscillator has a single equilibrium point and a corresponding one-well potential. The associated cable belongs to the class of initially prestressed cables which never undergo compression and thus actually only show one stationary point [20].

The presence of both even and odd nonlinear terms in the oscillator equation (1) gives rise to a rich and complex set of responses [17, 21]. Indeed, the model exhibits many fundamental characteristics of the behavior of nonlinear dynamical systems: multiple coexisting periodic attractors, strange chaotic attractors occurring in various ranges of control parameter values, different routes to chaos from periodic solutions, as well as regular and fractal basin boundaries. A variety of phenomena is associated with the occurrence of the resonance conditions for the system, namely the primary one, and the order 1/2- and 1/3-subharmonic, and the order 2 and 3 superharmonic.

These features suggest the oscillator should be classed as a further—simple but reach—prototypical model, besides the classical ones widely considered in the literature [2, 3, 12], to describe bifurcations from regular to chaotic behavior.

In the next sections, the main practical problems concerned with the understanding and description of bifurcation and chaos are examined with reference to such an oscillator. Problems of *diagnosis* are concerned with the implementation and the correct use of dynamic measures to reliably identify them. Problems of *scenario* refer to the construction of behavior charts in control parameter space and to the construction of basins of attraction in initial condition space. Problems of *prediction* are concerned with the determination of critical bounds, of technical interest, for the occurrence of bifurcations and chaos in regions of regular nonlinear response.

More specifically, the Helmholtz and Duffing oscillator is treated as a model system:

(i) to check the reliability and computational efficiency both of some quantitative measures for identifying chaos and of numerical procedures (continuation and cell mapping) for obtaining the system response;
(ii) to examine the bifurcation predictive capability of the stability analysis of periodic solutions obtained by either analytical or numerical techniques;
(iii) to interpret response unpredictability and chaos in light of geometrical description of the global attractor structure and of the attractor and basin bifurcations, by examination of the invariant manifolds of the saddle points of the mappings corresponding to various periodic solutions.

2. Behaviour Chart and Characterization of Chaotic Response

Analysis of chaotic response of a given system relies primarily on the results of numerical simulations of the relevant dynamic equation. Indeed, this is the method commonly followed in the literature, even though the present state of knowledge in this field is likely to suggest different initial steps to shed light on the possible chaotic behaviour of a system. In any case, however, behavior charts in a control parameter space have been systematically obtained for just a few oscillators [10, 12, 17].

For the Helmholtz and Duffing oscillator, the regions of response obtained in the forcing parameter space (Ω, P) through a point-by-point time integration search with zero initial conditions and a fixed damping value $(\mu = 0.1)$ are shown in Fig. 3. The degree of periodicity of the response was deduced mainly on the basis of the Poincaré map but quantitative measures - such as frequency response spectrum, Lyapunov exponents and fractal dimensions - were calculated in many specific situations, too.

Obtaining reliable results with each measure requires preliminary calibration of some computational parameters. Indeed, it is essential to distinguish between intrinsic sensitivity of potentially chaotic systems to variations of i.c. or of control parameters and spurious results produced by improper selection of some computational quantities. Accordingly, considerable care was taken in choosing general parameters for numerical simulations — the length of the

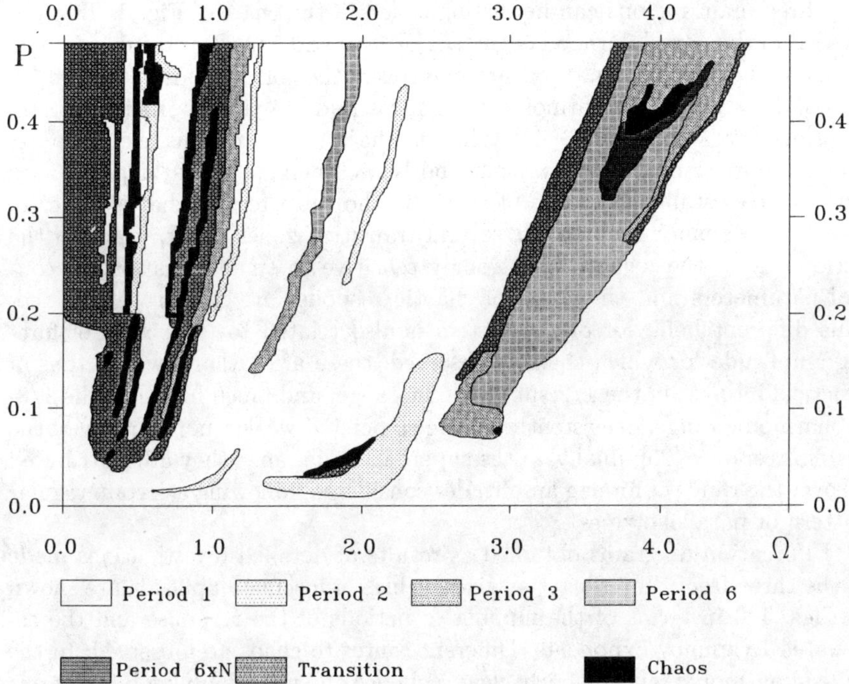

Fig. 3. Periodic and chaotic response in control space (P, Ω)

initial transient of response, the total time integration length and the time step increment - to be reasonably sure that the response obtained is really a steady one and that chaos, if it occurs, actually pertains to the differential equation of motion. Useful suggestions for proper selection were furnished by the requirement to obtain well stabilized values of some global measures of the system dynamics, like the Lyapunov exponents and the dimension of the attractor. Of course, use of these measures required computational care, too. As for the calculation of the exponents by a classical algorithm [22], attention was paid to the selection of both the step of reorthonormalization of the vector identifying a nearby trajectory and of the number of forcing periods used to obtain a reliable characterization of response [17]. The former was given a value of one, though the values suggested in the literature for other systems are no greater than ten. The latter depends on several factors. On the other hand, for reliable calculation of the correlation dimension of the $2D$ attractor, attention was paid to the dimension of the boxes with which to cover it, which has to be properly correlated with the density of the points in the map [16, 23, 24]. All this care is due to the conviction that a statement of "chaotically" must be supported by consistent indications obtained by a number of different measures.

Three main regions can be distinguished in the chart of Fig. 3. Two are located in the neighbourhood of the $1/2(\Omega \cong 2)$ and $1/3(\Omega \cong 3)$ subharmonic resonances of the system, the third one covers the zones of order $3(\Omega \cong 1/3)$ and $2(\Omega \cong 1/2)$ superharmonic resonances and extends approximately up to primary resonance ($\Omega \cong 1$). While in the first two regions the response, though quite complex, shows some fairly well-defined zones of periodicity with clearly established zones of chaos, in the third region, that is richer in terms of resonance frequencies, several transition zones occur, in which the periodicity of the response is strongly sensitive to small variations of control parameters and the zones of chaotic response are not so well-defined. This different behavior of the system is also related to the values of forcing amplitude for which chaos is observed: these are rather low and thus of practical interest in the 1/2-subharmonic range, and much higher in the 1/3-subharmonic range, consistent with the associated weaker nonlinearity of the system response [25]; finally, in the superharmonic range they are distributed all over the range of forcing amplitudes considered, following a certain regular pattern of parallel stripes.

Bifurcation diagrams obtained as results of detailed investigations made in the three frequency ranges at given values of forcing amplitude are shown in Figs. 4–6 in terms of the number of periods of the response and the associated Lyapunov exponents. Different routes to chaos are observed. In the 1/2-subharmonic range, a fairly clear sequence of period doubling bifurcations with decreasing frequency and sudden transition with increasing frequency occurs. In the 1/3-subharmonic, the transition to chaos from the left occurs via a basic period $6T$ response and responses with a period multiple of $6T$ originating from the former, whereas transition from the right is smoother and characterized mainly by a sequence of period doubled solutions somewhat dirtied by additional period multiples other than $6T$ responses. In the superharmonic range, at least at the forcing amplitude value considered, narrow zones of chaos exist, the transition to which seems to occur with a fairly clear period doubling sequence when decreasing the frequency, while it is preferably of sudden type when increasing the frequency.

The chaotic attractors obtained in the three zones are shown in Fig. 7. Remarkable differences occur amongst them. In the 1/2-subharmonic range, chaos is quite well-established as denoted by all the measures considered; in the 1/3-subharmonic, the strange attractor exhibits 6 independent bundles originating from the period $6T$ solution around which the motion fluctuates and, correspondingly, lower values of its dimension are obtained; finally, the attractor in the superharmonic range resembles the first one but has a much thinner structure.

Reliable calculation of quantitative measures of chaos underwent major difficulties in the 1/3- than in the 1/2-subharmonic range, which is likely to be due to the lower "chaoticity" of the response [17]. The exponents were found to stabilize after a few hundred forcing periods where they have rather high

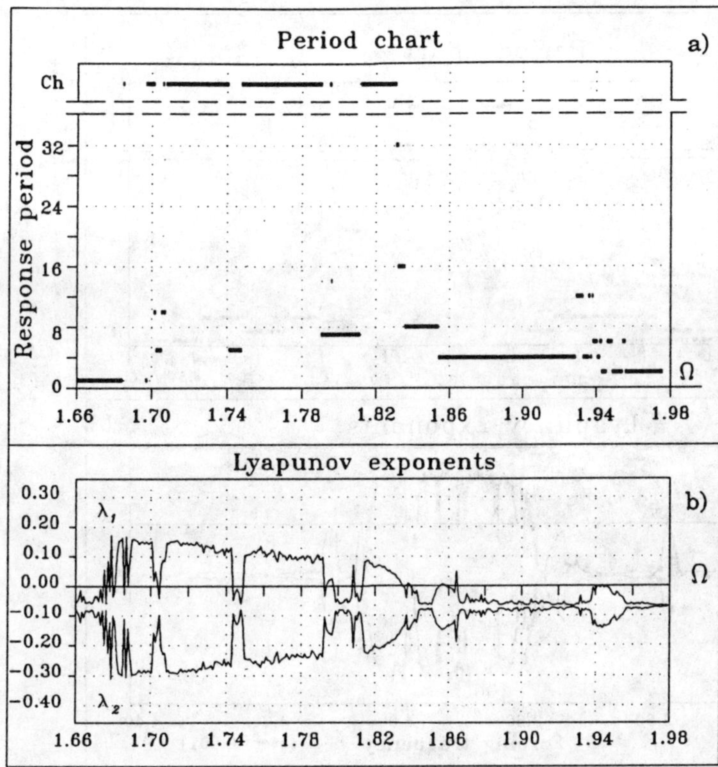

Fig. 4. Bifurcation diagram in the neighbourhood of 1/2-subharmonic resonance ($P = 0.04$)

positive values corresponding to evident chaotic response, as in the latter range; furthermore, a time step increment $h = T/200$ in the integration algorithm was enough for their reliable calculation. Instead, more than a thousand forcing periods were necessary to obtain stable values and correctly characterize chaos or high-period response where the exponents have lower positive values, such as in the 1/3-subharmonic range. An even higher number had to be used in both cases at the boundaries of chaos zones, where the above mentioned value of time step increment was shown to be too large, too.

Major problems also occurred in the 1/3-subharmonic range when calculating the correlation dimension of the attractor: they are likely to be due to thinness of its bundles.

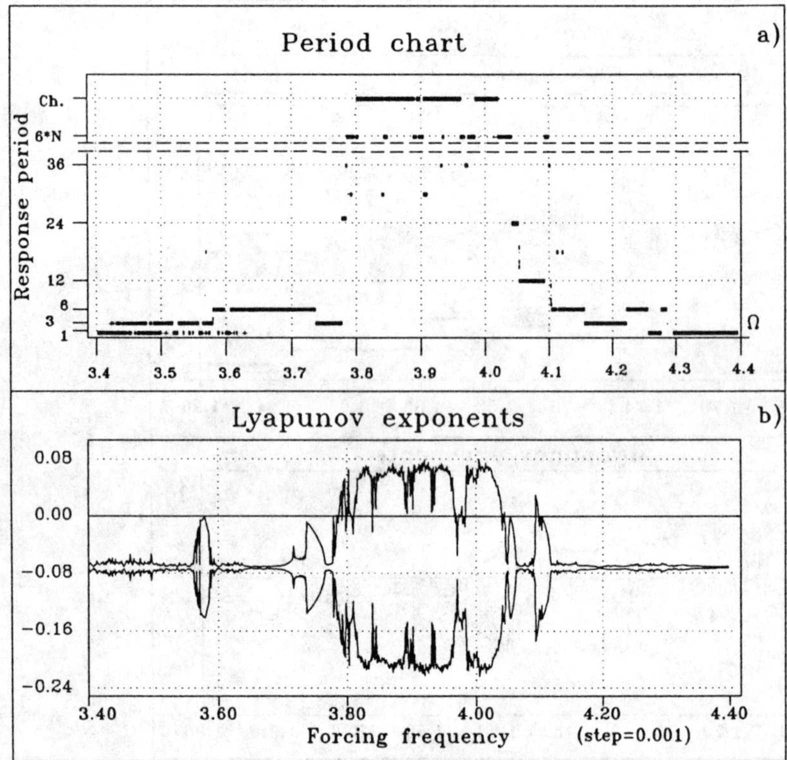

Fig. 5. Bifurcation diagram in the neighbourhood of 1/3-subharmonic resonance ($P = 0.40$)

3. Prediction of Local Bifurcations of Regular Solutions

Characterization of a response of a dynamic system by point-by-point computer simulations and associated subsequent calculations of specific measures is undoubtedly an essential step for the understanding of chaotic behavior. Indeed, the amount of information which can be obtained using these tools is very high, as outlined in the previous section.

However, from both a theoretical and an application point of view, there is a strong interest in a-priori evaluations of the possibility that the system will exhibit irregular and chaotic phenomena. Theoretically, the interest lies mainly in formulating "exact" criteria for loss of regular behavior and the onset of persistent chaos. Instead, from a technical point of view, the simple assessment of unpredictability of the response is often more important than its precise characterization as chaotic. E.g., in engineering applications, it can be sufficient to reliably predict the conditions leading a system designed

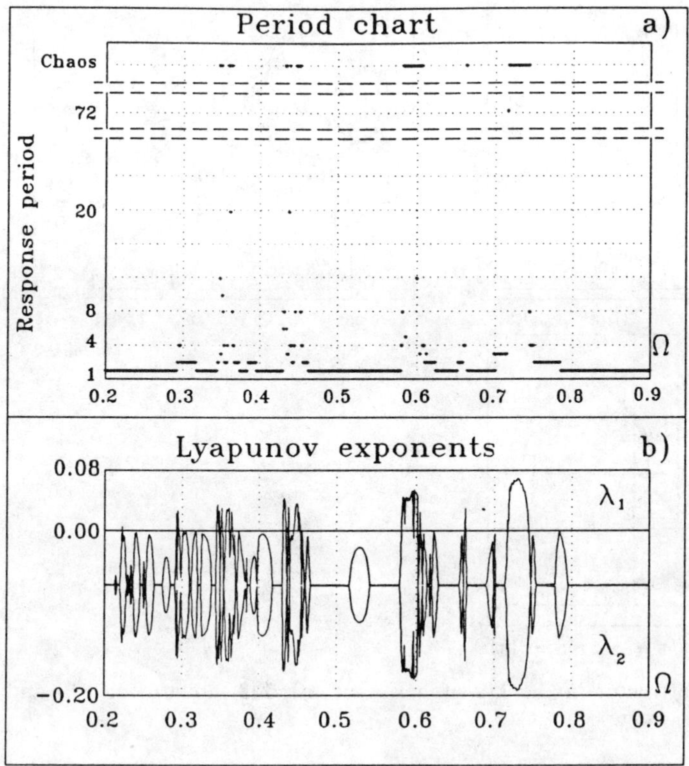

Fig. 6. Bifurcation diagram in the neighbourhood of superharmonic resonances ($P = 0.10$)

to work in a regime of regular dynamics to undergo irregular and possibly chaotic motions, too, which have to be controlled, quenched or even used! In light of this the circumstance that all theoretical criteria developed up to now just give necessary conditions for chaos is not a great drawback, since even the possibility of transient chaos is of interest in several applications.

It is thus important to examine the possibility of predicting disappearance of regular motions and of locating limited regions in the control parameter space, where one might expect transient or persistent chaos using extended and accurate — but localized — computational tools. Prediction of bifurcations — possible precursors to chaos — is reliably furnished by the stability analysis of periodic solutions of a system, obtained by one of the techniques described in the literature.

Classical perturbation or harmonic balance techniques can be used with some success. Low- or high-order approximations can be developed [5, 7, 9, 10, 26], the latter being able, in principle, to furnish better results usually as far as stability evaluations are concerned. Nevertheless, since the use of too

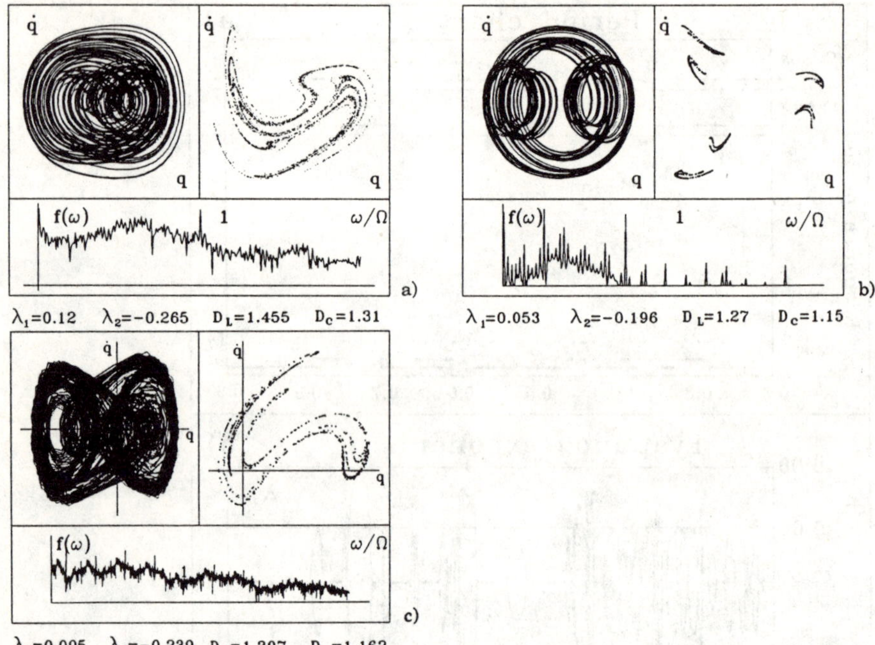

Fig. 7. Chaotic attractor in the 1/2-subharmonic (a), 1/3-subharmonic (b) and superharmonic (c) range

rich approximations can be inconsistent with the roughness or uncertainty of the simple model considered, it is chosen here to examine the bifurcation predictive capability of low-order approximations properly selected in the various regions of resonance of the system. The following expansions are considered:

$$\begin{aligned} q(t) &= a_0 + a_{1f}\cos(\Omega t+\psi), \\ q(t) &= a_0 + a_1\cos(\Omega t+\vartheta) + a_{1/2}\cos(\Omega t/2+\varphi), \\ q(t) &= a_0 + a_1\cos(\Omega t+\vartheta) + a_{1/3}\cos(\Omega t/3+\varphi), \\ q(t) &= a_0 + a_1\cos(\Omega t+\vartheta) + a_2\cos(2\Omega t+\varphi) + a_3\cos(3\Omega t+\psi). \end{aligned} \qquad (3a-d)$$

Consistent with the asymmetry of equation (1), all of them are asymmetric. This is of great importance with regard to the possible development of chaos. Indeed, according to the Floquet theory [27], jump or period doubling bifurcation can occur for an asymmetric solution when varying a control parameter, and the latter is the relevant starting point of a classical bifurcation process which could lead to chaos. From this point of view, the difference of a symmetric solution typical for Duffing oscillators, which must undergo symmetry breaking prior to the possibility of a period doubling bifurcation, is noted well.

The simplest period T expansion (3a) is used as the reference fundamental solution, whereas the simplest period $2T$ and $3T$ expansions (3b) and (3c) are used in the neighbourhood of the 1/2- and 1/3-subharmonic resonances respectively, and the improved period T expansion (3d) is used in the range of superharmonic resonances.

By applying the harmonic balance method, four sets of algebraic nonlinear equations are derived and then solved numerically for the unknown amplitudes and phases to obtain the resonance curves for each solution [21]. The relevant stability analysis is carried out repeatedly with a varying control parameter by calculating the eigenvalues of the monodromy matrix associated with the linearized variational equation [9–28].

Our attention is focused on the neighbourhoods of the two subharmonic resonance conditions $\Omega \cong 2$ and $\Omega \cong 3$, considering in each case a value of the forcing amplitude for which persistent chaos was observed in the computer simulations. The relevant frequency-response curves are shown in Figs. 8a and 8b, respectively, in terms of amplitude of the various harmonics, with the unstable parts plotted in a thin line. Apart from the unstable hysteresis branch, it is worth commenting on the stability features of the lower branch of the period T solution, which are different in the two cases. It the neighbourhood of 1/2-subharmonic, this branch becomes unstable by period doubling bifurcations: correspondingly, stable period $2T$ solution arises and then loses stability on its upper branch by period doubling bifurcation again. Instead, in the neighbourhood of 1/3-subharmonic, the lower branch of period T solution is practically always stable, but stable period $3T$ solution also arises at a certain frequency value and then becomes unstable by period doubling bifurcation. Thus, besides the upper period T solution which remains stable in both cases, a richer set of coexisting regular solutions seems to occur in the second case in some frequency ranges, whilst in the first case stronger suggestion of a possible sequence of period doubling bifurcations is obtained. Whether it actually occurs or not and how many stable solutions coexist in the two unstable zones, can only be examined by considering richer approximations allowing for solutions of a higher period, or by constructing bifurcation charts and basins of attraction using adapted computer simulations.

Bifurcation diagrams at the same values of forcing amplitude have already been discussed (Figs. 4 and 5). They show that a cascade of period doubling bifurcations up to chaos actually occurs with decreasing frequency - though plainer in the 1/2- than in 1/3-subharmonic case, while the situation is different and a bit more complicated with increasing frequency. In any case, one must remember that those diagrams lack a certain generality since they refer to a fixed set of initial conditions (zero values). As to their influence, an already richer picture is given in Figs. 9a, b, where the predictions obtained for the harmonic and subharmonic components with the chosen approximations are compared with the simulation results obtained with either

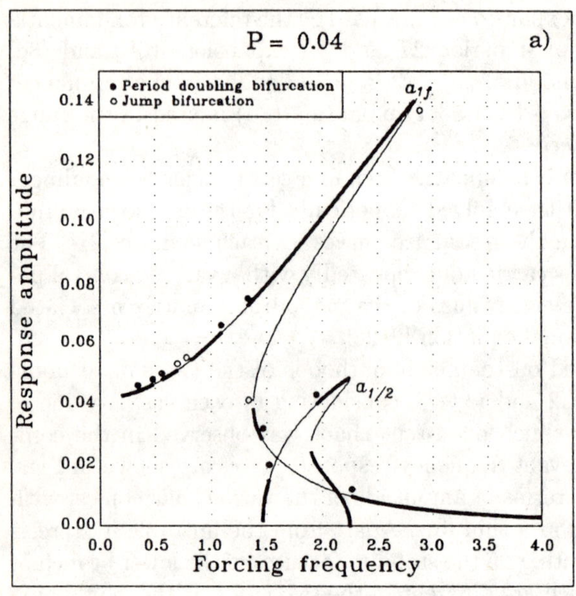

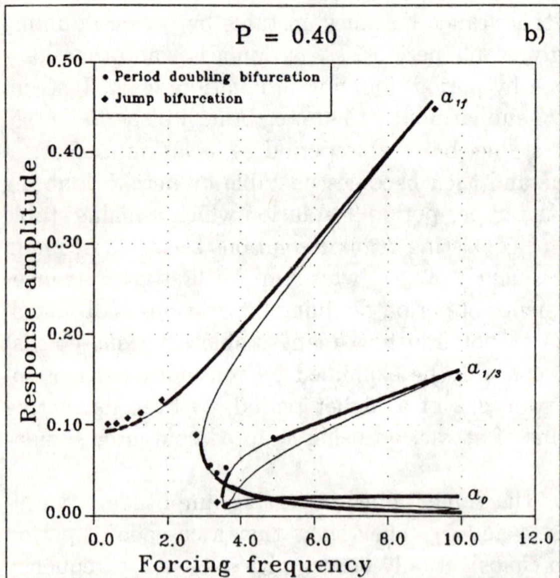

Fig. 8. Approximate frequency-response curves in the 1/2 (a) and 1/3 (b) subharmonic range

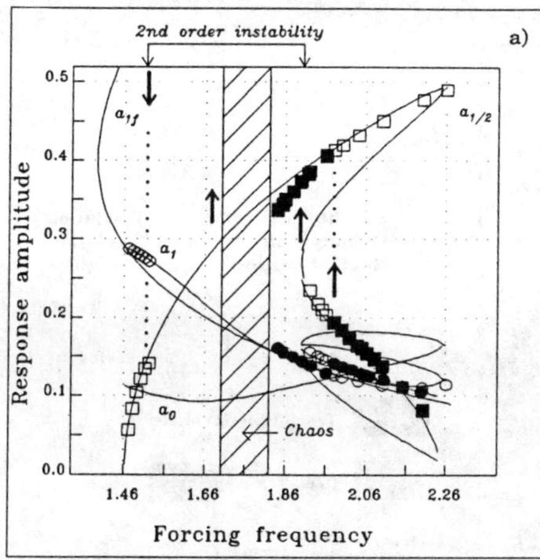

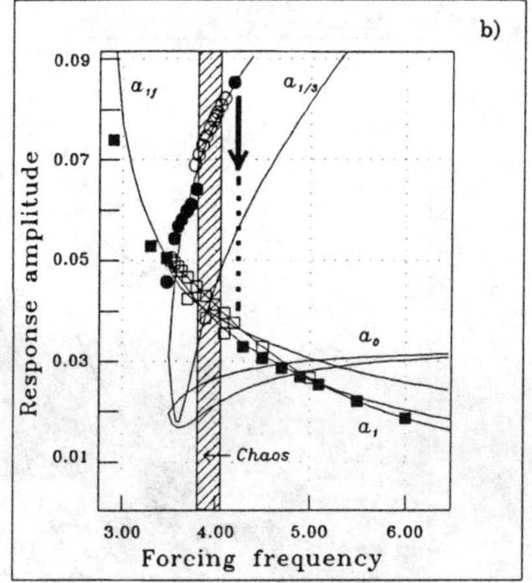

Fig. 9. Comparison between analytical and numerical results in (a) the 1/2 and (b) 1/3 subharmonic range

zero initial conditions (black dots) or the ones corresponding to the approximations themselves via equation (3b) or (3c) (white and black dots). Besides the general satisfactory agreement of the results and the location of the two chaos zones inside the unstable regions, one can observe a variety of jumps occurring in dependence upon the initial conditions. Moreover, in the 1/2-subharmonic resonance, the upper period T solution obviously plays a strong role, to which all considered conditions are attracted left of the chaos zone; in the 1/3-subharmonic resonance, the occurrence of both chaos and period $3T$ solutions hints at another presumably rich set of coexisting responses.

The bifurcation predictive capability of stability numerical analysis of simple periodic solutions is illustrated in more general terms in a control parameter space in Figs. 10 and 11, where the stability limit curves of the considered periodic solutions are superimposed upon the regions of response obtained through computer simulations around the 1/2- and 1/3-subharmonic resonance, respectively.

In the former case, the following kinds of response are obtained: lower period T in zone 1; lower period $2T$, that is always stable, in zone 2; upper period $2T$ in zone 3, where lower period $2T$ does not exist; upper period T, which is always stable, in zone 4. Thus one can observe: satisfactory prediction of the two zones where lower period T or upper period 2T motions do not occur by means of the corresponding stability limit curves; good prediction of the zone of possible existence of lower period $2T$ motions; near delimitation on the right of the zone where chaotic or highly periodic motions are obtained by the limit curve of vertical tangency of lower period $2T$ and the right stability limit curve of upper period 2T. In any case, which kinds of motion actually occur within the various zones depends considerably on the initial

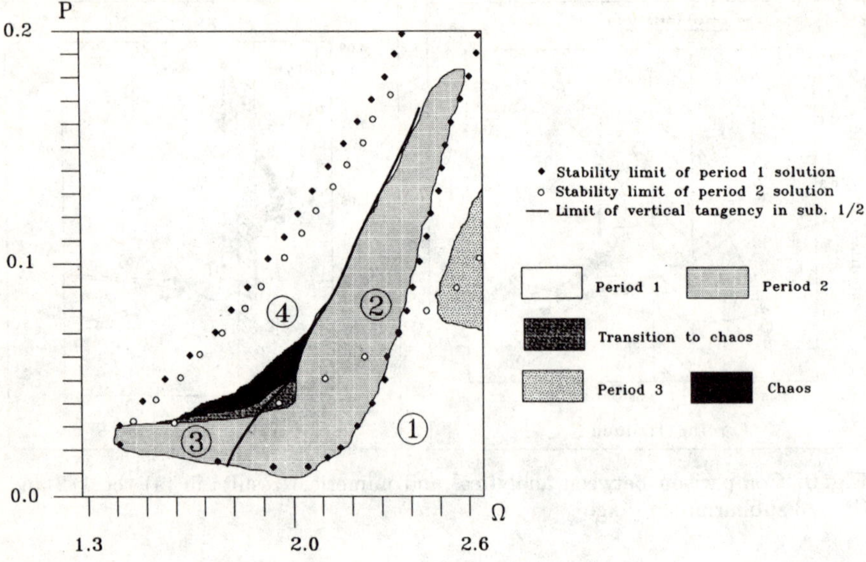

Fig. 10. Unstable regions of period T and $2T$ approximate solutions. Comparison with computer simulation results

conditions considered, which can entail shifts of the boundaries of some region of numerical response with respect to those obtained with the zero ones.

In the 1/3-subharmonic range, the picture of numerical response is more complex but for the approximate solution considered, the instability predictions are concerned only with the upper period $3T$ solution. A shift of the approximate unstable region towards higher frequencies is observed when compared to the region where periods other than $3T$ are found numerically. But the overall picture is very similar, with a wedge of stable solutions entering the upper part of the region. Moreover, one must again remember that the numerical results refer only to zero i.c., which seems to play a very special role in this case. Indeed, not all the conditions lead to chaos (see Fig. 9b) and some of them lead to period $3T$ motions at high frequencies, where the approximate period $3T$ solution is again stable.

Previous results show that simple and properly chosen approximate solutions can satisfactorily delimitate regions where complex and chaotic motions can occur. They also suggest strong sensitivity of response to the initial conditions and complement local bifurcation predictions and behavior numerical charts with systematic construction of basins of attraction. In particular, the extent of the chaotic basins has to be checked. This will be discussed in the next section in the framework of a global bifurcation description of the system response. Still in view of such a description, which relies on the construction of invariant manifolds of unstable periodic motions in control-phase

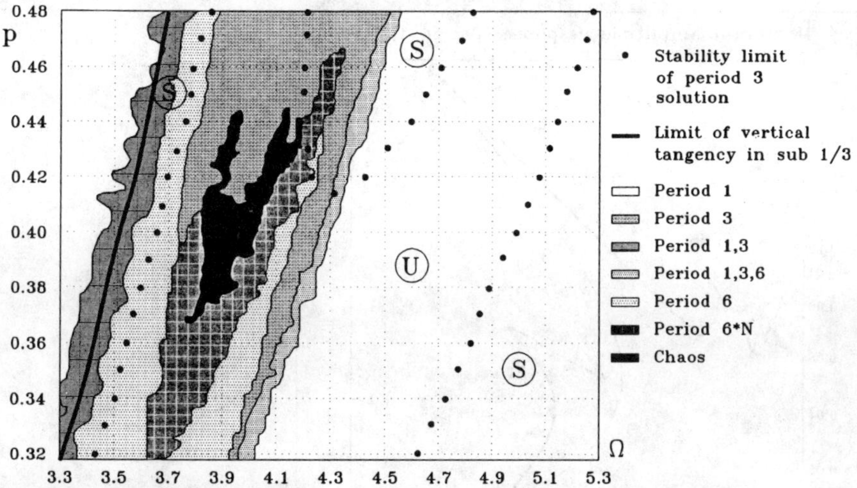

Fig. 11. Unstable region of the period $3T$ approximate solution. Comparison with computer simulation results

portraits, it is worth commenting briefly on a completely numerical technique for obtaining the system response.

In terms of the discrete dynamical systems theory [2, 29], the period mT solution $x = x(t; t_0, x_0, \dot{x}_0)$, $\dot{x} = \dot{x}(t; t_0, x_0, \dot{x}_0)$ of the flow (1), m being 1,2,3,... and zero denoting the initial conditions, is equivalent to the fixed points $\bar{p}_i(x_i, \dot{x}_i)$, $i = 1, 2, \ldots, m$, of the mapping $f : \mathbb{R}^2 \to \mathbb{R}^2$ of the phase plane $(x, \dot{x})$ into itself iterated m times (f^m). Each point is called an m-periodic point [12]. They are calculated by solving the equation $\bar{p}_i = f^m(\bar{p}_i)$ using the Newton-Raphson procedure. A path — following (or continuation [28]) algorithm with the fixed points of the Poincaré map as a control parameter is varied and can be used to obtain the response curves.

The results obtained in the 1/2- and 1/3-subharmonic range are reported in Figs. 12a,b in terms of the maximum amplitude of response. They show good general agreement with those in Figs. 6a,b obtained with the harmonic balance approximations. Obtaining response curves of solutions of period $T, 2T, \ldots$, as in Fig. 8a, corresponds here to finding the fixed points of the mappings $f, f^2, \ldots$. Solutions of higher periodicity were looked for, since previous analysis reported of their existence and importance, but their determination is generally both computationally expensive and difficult with regard to the selection of suitable triggering values. The picture of responses obtained is quite complex mainly in the 1/3-subharmonic range, though many of them are unstable in large frequency ranges and thus very hard to be found numerically.

Before closing this section, it is worth observing that the path — following algorithm, besides being a more modern technique to obtain response curves,

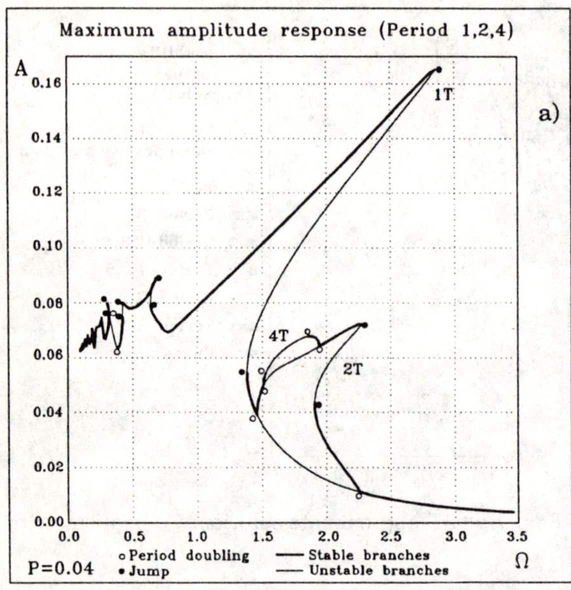

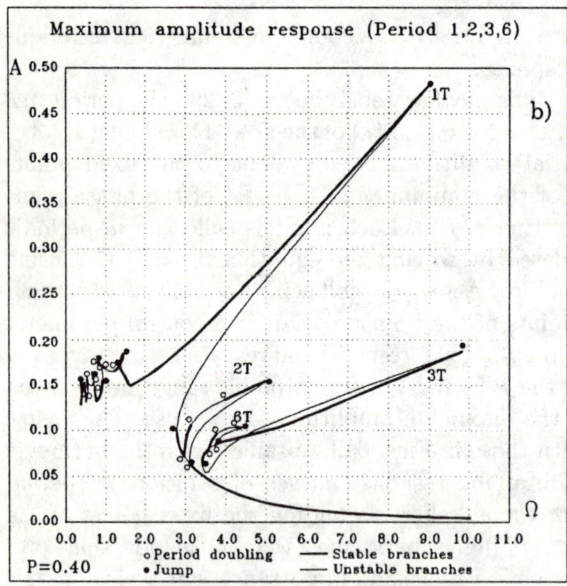

Fig. 12. Numerical frequency-response curves in the 1/2 (a) and 1/3 (b) subharmonic range

proves to be useful for directly furnishing the fixed points of the motion, which play a fundamental role in its geometrical description, as will be seen in the next section. Fixed points corresponding to a stable periodic solution are sinks (S), unstable fixed points are direct (D) or inverse (I) saddles which are established by fold (jump) or flip (period doubling) bifurcations respectively [12], when varying a control parameter.

4. Geometrical Description of System Response Using Attractor-Basin Portraits and Invariant Manifolds

A geometrical description of the system's global dynamics helps us understand the complicated features of nonlinear and chaotic behavior. First, attractor-basin phase portraits are constructed, in which periodic and/or chaotic coexisting attractors are superimposed on the basins in the i.c. space obtained by systematic computer simulations or by a cell mapping technique [30]. Here the basins are obtained with a simple but efficient algorithm by the following steps. (i) A proper region of the i.c. plane is subdivided into square boxes using a grid of points. (ii) Starting from the centre of a box, the fixed points of the relevant stroboscopic trajectory under a mapping of given period are obtained and all boxes crossed by the trajectory itself are labelled accordingly. (iii) The procedure is repeated sequentially for all boxes, discarding those already labelled. The periodic points located inside the basins are sinks (S) while the ones in the basin boundaries are saddles (D or I), with subscripts referring to the order of mapping. The following symbols are used in the figures: ($\cdot$) for sinks, ($\bullet$) and ($\circ$) for direct and inverse saddles respectively.

The attractor-basin phase portraits are then examined in the light of the global dynamics of the system, described by the invariant manifolds of the mapping's saddle points corresponding to various unstable periodic solutions. The manifolds are obtained by forward and backward integrations from the saddles themselves and are superimposed on the basins in the following figures. Analysis of the evolution of basins/manifolds with a varying control parameter in certain sagnificant ranges enables a deeper understanding of the global attractor structure as well as attractor and basin bifurcations.

It is worth examining a collection of attractor-basin portraits obtained in the 1/2-subharmonic range with decreasing frequency [31].

To the right of the value $\Omega = 2.26$, only the two basins of the upper (light grey) and lower (grey) period T solutions spiralling into each other occur. At that frequency value (Fig. 13a) a local fold bifurcation occurs (see Fig. 12a), which causes very thin tongues (white) corresponding to the just established upper period $2T$ solution to suddenly appear within the basin of lower period T. At about the same Ω value a local flip bifurcation occurs as well: accordingly, the sink S_1^1 in the grey basin becomes inversely unstable

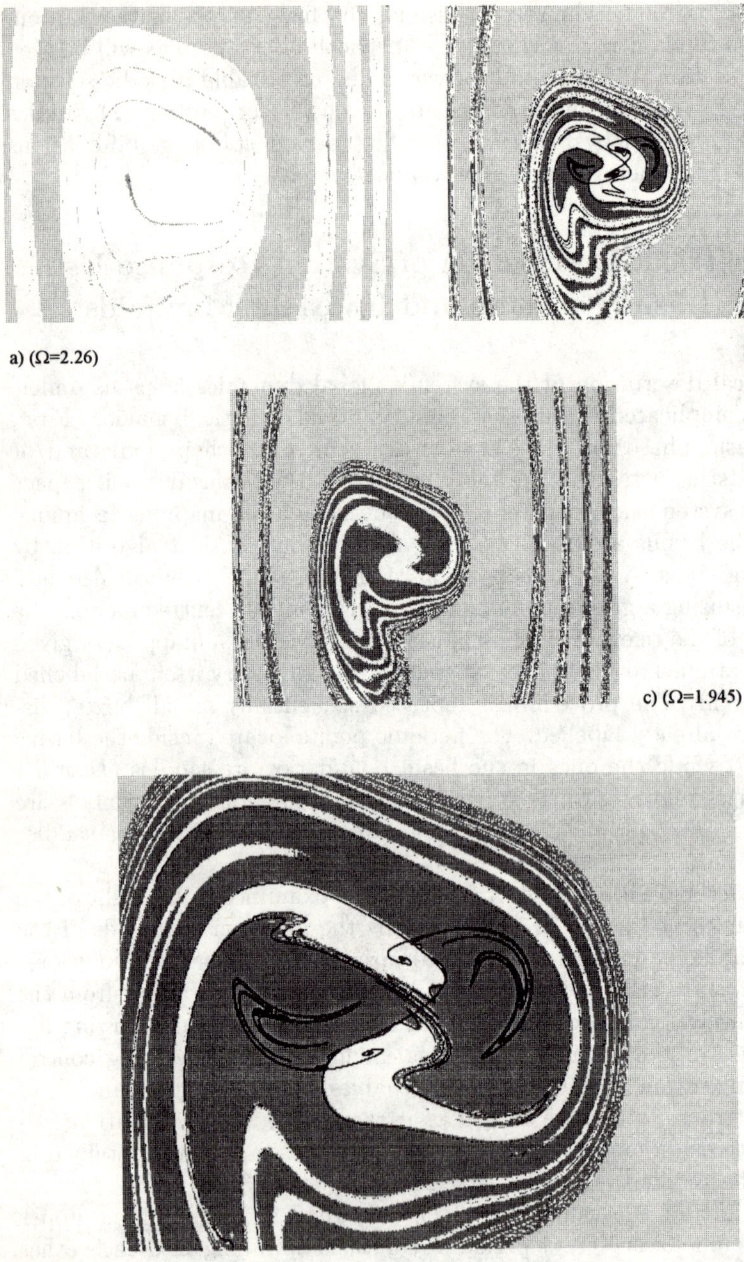

a) (Ω=2.26)

b) (Ω=2.00)

c) (Ω=1.945)

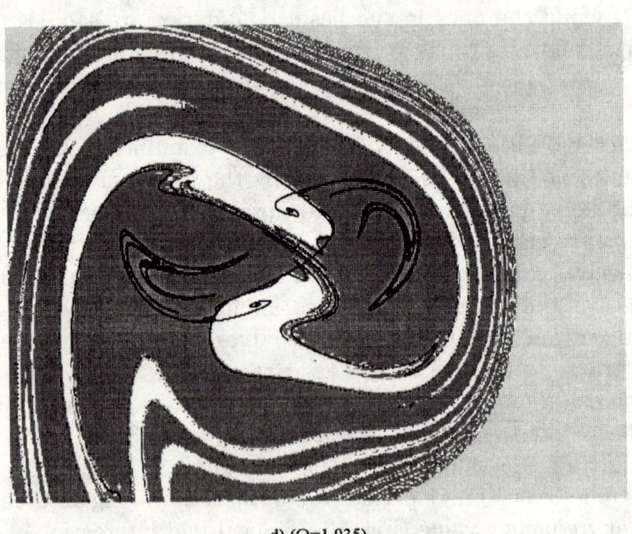

d) (Ω=1.935)

Fig. 13. Attractor-basin portraits. Infinite folding of stable manifolds of D_2 and unstable manifold of I_1 (Fig. d)

($S_1^1 \rightarrow I_1$) and two new sinks corresponding to the bifurcated lower period $2T$ solution appear in this basin. In Fig. 13b ($\Omega = 2.00$) the stable and unstable manifolds of the two direct saddles D_2 of f^2, which are located in the boundary separating the two period $2T$ basins, and those (inside the grey basin) of the inverse saddle I_1 of f are superimposed on the basins of attraction. The upper period $2T$ basin (white) increases in size with the decreasing control parameter and appears disjoined in two subdomains whose tails, however, ravel out becoming thinner and thinner as they accumulate along the stable manifolds of the direct saddles. Theoretically, these should coincide with the basin boundary between the two period $2T$ solutions; the practical agreement is generally fairly good near the saddles. The lower period $2T$ basin decreases in size and appears joined, but it is actually separated in two subdomains, too, by the stable manifold of the inverse saddle I_1. Four distinct subdomains were obtained if each of the relevant sinks were determined as the unique fixed point of a period $2T$ mapping (two-periodic point).

At the frequency value $\Omega = 1.935$ (Fig. 13d), the four subdomains are approximately equally disjoined. In the neighbourhood of I_1, repeated folding of the stable manifolds of the two direct saddles of f^2 is observed. They intersect an infinite number of times with the unstable manifold of the inverse saddle of f (heteroclinic tangle [2, 3, 29]), which entails locally fractal basin boundaries between periodic solutions. Point I_1 is located at about the position where the previously disjoined subdomains will join and the previously joined ones will disjoin. Notwithstanding the fractal basin boundary, it is not badly predicted by the stable manifolds of the saddles. The joining basin (white) corresponds to the period $4T$ solution to which the upper period $2T$ in Fig. 12a has locally bifurcated at this frequency value: this is also deduced from Fig. 4a, though relevant to a fixed set of i.c. The latter figure also shows the occurrence of further solutions of higher periodicity, whose narrow basins were actually detected inside the white region at some frequency values (Fig. 13c). The disjoining basin (grey) vanishes at the frequency value where a fold bifurcation occurs in Fig. 12a, causing the lower period $2T$ solution to suddenly disappear.

With decreasing frequency, the solution in the white basin evolves towards chaos with successive period doublings. Fig. 14a shows a steady-state chaotic attractor superimposed on the basin ($\Omega = 1.77$). It contains all the infinite inverse saddles ($I_1, I_2, I_4, \ldots$) corresponding to the periodic solutions which became progressively unstable through successive period doublings. The stable and unstable manifolds of the inverse saddle I_1 of the mapping f are shown in Fig. 14b; they have an infinite number of homoclinic intersections. The same occurs for the invariant manifolds of the two saddles I_2 of the mapping f^2, as shown in Fig. 14c. The chaotic attractor in Fig. 14a is identical with the closure of the unstable manifold of the inverse saddle of f, and with that of the inverse saddles of f^2, and so on. It is thus possible to

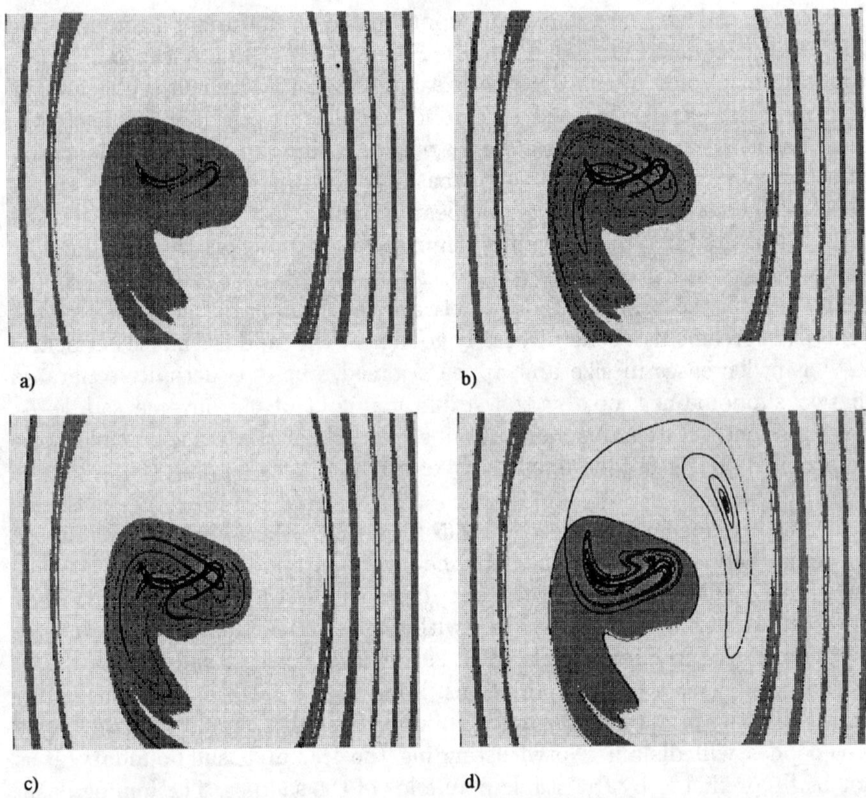

Fig. 14. Strange attractor (a), stable and unstable manifolds of saddles I_1 (b), I_2 (c), D_1 (d). ($\Omega = 1.77$)

have suggestions about a sequence of period doubling bifurcations progressing to chaos by simply comparing the invariant manifolds of low inverse periodic saddles systematically obtained by forward and backward integrations of limited length, whilst the control parameter is varied. The chaotic attractor can also be obtained, after transients have died away, by constructing the lower unstable branch of the direct saddle D_1 of f in the basin boundary that separates the light grey region of the upper period T solution from the chaotic one (Fig. 14d). Such a smooth boundary, in turn, is satisfactorily matched by the stable manifold of the saddle itself.

As the frequency value is further decreased, the attractor increases in size and approaches the basin boundary. Homoclinic tangency of the two manifolds of the direct saddle of f occurs at about $\Omega = 1.705$ and causes the steady-state chaotic attractor to suddenly disappear by colliding with the saddle point and its stable branch (Fig. 15a,b). This global bifurcation, after which the basin of upper period T solution occupies the whole i.c.

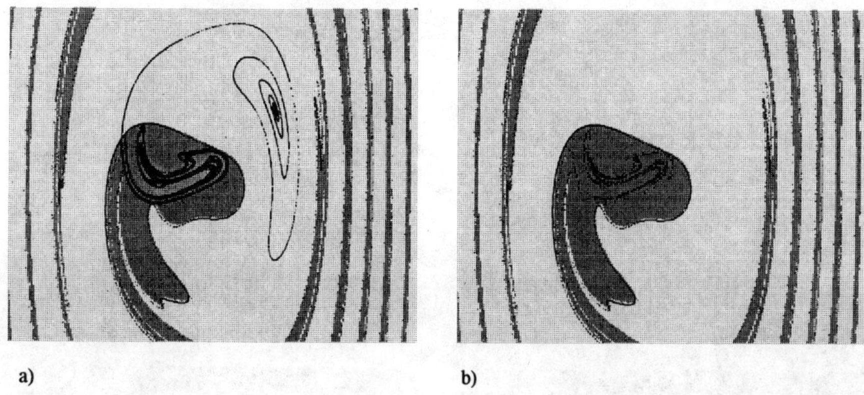

Fig. 15. Homoclinic tangency of manifolds of D_1 (a), collision of attractor with D_1 and its stable branch (b). ($\Omega = 1.705$)

space, is of the type commonly referred to as boundary crisis or blue sky catastrophe bifurcation in the literature [2, 12, 32]. Actually, such a threshold of catastrophic bifurcation is "spread" over an extremely narrow range of forcing frequency values [16], so that the left boundary of the chaos zone in Fig. 10 is likely to have marked fractal character. Correspondingly, strong mixing of the boundary between the period T and chaotic basins occurs in the i.c. space, though it is clearly observable only through a very localized window near the boundary (Fig. 16a). Of course, in this case, it becomes impossible to locate the basin boundary by numerical experiments to determine the stable manifold of the direct saddle of f, which has been infinitely stretched and folded by the homoclinic structure.

To the left of the bifurcation threshold, there is a rather large zone of frequency values where any i.c. leads to the upper period T solution: indeed, this is the only stable one, as observed in Sect. 3. It is still worth, however, examining the pattern of the invariant manifolds of the direct saddle (Fig. 16b, $\Omega = 1.62$). The stable branches delimitate a basin with the same shape as that previously identified as being steady chaotic. The lower unstable branch escapes to the upper sink S_1^{u} after remaining in that basin for a rather long time. This means the formerly chaotic topological structure still has notable attractive properties, however it is no longer trapped inside the stable manifold and is thus only transient.

Afterwards, markedly fractal tongues arise within that attractive, lessening, region. They correspond to a multiple of the two high periodicity solutions undergoing a reverse sequence of flip bifurcations up to period $2T$ in an extremely narrow frequency range, as the relevant unstable solutions become stable again (see also Fig. 12a). The invariant branches of the direct saddle in the basin boundary intersect an infinite number of times; this homoclinic structure is strictly associated with the fractal character of the

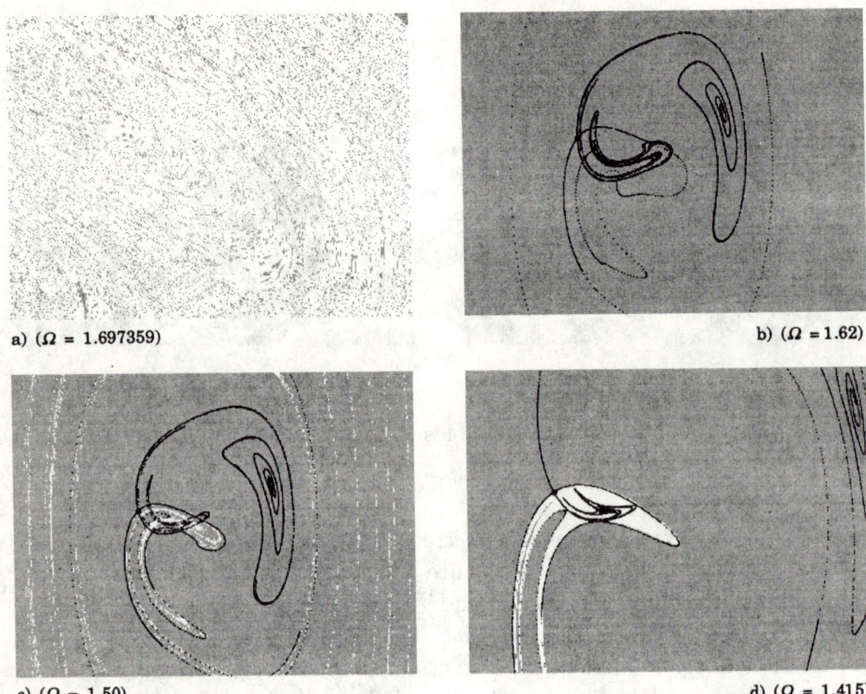

Fig. 16. Strong mixing of basin boundary (a). Manifolds of saddle D_1: fractal (c) and smooth (d) basin boundary

boundary itself (Fig. 16c, $\Omega = 1.50$). When this becomes smooth, the unstable manifold touches the stable one and then remains trapped inside the basin of lower period T (Fig. 16d, $\Omega = 1.415$) again. The occurrence of this global behavior in connection with basins of regular response shows that the homoclinic tangency merely represents a threshold of unpredictability of the response itself.

Examination of attractor-basin phase portraits and global invariant manifold structures in the ranges of 1/3-subharmonic and of superharmonic resonances further stresses the richness and variety of the bifurcation and chaos features of the Helmholtz and Duffing oscillator. These analysis would go beyond the scope of this work. Nevertheless, the results presented extensively show how deep an understanding of the global dynamics of a system can be achieved through the combined use of numerical techniques and geometrical interpretation.

5. Conclusions

A Helmholtz and Duffing oscillator, which represents a problem of interest in the field of elastic nonlinear structural dynamics, was considered as a prototypical model to describe the main features of bifurcation and chaos phenomena for single-degree-of-freedom systems. Indeed, it exhibits a wide range of regular and chaotic motions, as well as an extensive range of local and global bifurcations and routes to chaos.

The picture of response was discussed in three main steps. Computer simulations allow us to calculate reliable measures of chaotic response and to obtain behavior charts in control parameter space and basins of attraction in initial condition space. Stability analysis of regular solutions obtained either by purely numerical or by a combination of analytical and numerical techniques allow us to predict local bifurcations and their siginificance as possible precursors to chaos. Geometrical analysis throws light on the global features of the attractor and basin bifurcations, and gives us a better insight into situations of response unpredictability, transient chaos and persistent chaos.

Considering the present point of development of the bifurcation and chaos theory, all three steps have to be considered as standard steps in the analysis of a system susceptible of chaotic behavior. However, in engineering applications, owing to the notable computational effort necessary for a thorough description of response, their sequence will be mostly aimed at clarifying simple unpredictability with respect to the expected behavior and possible transient chaos.

Thus, after developing a fairly realistic but tractable mathematical model of a mechanical or structural system — which is likely to be low-dimensional but have more than one-degree-of-freedom — the sagnificant zones of nonlinear dynamic behavior and the relevant regular responses will be determined using numerical, analytical or a mixture of these techniques. Prediction of local bifurcations of those responses are essential for localizing regions of irregular and complex behavior. For this purpose, accurate and chaos-oriented computer simulations can be performed to characterize the kind of response occurring in dependence of various control parameters, but only the geometrical interpretation will throw some light on the actual features of the model global dynamics.

Acknowledgements. This work was partially supported with MURST 40% Grants. The author is indebted to Drs. F. Benedettini and A. Salvatori for their help with computations and for their participance in some useful discussions.

References

1. P. Holmes: A nonlinear oscillator with a strange attractor. Phil. Trans. Royal Soc. London **A 292**, 419–448 (1979)
2. J.M.T. Thompson, H.B. Stewart: Nonlinear dynamics and chaos. Wiley, Chichester 1986
3. F.C. Moon: Chaotic vibrations. Wiley, New York 1987
4. B.H. Tongue: Characteristics of numerical simulations of chaotic systems. ASME J. Appl. Mech. **54**, 695–699 (1987)
5. W. Szemplinska-Stupnicka: Secondary resonance and approximate models of routes to chaotic motion in non-linear oscillators. J. Sound Vibrat. **113**, 155–172 (1987)
6. T. Fang, E.H. Dowell: Numerical simulations of periodic and chaotic responses in a stable Duffing system. Int. J. Non-Linear Mech. **22**, 401–425 (1987)
7. R. Van Dooren: On the transition from regular to chaotic behaviour in the Duffing oscillator. J. Sound Vibrat. **12**, 327–339 (1988)
8. S.T. Ariaratnam, Wei Chan Xie, E.R. Vrscay: Chaotic motion under parametric excitation. Dyn. Stab. Syst. **4**, 111–130 (1989)
9. A.H. Nayfeh, N.E. Sanchez: Bifurcations in a forced softening Duffing oscillator. Int. J. Non-Linear Mech. **24**, 483–497 (1989)
10. A.Y.T. Leung, T.C. Fung: Construction of chaotic regions. J. Sound Vibrat. **131**, 445–455 (1989)
11. G.X. Li, F.C. Moon: Criteria for chaos of a three-well potential oscillator with homoclinic and heteroclinic orbits. J. Sound Vibrat. **136**, 17–34 (1990)
12. Y. Ueda: Survey of regular and chaotic phenomena in the forced Duffing oscillator. Chaos, Solits. & Fracts. **1**, 199–231 (1991)
13. L.D. Zavodney, A.H. Nayfeh: The response of a single-degree-of-freedom system with quadratic and cubic non-linearities to a fundamental parametric resonance. J. Sound Vibrat. **120**, 63–93 (1988)
14. W. Szemplinska-Stupnicka, R.H. Plaut, J.C. Hsieh: Period doubling and chaos in unsymmetric structures under parametric excitation. ASME J. Appl. Mech. **56**, 947–952 (1989)
15. W. Szemplinska-Stupnicka, P. Niezgodzki: The approximate approach to chaos phenomena in oscillators having single equilibrium position. J. Sound Vibrat. **141**, 181–192 (1990)
16. F. Benedettini, G. Rega: Numerical simulation of chaotic dynamics in a model of elastic cable. Nonlinear Dyn. **1**, 23–38 (1990)
17. G. Rega, F. Benedettini, A. Salvatori: Periodic and chaotic motions of an unsymmetrical oscillator in nonlinear structural dynamics. Chaos., Solits. & Fracts. **1**, 39–54 (1991)
18. G. Rega: From regular nonlinear oscillations to chaotic motions in structural dynamics. In Nonlinear Problems in Engng. (Eds. C. Carmignani and G. Maino), 195–209, World Scientifc, Singapore (1991)
19. F. Benedettini, G. Rega: Nonlinear dynamics of an elastic clable under planar excitation. Int. J. Non-Linear Mech. **22**, 497–509 (1987)
20. G. Rega, F. Vestroni, F. Benedettini: Parametric analysis of large amplitude free vibrations of a suspended cable. Int. J. Solids Struct. **20**, 95–105 (1984)
21. F. Benedettini, G. Rega, A. Salvatori: Prediction of bifurcations and chaos for an asymmetric elastic oscillator. Chaos, Solits & Fracts. **2**, 303–321 (1992)
22. A. Wolf, J. Swift, H.L. Swinney, J. Vastano: Determining Lyapunov exponents from a time series. Physica **16D**, 285–317 (1985)
23. P. Grassberger, I. Procaccia: Characterization of strange attractors. Phys. Rev. Lett. **50**, 346–349 (1983)

24. F.C. Moon, G.X. Li: The fractal dimension of the two-well potential strange attractor. Physica **17D**, 99–108 (1985)
25. G. Rega, F. Benedettini: Planar nonlinear oscillations of elastic cables under subharmonic resonance conditions. J. Sound Vibrat. **132**, 367–381 (1989)
26. H. Tamura, T. Kondou, A. Sueoka, N. Ueda: Higher approximate solutions of the Duffing equation. Bull. JSME **29**, 3075–3082 (1986)
27. C. Hayashi: Nonlinear oscillations in physical system. McGraw-Hill, New York 1964
28. R. Seydel: From equilibrium to chaos. Elsevier, New York 1988
29. J. Guckenheimer, P. Holmes: Nonlinear oscillations dynamical system, and bifurcation of vector fields. Springer-Verlag, New York 1983
30. C.S. Hsu: Cell to cell mapping. Springer-Verlag, New York 1987
31. G. Rega, A. Salvatori, F. Benedettini: Numerical and Geometrical Analysis of Bifurcation and Chaos for an Asymmetric Nonlinear Oscillator. Nonlinear Dyn. (in press), (1994)
32. C. Grebogi, E. Ott, J.A. Yorke: Crises, sudden changes in chaotic attractors, and transient chaos. Physica **7D**, 181–200 (1983)

Bifurcations and Chaotic Motions in Resonantly Excited Structures

S.I. Chang, A.K. Bajaj and P. Davies

School of Mechanical Engineering
Purdue University
West Lafayette, IN 47907–1288, U.S.A.

Abstract

The focus of this chapter is a discussion of the behavior of multi-degree-of-freedom models of structures with nonlinearities. While an overview of the research conducted in this area is given, the latter part of the chapter is devoted to a study of the response of weakly nonlinear multi-degree-of-freedom models under harmonic excitation. These models were derived from the von Karman equations that describe the behavior of a thin rectangular plate under initial tension. The types of behavior that result from internal resonances are of particular interest, whereby one mode is driven directly but other modes are excited by the nonlinear coupling between the modes. Energy sharing between the directly driven mode and the other modes leads to an amplitude-modulated coupled-mode response that can become chaotic. The approach is to develop models of the slowly varying amplitude and phase of the nonlinear response of the interacting modes through averaging. These equations are studied using the local bifurcation theory for their steady-state solutions. Various bifurcation points are identified in order to understand which types of solutions are possible for a given set of excitation conditions and model parameter values. It is shown that the response of the plate is qualitatively distinct and depends on the mode which is directly excited by the external loading.

1. Introduction

The nonlinear behavior of structures, including that of strings, beams, arches, plates and shells, has been studied, both statically and dynamically, for a long time [1, 2]. The source of the nonlinearity may be geometric, inertial, due to material properties, or may arise due to damping mechanisms or boundary conditions. The geometric nonlinearity may be induced by nonlinear stretching or by curvature effects that are significant when deformations are large. The membrane forces, induced by stretching, accompany the transverse motion of the structure, if the boundary is constrained against movement in the longitudinal direction. This longitudinal stretching leads to a nonlinear relationship between the strain and the displacement. Therefore, when large amplitude vibrations of a structure are studied, this nonlinear geometric effect needs to be considered. Nonlinear inertial effects, caused by the presence of concentrated or distributed masses, also couple the transverse and in-plane motion. Material nonlinearity in structures arises when the relationship between the stress and the strain is nonlinear.

When the structure has commensurable or nearly commensurable natural frequencies of small integer ratios, the interaction among the modes through *internal resonances* can be of significance depending on the type of nonlinearity present in the system. The nonlinear system behavior can be classified into several cases [1, 3] depending on the presence or absence of internal and/or external resonances. Systems which possess internal, as well as external resonances, are found to exhibit interesting responses, arising due to the exchange of energy between the modes in internal resonance by means of external resonance. Trough the external resonance, energy can be fed to one or many modes in internal resonance. Even when only one mode is directly excited, the system can exhibit the so called "coupled-mode response" (in contrast to the "single-mode response") due to the exchange of energy between the various modes. This modal coupling in the structure is caused by the nonlinearities present. A classical example of this behavior is the stretched nonlinear string, which exhibits non-planar whirling motions even when the resonant harmonic excitation is restricted to one plane [4–6] . Higher levels of excitation are usually necessary to destabilize the single-mode response into a coupled-mode response. Some studies [7–9] have shown that this coupled-mode response can be in the form of traveling waves, and that the steady-state traveling waves become amplitude-modulated, period-doubled and finally chaotic as a system parameter is varied.

The literature on nonlinear structural vibrations is quite extensive with both single and multiple degree-of freedom models having been investigated. Studies with single degree-of freedom models include weakly nonlinear [1] as well as global or strongly nonlinear responses [10, 11], whereas, the studies with multiple degree-of freedom models are mostly restricted to the weakly nonlinear behavior involving modal interactions. The methods used for analyzing the weakly nonlinear responses, usually consisting of perturbation and asymptotic techniques [1], are quite distinct form the geometric-perturbation methods [12–14] used in producing the strongly nonlinear behavior including the existence of chaotic dynamics. Applications of the geometric-perturbation methods in the study of the global behavior of structural systems include the works of Holmes and Marsden [11] and Yagasaki [15] for one-mode responses, and the works of Yang and Sethna [7, 16] for coupled-mode dynamics. The work described in this chapter is limited mostly to coupled-mode responses and only a brief remark will be made about the application of global analysis techniques in proving the existence of chaos in these averaged systems.

The coupled-mode behavior of resonantly excited structures has received considerable attention in recent years although classical studies [17–19] conducted two to three decades ago do exist. The authors of more recent studies were able to utilize developments in the local bifurcation theory. In the following section, recent studies on nonlinear behavior of basic structural elements are first reviewed and then briefly discussed. The discussion here is restricted to works which use the local bifurcation theory. In subsequent sections, the nonlinear flexural motions of rectangular plates are considered and some new results on the coupled-mode dynamics under resonant harmonic excitation are presented.

2. Nonlinear Structural Members

In this section the recent literature on the nonlinear response of structural elements will be reviewed. The systems discussed include strings, beams, shells and rings, and plates. They are subjected to either external or parametric harmonic excitations. The governing partial differential equations for the continuous systems are transformed to a nonlinear temporal set of ordinary differential equations by the Galerkin procedure in most studies. The finite set of second-order ordinary differential equations are then analyzed using the asymptotic methods of multiple time scales or averaging, to obtain a first-order approximate solution for the case of weakly nonlinear systems. The form of the amplitude equations, obtained by the asymptotic method, depends on the types of "internal" and "combination" resonances and on the nature of the nonlinearities in the structure. The responses of specific structures are discussed below.

2.1 Strings

The prediction of the nonlinear response of stretched strings to harmonic excitations is a classic problem. Numerous studies have been carried out, both analytical and experimental ones, the results of which are summarized in Nayfeh and Mook [1]. This system, and in particular the non-planar response, has received considerable attention in recent years beginning with the work of Miles [20] who reviewed the problem of stretched string vibration using the local bifurcation theory.

Following Miles' work, Bajaj and Johnson [4, 21, 22], took up the problem of the non-planar response of a string and showed that periodically and chaotically modulated responses are possible for some parameter values.

Numerical investigations with the single-mode truncation of the nonautonomous string system shows good correspondence, even between chaotic solutions of the averaged system and those of the original system.

Tufillaro [5] adopted a very simple single-mode model for string vibrations. Bajaj and Johnson had started with the first order continuum model, which accounts for axial motions, but considers the longitudinal wave speed to be much higher than the transverse wave speed, an assumption valid for metallic strings. Tufillaro's model is capable of showing nonlinear phenomena including hysteresis, periodic, quasiperiodic, and chaotic motions. Chaotic vibrations were predicted for an experimentally accessible regime. Molteno and Tufillaro [23] reported experimental results that show good qualitative agreement of the theoretical and numerical results described by Johnson and Bajaj [4] for a torus doubling transition to chaos. They observed the following bifurcation sequence as the frequency was varied: periodic → quasi-periodic → chaotic → quasi-periodic → periodic and, at lower forcing amplitude, periodic → quasi-periodic → periodic. They also observed torus mergings for excitation frequencies near the second harmonic with the bifurcation sequence: period one period two → two separate tori → torus merging → torus doubling.

O'Reilly and Holmes [6] also reported both experimental and theoretical results for the nonlinear motions of a stretched string. They observed multiple periodic motions, planar and non-planar, as well as quasi-periodic whirling and irregularly processing oscillations. In a more recent work, O'Reilly [24] showed how Silnikov's analysis can be used to predict the existence of chaotic behavior in the averaged equations when the string exhibits non-planar or whirling motion. This analysis requires the existence of a homoclinic orbit which was observed in the numerical work of both Bajaj and Johnson [22] and O'Reilly and Holmes [6].

2.2 Beams

Studies of the nonlinear resonant response of inextensible as well as extensible elastic beams have been widely reported in the literature. The most studied cases are coupled-mode responses with 1:1 and 1:3 internal resonances among the participating modes.

Crespo da Silva and Glynn [25, 26] investigated the nonlinear, non-planar oscillations of inextensional elastic beams. They derived a set of equations of motion for inextensional elastic beams which model the flexure about two principal axes, account for torsion and retain the order-three nonlinear inertia and curvature terms [25]. Crespo da Silva [27] reported that, for a certain range of excitation frequency and for particular values of parameters, no stable steady-state response, neither planar nor non-planar, exists, and the tip of the cantilever moves about in space without reaching a steady-state.

Maewal [28] studied the problem of resonant motions of simply-supported elastic beams whose cross-sections are invariant with respect to planar rotation of ninety degrees. He showed that, for certain values of the excitation frequency, the response of the beam may be chaotically modulated and that one of the Lyapunov exponents is positive for the cases of chaotic responses.

Nayfeh and Pai [29] investigated the planar and the non-planar responses of a fixed-free beam, with a 1:1 internal resonance between in-plane and out-of plane flexural modes, to a principal parametric excitation.

Restuccio, Krousgrill and Bajaj [30] investigated the nonlinear response of a clamped-clamped/sliding inextensional elastic beam subjected to a harmonic axial load. The amplitude equations for the two-mode approximation were analyzed for steady-state and periodic solutions arising from Hopf bifurcations. Depending on the amplitude of excitation, the damping and the ratio of principal flexural rigidities, various qualitatively different frequency response diagrams are found and limit cycles and chaotic motions were found.

2.3 Cylindrical Shells and Rings

Some of the earliest studies on the coupled-mode responses of rings and shells are the works of Evensen [18] and Chen and Babcock [19] who reported experimental observations of coupling and modal interaction between in-plane and out-of-plane modes in a thin elastic ring, and between two flexural modes in a finite circular cylinder. More recently 1:1 internal resonance between the appropriate modes. In more recent years Maewal [31, 32] investigated the two related problems in order to study the influence of gyroscopic forces on nonlinear harmonic oscillations of rotationally symmetric shell structures. The results show the existence of a secondary bifurcation, and the response on the secondary branch is found to be similar to that of standing waves which do not appear in the linear free vibration model of the system.

Maewal [32] showed that the evolution equations of Miles [20, 33, 34] for the amplitudes of the two modes in 1:1 internal resonance also appear in studies of the nonlinear dynamics of axisymmetric elastic shells. Results of numerical integration of the evolution equations for a ring and a cylindrical shell indicate that both of these elastic structures can exhibit chaotically modulated behavior for certain values of damping and excitation frequency. He pointed out that the equations are very similar to those for a spherical pendulum, a stretched elastic string, an elastic beam, and surface waves in a cylindrical container. In fact, they form a two-parameter family of equations valid for any system with O(2) symmetry (Bajaj and Johnson [22]).

Maganty and Bickford [35] derived a set of geometrically nonlinear equations of motion that describe the behavior of a thin circular ring. The resulting equations for free oscillations are analyzed using a single bending mode approximation for both the in-plane and the out-of-plane motions. The results for the resonant case indicate the presence of unsteady oscillations with an exchange of energy between the in-plane and the out-of-plane modes. The equations of free oscillations are then extended to include the effect of the in-plane and the out-of-plane excitations [36]. The response due to the in-plane excitation exhibits unsteady motions with an exchange of energy between the in-plane and the out-of-plane modes.

Nayfeh and Raouf [37] investigated the nonlinear forced response of infinitely long circular cylindrical shells, in the presence of 2:1 (autoparametric) internal resonances between a flexural mode and a breathing mode. A saturation phenomenon was found to exist when the excitation frequency is close to the natural frequency of the breathing mode. Results of numerical investigations showed that Hopf bifurcations occur, yielding amplitude-and phase-modulated motion. The amplitudes and phases experience a cascade of period-doubling bifurcations ending in a chaotic response. Raouf and Nayfeh [38] also studied the case of 1:1 (autoparametric) internal resonance between a flexural mode and its companion mode. They found results similar to those in the case of 2:1 internal resonance. Numerical integration of the amplitude equations in the frequency range between the two Hopf bifurcation points showed two branches of attractors which exhibited types of behavior similar to those found in a string (Johnson and Bajaj [4]). Nayfeh, Raouf and Nayfeh [39] studied the response of infinitely long, circular, cylindrical shells to subharmonic radial excitation, of order one-half, in the presence of a 2:1 internal resonance. Amplitude- and phase-modulated solutions were found to exist. Some limit cycle solutions, corresponding to modulated solutions, underwent symmetry-breaking bifurcations, whereas some others underwent cyclic-fold bifurcations. Some cyclic-folds were also found to result in a transition to chaos.

2.4 Plates

The nonlinear response of thin as well as thick plates has been the subject of extensive studies and many recent reviews exist on the subject [40]. Sridhar, Mook and Nayfeh [8, 9] used the dynamic analogue of the von Karman equations to study the forced response of the plate. They analyzed symmetric as well as asymmetric vibrations, and traveling waves in a clamped circular plate subjected to harmonic excitations, when the frequency of excitation is close to of the natural frequencies. For the symmetric responses in the presence of an internal resonance among the first three modes, when more than one mode is directly excited, the lower modes can dominate the response even when the frequency of the excitation is close to that of the highest mode. When the response is asymmetric, they found that in the absence of internal resonance, or when the frequency of excitation is close to one of the lower frequencies involved in internal resonance, the steady-state response can only be of standing wave form. Hadian and Nayfeh [41] showed that in the case of a symmetric response, a multi-mode motion loses its stability through a Hopf bifurcation, resulting in periodically- or chaotically-modulated motions of the plate.

Yang and Sethna [16] studied nonlinear flexural vibrations of nearly square plates subjected to parametric in-plane excitations. Local bifurcation analysis of the amplitude equations showed that the system is capable of extremely complex standing as well as traveling wave motions including periodic, almost-periodic and chaotic oscillations. These motions were physically interpreted in terms of the rotations of the nodal patterns. A global bifurcation analysis, based on a Melnikov type theory for two degrees-of-freedom

Hamiltonian systems, was also undertaken. It showed the existence of heteroclinic loops which, when they break, lead to Smale horseshoes and chaotic behavior on an extremely long time scale. Yang and Sethna [7] carried out a similar analysis of plate motions with harmonic excitations normal to the midplane of the plates.

Previously, Yasuda and Torii [42] had also studied the response of square membranes to transverse harmonic excitations which can lead to a coupled-mode response arising from 1:1 internal resonance. Following the analytical and experimental work of Yasuda and Asano [43], in which they analytically predicted as well as experimentally observed amplitude-modulated motions, Chang et al. [44] investigated nonlinear flexural vibrations of rectangular plates with uniform stretching, subject to excitations normal to the mid-plane. For low damping levels, the presence of a Hopf bifurcation in the multi-mode response leads to complicated amplitude-modulated dynamics including period-doubling bifurcations, chaos, coexistence of multiple chaotic motions, and crisis, whereby the chaotic attractor suddenly disappears and the plate resumes small amplitude harmonic motions in a single-mode.

3. Resonant Motions of Rectangular Plates with Internal and External Resonances

In this section, results of an investigation into the dynamic response of a rectangular plate to harmonic excitations are presented. The von Karman plate equations, accounting for membrane forces, are first reduced to a set of second-order nonlinear modal equations using the Galerkin procedure. The method of averaging is then utilized to transform the modal equations to a set of first-order ordinary differential equations representing the slow-time evolution of amplitudes of harmonic motion of the interacting modes. These amplitude or averaged equations are a generalization of those that describe the motion of square plates [7] and a membrane [43], and when the additional restriction of circular symmetry is imposed they have arisen in the study of resonant motion of a spherical pendulum [34], a stretched string [4], and forced response of axisymmetric shells [32] and beams [28]. The amplitude equations for the rectangular plate depend on three nonlinear coefficients, in contrast to the two independent nonlinear coefficients found in the above mentioned studies.

The conditions for various external and internal resonances are identified and only the case for primary external resonance with 1:1 internal resonance is studied extensively. The amplitude equations are analyzed for steady-state constant solutions and their various local bifurcations as a function of the excitation frequency, amplitude and modal damping. Dynamic solutions created by local and global bifurcations are studied numerically using AUTO

[45], a bifurcation analysis and two-parameter continuation computer software package, and by direct time integration of the amplitude equations.

3.1 Equations of Motion

Consider a rectangular plate of thickness h, and edge lengths a and b. Let $Oxyz$ be a Cartesian coordinate system with Oxy in the midplane of the plate and the origin at a corner. The plate is subjected to a uniform stretching force N_0 (in x-and y-directions). Under these conditions, the von Karman-type equations of motion for the plate, in nondimensional form, are as follows:

$$\begin{aligned} w_{,u} - \frac{1}{\pi^2}(w_{,xx} + \kappa^2 w_{,yy}) + D(w_{,xxx} + 2\kappa^2 w_{,xxyy} + \kappa^4 w_{,yyyy}) \\ = \varepsilon(F_{,yy}w_{,xx} - 2F_{,xy}w_{,xy} + F_{,xx}w_{,yy}) - cw_{,t} + q\ , \end{aligned} \tag{1}$$

$$F_{,xxxx} + 2\kappa^2 F_{,xxyy} + \kappa^4 F_{,yyyy} = w_{,xy}^2 - w_{,xx}w_{,yy}\ , \tag{2}$$

where $w(x,y,t)$, $F(x,y,t)$ and $q(x,y,t)$ are the nondimensional transverse deflection, the stress function, and the external force normal to the plate, respectively. The dimensionless parameters ε, κ, D and c represent the thickness parameter, the aspect ratio, the ratio of bending stiffness to uniform stretching force and the damping coefficient, respectively. Furthermore, the subscript x, y or t denotes a partial differentiation with respect to that nondimensional variable. In equations (1) and (2) the following transformations for the variables and parameters have been used:

$$\begin{aligned} &x = \frac{\bar{x}}{a},\ y = \frac{\bar{y}}{b},\ \kappa = \frac{a}{b},\ t = \frac{\pi}{a}\sqrt{\frac{N_0}{\rho h}}\,\bar{t},\ w = \frac{\bar{w}}{h},\ c = \frac{a}{\pi}\sqrt{\frac{1}{\rho h N_0}}\,\bar{c}, \\ &q = \frac{a^2}{\pi^2 h N_0}\bar{q},\quad F = \frac{\bar{F}}{Eh^3\kappa^2},\quad \varepsilon = \frac{Eh^3\kappa^2}{N_0\pi^2 b^2},\quad D = \frac{\bar{D}}{\pi^2 N_0 a^2}\ . \end{aligned} \tag{3}$$

Here the variables and parameters with an overbar represent the physical quantities.

The boundary conditions considered here are that all the edges are simply supported and immovable. The transverse displacement w then satisfies

$$w = w_{,xx} = 0 \text{ at } x = 0.1, \text{ and } w = w_{,yy} = 0 \text{ at } y = 0.1\ . \tag{4}$$

The in-plane boundary conditions of $u = v = 0$ along the four sides of the plate, where u and v are the in-plane displacements in the x and y direction, respectively, can only be satisfied by the average [43]. These conditions, expressed in terms of the stress function F, and expressed in nondimensional form, are as follows:

$$\int_0^1\int_0^1 \left(\kappa^4 F_{,yy} - \nu\kappa^2 F_{,xx} - \frac{1}{2}w_{,x}^2\right) dxdy = 0\ ,$$

$$\int_0^1 \int_0^1 \left(F_{,xx} - \nu\kappa^2 F_{,yy} - \frac{1}{2} w_{,y}^2 \right) dxdy = 0 \,, \tag{5}$$

$$\int_0^1 \int_0^1 \left(2(1+\nu)\kappa^2 F_{,xy} + w_{,x} w_{,y} \right) dxdy = 0 \,.$$

where ν is the Poisson's ratio.

To investigate the nonlinear dynamical response of the plate, we use the Galerkin technique. Thus, the transverse deflection w can, in general, be chosen as

$$w(x,y,t) = \sum_m^\infty \sum_n^\infty W_{mn}(t) \phi_m(x) \psi_n(y) \,, \tag{6}$$

where $\phi_m(x)$, $\psi_n(y)$ are groups of comparison functions satisfying the appropriate boundary conditions. For simply supported boundary conditions on the four sides, we can choose

$$\phi_m(x) = \sin m\pi x, \qquad \psi_n(y) = \sin n\pi y \,, \tag{7}$$

which together define the shape of an (m,n) mode of the plate.

Thus, the motion consists of a linear combination of an infinite number of spatial modes. The modal amplitudes W_{mn} are a function of time, and the nonlinear terms in the system determine their time evolution. Substituting equations (6) and (7) into equation (2), the solution for the resulting linear partial differential equation in the stress function F can be written as

$$F(x,y,t) = F^h(x,y,t) + F^p(x,y,t) \,, \tag{8}$$

where F^h is the homogeneous solution that includes the effect of in-plane stretching forces independent of the transverse deflection, and F^p is the particular solution that includes the effect of out-of-plane boundary conditions. The particular solution F^p can be shown to be

$$\begin{aligned} F^p(x,y,t) = \sum_{m,n,r,s}^\infty \big(& a_{mnrs} \cos(m-r)\pi x \cos(n-s)\pi y + \\ & b_{mnrs} \cos(m-r)\pi x \cos(n+s)\pi y + \\ & c_{mnrs} \cos(m+r)\pi x \cos(n-s)\pi y + \\ & d_{mnrs} \cos(m+r)\pi x \cos(n+s)\pi y \big) W_{mn} W_{rs} \,, \end{aligned} \tag{9}$$

where a_{mnrs}, b_{mnrs}, c_{mnrs} and d_{mnrs} are functions of the mode numbers (m,n), (r,s) and of the aspect ratio κ. Their expressions are given in the Appendix. For F^h to satisfy the boundary conditions, equations (5), the homogeneous solution F^h can be assumed to be

$$F^h(x,y,t) = \frac{1}{2}N_{x0}y^2 + \frac{1}{2}N_{y0}x^2 + N_{xy0}xy \ . \tag{10}$$

Substituting $F = F^p + F^h$ into the in-plane boundary conditions, equations (5), and carrying through the algebra, the time dependent functions N_{x0}, N_{y0}, and N_{xy0} turn out to be

$$\begin{aligned} N_{x0} &= \sum_{m,n}^{\infty} S_{x0mn} W_{mn}^2 \ , \qquad N_{y0} = \sum_{m,n}^{\infty} S_{y0mn} W_{mn}^2 \ , \\ N_{xy0} &= \sum_{m,n,r,s}^{\infty} S_{xy0mnrs} W_{mn} W_{rs} \ , \end{aligned} \tag{11}$$

where S_{x0mn}, S_{y0mn} and $S_{xy0mnrs}$ are functions of the mode numbers (m,n), (r,s), the Poisson ratio ν, and the aspect ratio κ. Their expressions are also given in the Appendix.

Substituting the solution for F obtained above, and equations (6) and (7), into equation (1), multiplying by $\sin k\pi x \sin l\pi y$, and integrating over the domain of the plate, we get the following discretized equations of motion:

$$\begin{aligned} \ddot{W}_{kl} + \Omega_{kl}^2 + c\dot{W}_{kl} + \sum_{m,n,r,s,i,j} L_{mnrsijkl} W_{mn} W_{rs} W_{ij} &= q_{kl} \\ (k,l = 1,2,\ldots) \ , \end{aligned} \tag{12}$$

where Ω_{kl} is the nondimensional natural frequency of the (k,l) mode, q_{kl} is the contribution of the transverse excitation q to the (k,l) mode, and $L_{mnrsijkl}$ are coefficients for the nonlinear terms of the (k,l) mode. Their expressions are given in the Appendix. For brevity of expressions, and with an N-mode approximation, we can write equations (12) as

$$\begin{aligned} \ddot{X}_m + \Omega_m^2 X_m + c\dot{X}_m + \sum_{i,j,k} \bar{L}_{ijklm} X_i X_j X_k &= Q_m \cos\omega t, \\ m = 1,2,\ldots,N \ , \end{aligned} \tag{13}$$

where we have introduced harmonic forces, $q_m = Q_m \cos\omega t$, and X_i, $i = 1,2,\ldots,N$ are the amplitudes of the N modes used in the approximation.

3.2 Averaged Equations

When the nonlinear response of equations (13) is small, the method of averaging [1, 14, 46] can be used effectively. The plate response depends critically on the modes involved in internal and external resonances, that is, the response depends on the excitation frequency ω, and the natural frequencies Ω_{kl} or Ω_m. The natural frequencies of the plate, with expressions given in equation (A8), depend on the mode combination, (k,l), the aspect ratio κ, and the nondimensional bending stiffness parameter D. As the aspect ratio and the stiffness parameter are varied, many different mode combinations undergo various internal resonances. Now one would like to show that only a small number of modes, which are in internal/combination resonance with the external excitation frequency, contribute to the response in the first approximation. This was shown by Bajaj and Johnson [22] in the case of a stretched string, and by Holmes [47] in the case of surface waves in a cylindrical container. Similar results can be expected in the case of a rectangular plate, although the analysis is complicated by the fact that the natural frequencies, and the relationships between them, change as a function of the aspect ratio, and hence the N modes that should be considered in the approximation also vary as a function of the aspect ratio. Thus, if we restrict the discussion to a four-mode approximation, say at $\kappa \cong 1.633$, the four lowest relevant modes are the (1,1), (2,1), (1,2) and (3,1) modes of the plate. Note that Ω_{12} and Ω_{31} are nearly equal for $\kappa = 1.633$ so that there is 1:1 internal resonance between the (1,2) and (3,1) modes. An analysis along the line of Bajaj and Johnson [22] then shows that, for weak excitation with primary resonance ($\omega \simeq \Omega_{12} \simeq \Omega_{31}$) the response is essentially (to $\mathrm{O}(\varepsilon)$) determined by the (1,2) and (3,1) modes.

Bearing the aforementioned results in mind, we analyze a two-mode approximation of the plate system ($N = 2$ in equations (13)) where the two modes of interest are in 1:1 internal resonance. The general discussion and results will be valid for all two-mode pairs in 1:1 resonance, irrespective of the aspect ratio κ of the plate. Specific numerical results will be mostly limited to the $\kappa = 1.633$ case.

As for external resonances, there are various possibilities depending on the strength of the external excitation. These include primary resonance ($\omega \approx \Omega_1, \omega \approx \Omega_2$), subharmonic resonances ($\omega \approx 3\Omega_1, \omega \approx 3\Omega_2$), superharmonic resonances ($3\omega \approx \Omega_1, 3\omega \approx \Omega_2$) and various combination resonances:

$$\omega \approx \pm 2\Omega_1 \pm \Omega_2 , \qquad \omega \approx \Omega_1 \pm \Omega_2 , \qquad 2\omega \approx \pm\Omega_1 \pm \Omega_2 . \tag{14}$$

The present work concentrates on the case of primary external resonance with 1:1 internal resonance, and the two-mode approximation can then be written as:

$$\begin{aligned} \ddot{X}_1 + \Omega_1^2 X_1 &= \varepsilon(A_1 X_1^2 + A_2 X_2^2)X_1 - c\dot{X}_1 + Q_1 \cos\omega t , \\ \ddot{X}_2 + \Omega_2^2 X_2 &= \varepsilon(A_2 X_1^2 + A_3 X_2^2)X_2 - c\dot{X}_2 + Q_2 \cos\omega t , \end{aligned} \tag{15}$$

where A_1, A_2 and A_3 are the constant non-linear coefficients determined for the specific mode combinations, and Ω_1 and Ω_2 are the corresponding natural frequencies of the two linear modes. Here X_1 is the amplitude of some (m,n) mode and X_2 is the amplitude of some other (r,s) mode which is in 1:1 internal resonance with the (m,n) mode. The expressions for A_1, A_2 and A_3 in terms of the mode numbers (m,n) and (r,s) and the aspect ratio κ, are already given in the Appendix, and in Chang et al. [44]. Note that these equations possess the reflection symmetry of $Z_2 \times Z_2$ in the absence of external forcing, that is, they remain unchanged by the transformations $(X_1, X_2) \to (-X_1, X_2)$ and $(X_1, X_2) \to (X_1, -X_2)$. The external excitation breaks this symmetry property partially or completely, depending on the amplitudes Q_1 and Q_2.

Let

$$X_i = R_i \cos(\omega t - \gamma_i) = u_i \cos \omega t + v_i \sin \omega t\,, \qquad i = 1, 2\,. \tag{16}$$

Then, by using a variation of constants procedure and the method of averaging [1, 14], and noting that the excitation frequency ω is nearly equal to the two close natural frequencies, equations (15) result in the following averaged equations for the amplitudes R_i and the phases γ_i:

$$\begin{aligned}
\dot{R}_1 &= -\frac{c}{2} R_1 + \frac{Q_1}{2\omega} \sin \gamma_1 + \frac{\varepsilon A_2}{8\omega} R_2^2 R_1 \sin 2(\gamma_1 - \gamma_2)\,,\\
\dot{\gamma}_1 &= \frac{\omega^2 - \Omega_1^2}{2\omega} + \frac{Q_1}{2\omega R_1} \cos \gamma_1 + \frac{3\varepsilon A_1}{8\omega} R_1^2 + \frac{\varepsilon A_2}{8\omega} R_2^2 \left(2 + \cos 2(\gamma_1 - \gamma_2)\right)\,,\\
\dot{R}_2 &= -\frac{c}{2} R_2 + \frac{Q_2}{2\omega} \sin \gamma_2 + \frac{\varepsilon A_2}{8\omega} R_1^2 R_2 \sin 2(\gamma_2 - \gamma_1)\,,\\
\dot{\gamma}_2 &= \frac{\omega^2 - \Omega_2^2}{2\omega} + \frac{Q_2}{2\omega R_2} \cos \gamma_2 + \frac{3\varepsilon A_3}{8\omega} R_2^2 + \frac{\varepsilon A_2}{8\omega} R_1^2 \left(2 + \cos 2(\gamma_2 - \gamma_1)\right)\,.
\end{aligned} \tag{17}$$

These equations were also derived and studied by Yasuda and Asano in [43] for the case of a rectangular membrane. Given a specific value of the aspect ratio κ, and the degeneracy of two specific modes, the plate and the membrane have the same averaged or amplitude equations. The nonlinear coefficients A_1, A_2 and A_3 depend only on the mode combinations, the Poisson's ratio, and the form of the nonlinearity assumed (von Karman-type nonlinearities). The values of the natural frequencies Ω_1 and Ω_2 for the degeneracy of two specific plate modes are however different (equations (A8)) from these for the membrane.

In a general external loading case, the force amplitudes Q_1 and Q_2 are not zero. There can be special situations when one (both) of them is (are) zero depending on the spatial distribution of the loading and the mode numbers in internal resonance. Yasuda and Asano [43] presented results for $Q_1 = Q_2 = 10.0$. Here, we are much more interested in the situation when only one mode is externally excited and the second mode is driven due to its nonlinear coupling to the excited mode. Two such specific cases arise, that is when $Q_1 \neq 0$ and $Q_2 = 0$, or when $Q_1 = 0$ and $Q_2 \neq 0$. Due to the similar nature of the equations for (R_1, γ_1) and for (R_2, γ_2) the analytical expressions for various steady-state constant solutions turn out to be identical except for the role of the nonlinear coefficients A_1 and A_3. In view of the possible bifurcations and stability considerations, however, considerable qualitative as well as quantitative differences in the overall response can arise in the two cases. We describe these in the next section, where a local bifurcation analysis of equations (17) is carried out. In fact, it is shown that the qualitative behavior is strongly dependent on the nonlinear coefficients, and rectangular plates with two interacting modes in 1:1 resonance can be classified based on the nonlinear coefficients.

Finally, it is easy to see that the divergence of the averaged system (17), when expressed in Cartesian form (equations (26)), $\sum_{i=1}^{2}\left(\frac{\partial \dot{u}_i}{\partial u_i}+\frac{\partial \dot{v}_i}{\partial v_i}\right)$, is $-2c$ form which it follows that the volume in (u_1, v_1, u_2, v_2) space contracts and that every solution trajectory must ultimately be confined to a limiting subspace of dimension less than four. Furthermore, equations (17) can be combined to show that

$$\frac{dE}{dt}=-\frac{c}{q}E+\frac{Q_1}{2\omega}\left(\frac{R_1\sin\gamma_1}{E}\right)+\frac{Q_2}{2\omega}\left(\frac{R_2\sin\gamma_2}{E}\right), \tag{18}$$

where $E^2 = R_1^2 + R_2^2$. This has the implicit solution

$$E(t)=E(0)e^{-\frac{c}{2}t}+\int_0^t\left(\frac{Q_1}{2\omega}\left(\frac{R_1\sin\gamma_1}{E}\right)+\frac{Q_2}{2\omega}\left(\frac{R_2\sin\gamma_2}{E}\right)\right)e^{-\frac{c}{2}(t-\tau)}d\tau .$$

Noting that $|R_i \sin\gamma_i/E| \leq 1$, $i = 1, 2$, we obtain the inequality

$$\left|E(t)-E(0)e^{-\frac{c}{2}t}\right|\leq\frac{1}{c\omega}\left(1-e^{-\frac{c}{2}t}\right)(Q_1+Q_2)\,. \tag{19}$$

Thus, the steady-state solution is ultimately $(t \to \infty)$ bounded and confined to a hypersphere of radius $(Q_1 + Q_2)/c\omega$.

3.3 Steady-State Constant Solutions

As already discussed, we emphasize the cases when only one of the two modes is externally excited. First, let us consider the case when $Q_2 = 0$ and $Q_1 \neq 0$. Thus, the (m, n) mode is directly excited by an external harmonic force. There are two types of steady-state constant solutions. One set of solutions is characterized by the fact that $R_2 = 0$, i.e. the indirectly excited mode is absent. Then the only response is in the (m, n) mode with $R_1 \neq 0$ and this is called the single-mode solution. The other class of solutions corresponds to both R_1 and R_2 being nonzero and such motions are called coupled-mode responses. A similar situation exists when the (r, s) mode is directly excited and $Q_1 = 0$.

From equations (17), the steady-state constant solutions for single-mode motions ($\bar{R}_2 = 0$) are determined by

$$\frac{c}{2}\bar{R}_1-\frac{Q_1}{2\omega}\sin\bar{\gamma}_1=0,\qquad \frac{\omega^2-\Omega_1^2}{2\omega}\bar{R}_1+\frac{3\varepsilon A_1}{8\omega}\bar{R}_1^3+\frac{Q_1}{2\omega}\cos\bar{\gamma}_1=0\,, \tag{20}$$

where an overbar indicates the single-mode steady-state solutions. Combining the equations for $\bar{R}_1$ and $\bar{\gamma}_1$ results in the following polynomial in $\bar{R}_1$:

$$\bar{R}_1^6 + \frac{8(\omega^2 - \Omega_1^2)}{3\varepsilon A_1}\bar{R}_1^4 + \frac{16[\omega^2 c^2 + (\omega^2 - \Omega_1^2)^2]}{9\varepsilon^2 A_1^2}\bar{R}_1^2 - \frac{16 Q_1^2}{9\varepsilon^2 A_1^2} = 0\,. \tag{21}$$

Real roots of the equation (21), which is identical to those arising in the primary resonant response of the harmonically excited Duffing equation [1], determine the single-mode steady-state constant solutions.

Differentiating equation (21) with respect to $\bar{R}_1$ and setting $\frac{\partial\omega}{\partial\bar{R}_1} = 0$ gives, the saddle-node bifurcation points [12] or the points of vertical tangency for single-mode steady-state solutions:

$$(\bar{R}_1^2)_{\mathrm{SNS}} = \frac{4}{9\varepsilon A_1}\left[-2(\omega^2 - \Omega_1^2) \pm \sqrt{(\omega^2 - \Omega_1^2)^2 - 3c^2\omega^2}\right]. \tag{22}$$

Here the subscript SNS implies the saddle-node bifurcation for single-mode solutions. We will show later (when the stability of solutions in the single-mode branches is considered) that equation (22) also corresponds to the occurrence of zero eigenvalues.

The problem of finding steady-state constant solutions for the coupled-mode response ($R_1 \neq 0, R_2 \neq 0$) can also be formulated as that of finding the real roots of a polynomial of the 8th order in $\hat{R}_2^2$, where the hat indicates the coupled-mode steady-state solution. Due to its complexity, the polynomial expression in $\hat{R}_2$ has been determined using symbolic algebra programs (e.g. SMP, MACSYMA), and is not shown here. The corresponding expression for the coupled-mode steady-state solution $\hat{R}_1$ is given in terms of $\hat{R}_2$ by

$$\begin{aligned}\hat{R}_1^2 = -2\left[\frac{A_3}{A_2}\hat{R}_2^2 + \frac{4}{3}\frac{(\omega^2 - \Omega_2^2)}{\varepsilon A_2}\right] \pm \\ \pm\sqrt{\left[\frac{A_3}{A_2}\hat{R}_2^2 + \frac{4}{3}\frac{(\omega^2 - \Omega_2^2)}{\varepsilon A_2}\right]^2 - \frac{16}{3}\frac{\omega^2 c^2}{\varepsilon^2 A_2^2}}\,.\end{aligned} \tag{23}$$

When damping is absent, the equation governing the amplitude $\hat{R}_2$ is of the form

$$C_1\hat{R}_2^8 + C_2\hat{R}_2^6 + C_3\hat{R}_2^4 + C_4\hat{R}_2^2 + C_5 = 0\,, \tag{24}$$

where the coefficients of the polynomial are functions of the parameters A_1, A_2, A_3, ω, Ω_1, Ω_2, ε and Q_1. These expressions for coefficients C_i, $i = 1,2,3,4,5$, are given in the work of Chang et al. [44] .

Setting $R_2 = 0$ in equation (23), we can obtain the critical points for the onset of coupled-mode steady-state harmonic response. The condition for the occurrence of pitchfork bifurcation from the single-mode response is

$$\left(\hat{R}_1^2\right)_{\mathrm{PF}} = \frac{4}{3\varepsilon A_2}\left[-2(\omega^2 - \Omega_2^2) \pm \sqrt{(\omega^2 - \Omega_2^2)^2 - 3c^2\omega^2}\right], \tag{25}$$

where PF refers to a pitchfork bifurcation [12]. We will show later that equation (25) also corresponds to the occurrence of a zero eigenvalue.

It is clear form the polynomials (21) and (24) that, given the mode numbers (m, n) and (r, s), and the aspect ratio κ, the number of real solutions of the single-mode and the coupled-mode type depends on the physical parameters Ω_1, Ω_2, c, ω and Q_1. While the condition of $\kappa = 1.633$ fixes the two natural frequencies $\Omega_1 = \Omega_2$, any small deviations form the precise value of the aspect ratio lead to small mistuning in the internally resonant modes and thus $(\Omega_1^2 - \Omega_2^2)$ is an important "internal" mistuning parameter. The other frequency parameter is $(\omega^2 - \Omega_1^2)$ or $(\omega^2 - \Omega_2^2)$ which represents the "external" mistuning. Numerical values of the natural frequencies Ω_1 and Ω_2, as indicated earlier, also depend on the bending stiffness D and the Poisson's ratio ν. The nonlinear coefficients A_1, A_2, and A_3, however, only depend on the Poisson's ratio.

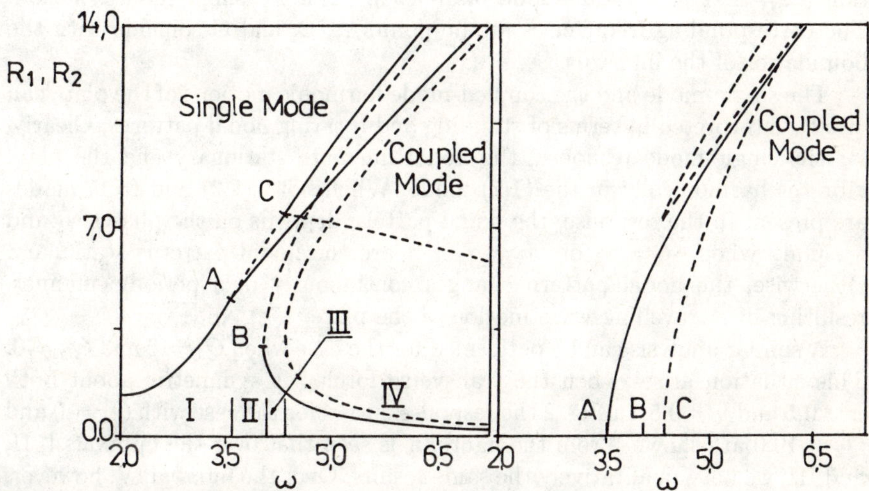

Fig. 1. Constant amplitude-response R_1, for the (1,2) plate mode, and R_2 for the (3,1) plate mode; $Q_1 = 10.0$, $Q_2 = 0.0$, $c = 0.0$

The various single-mode and coupled-mode steady-state constant solutions R_1 and R_2 as a function of the excitation frequency ω are shown in Fig. 1. These response curves are for (1,2) and (3,1) interacting modes with the damping $c = 0.0$, and force amplitudes $Q_1 = 10.0$ and $Q_2 = 0.0$. This situation arises when the loading is symmetric about $x = 0.5$ and is antisymmetric about $y = 0.5$. For all the numerical results presented in this work $\varepsilon = 6 \times 10^{-4}$, $\nu = 0.3$, $\Omega_1^2 = \Omega_2^2 = 35/3$, and $D = 0.0$. The nonlinear coefficients for the (1,2) and (3,1) modes are $A_1 = -326.27$, $A_2 = -274.79$ and $A_3 = -268.32$. The frequency axis is divided into 4 intervals, I, II, III, and IV, according to the nature of the solutions. Over the interval I, only one single-mode solution exists. Over the interval II, we have a stable coupled-mode solution and an unstable single-mode solution. Therefore, in the intervals I and II, the initial conditions are not critical to determining the final steady-state response. In the frequency intervals III and IV, a stable single-mode and a stable coupled-mode solution exists. In the frequency interval IV, two stable single-mode solutions and a stable coupled-mode solution exist. Thus, in intervals III and IV, the initial conditions are very important in determining the final steady-state response reached in any experiment or numerical simulation. Note also, that for every mixed-mode solution for a certain value of γ_2, there is another solution with phase angle $\gamma_2+\pi$ for the same amplitude R_2. Thus, the response curves really represent two coupled-mode solutions which are phase shifted by π radians.

The points A and C in Figure 1 are associated with equation (25), that is, the pitchfork bifurcation points, and the point B is associated with equation (22), that is, a saddle-node bifurcation point for single-mode solution. The corresponding frequencies at the points A, B, and C coincide with the boundaries of the intervals.

The single-mode and the coupled-mode harmonic motions of the plate can also be interpreted in terms of standing and rotating nodal patterns. Clearly, for the single-mode response, the nodal lines are stationary and the plate vibrates harmonically in the (1,2) mode. When both (1,2) and (3,1) modes are present in the response, the nodal pattern depends on the phases γ_1 and γ_2. Only when $\gamma_1 = \gamma_2$ or $\gamma_1 = \gamma_2 \pm \pi$ are the nodal patterns stationary. Otherwise, the nodal pattern changes continuously in a periodic manner, resulting in a travelling wave motion of the plate.

A similar analysis can be performed for the case when $Q_1 = 0$ and $Q_2 \neq 0$. This situation arises when the transverse forcing is symmetric about both $x = 0.5$ and $y = 0.5$. In Fig. 2 the response curves for the case with $Q_1 = 0$ and $Q_2 = 10.0$ are shown. From the figure, it is seen that over the intervals I, II, and III, we have qualitatively the same results. Over the interval IV, however, two solutions exist: one stable single-mode and one stable coupled-mode, whereas, there are two stable single-mode solutions and one stable coupled-mode solution for the case with $Q_1 = 10.0$ and $Q_2 = 0$. This qualitative difference arises because here one of the pitchfork bifurcations form the single-mode solutions occurs in the lower branch (point C), while in the earlier case both the pitchfork bifurcations occur only in the upper branch of the single-mode solutions. As is shown in Section 3.4, this is a consequence of the relative magnitude of the nonlinear coefficients A_i, $i = 1, 2, 3$. Further discussion of the qualitative differences between the responses for the two cases will be given following the stability analysis.

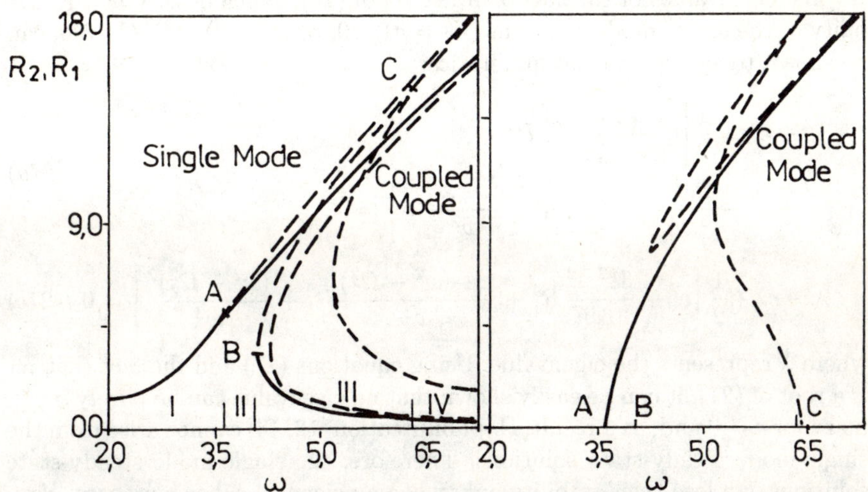

Fig. 2. Constant amplitude-response R_1, for the (1,2) plate mode, and R_2 for the (3,1) plate mode; $Q_1 = 0.0$, $Q_2 = 10.0$, $c = 0.0$

3.4 Stability Analysis of Constant Solutions

Stability analysis of steady-state constant solutions of the averaged equations is most readily accomplished with equations in Cartesian form. The polar form of the averaged equations (17), when transformed to Cartesian variables (u_i, v_i), $i = 1, 2$, are given by

$$
\begin{aligned}
\dot{u}_1 &= -\frac{c}{2}u_1 - \frac{\omega^2 - \Omega_1^2}{2\omega}v_1 - \frac{3\varepsilon A_1}{8\omega}v_1(u_1^2 + v_1^2) \\
&\quad + \frac{\varepsilon A_2}{8\omega}(-v_1 u_2^2 - 3v_1 v_2^2 - 2u_1 u_2 v_2)\,, \\
\dot{v}_1 &= -\frac{c}{2}v_1 + \frac{Q_1}{2\omega} + \frac{\omega^2 - \Omega_1^2}{2\omega}u_1 + \frac{3\varepsilon A_1}{8\omega}u_1(u_1^2 + v_1^2) \\
&\quad + \frac{\varepsilon A_2}{8\omega}(3u_1 u_2^2 + u_1 v_2^2 + 2v_1 u_2 v_2)\,, \\
\dot{u}_2 &= -\frac{c}{2}u_2 - \frac{\omega^2 - \Omega_2^2}{2\omega}v_2 - \frac{3\varepsilon A_3}{8\omega}v_2(u_2^2 + v_2^2) \\
&\quad + \frac{\varepsilon A_2}{8\omega}(-v_2 u_1^2 - 3v_2 v_1^2 - 2u_2 u_1 v_1)\,, \\
\dot{v}_2 &= -\frac{c}{2}v_2 + \frac{Q_2}{2\omega} + \frac{\omega^2 - \Omega_2^2}{2\omega}u_2 + \frac{3\varepsilon A_3}{8\omega}u_2(u_2^2 + v_2^2) \\
&\quad + \frac{\varepsilon A_2}{8\omega}(3u_2 u_1^2 + u_2 v_1^2 + 2v_2 u_1 v_1)\,.
\end{aligned}
\tag{26}
$$

The eigenvalues of the Jacobian matrix of (26), which determine the stability of the single-mode solutions ($u_2 = v_2 = 0$, or $R_2 = 0$) with $Q_2 = 0$, can be shown to satisfy the two quadratics:

$$\lambda^2 + c\lambda + \frac{1}{2}\left[c^2 + \frac{27\varepsilon^2 A_1^2}{16\omega^2}\bar{R}_1^4 + \frac{3\varepsilon A_1(\omega^2 - \Omega_1^2)}{\omega^2}\bar{R}_1^2 + \frac{(\omega^2 - \Omega_1^2)^2}{\omega^2}\right] = 0\,, \tag{27a}$$

$$\lambda^2 + c\lambda + \frac{1}{2}\left[c^2 + \frac{3\varepsilon^2 A_2^2}{16\omega^2}\bar{R}_1^4 + \frac{\varepsilon A_2(\omega^2 - \Omega_2^2)}{\omega^2}\bar{R}_1^2 + \frac{(\omega^2 - \Omega_2^2)^2}{\omega^2}\right] = 0\,, \tag{27b}$$

where λ represents the eigenvalue. Using equations (27) and the fact that $\bar{R}_1$ is a root of (21), it can be easily shown that no eigenvalue can be purely imaginary for $c \neq 0$ and, as a result, Hopf bifurcation [12, 14] cannot arise from the single-mode steady-state solutions. Therefore, the single-mode steady-state solutions can lose their stability only when an eigenvalue becomes zero. Vanishing of the constant term in equation (27a) is really equivalent to equation (22), the condition for a saddle-node bifurcation or a turning point. Similarly, the vanishing of the constant term in equation (27b) is equivalent to equation (25), the condition for a pitchfork bifurcation. It can thus be concluded that the single-mode steady-state constant solutions lose their stability either at the saddle-node bifurcation points or at the pitchfork bifurcation points.

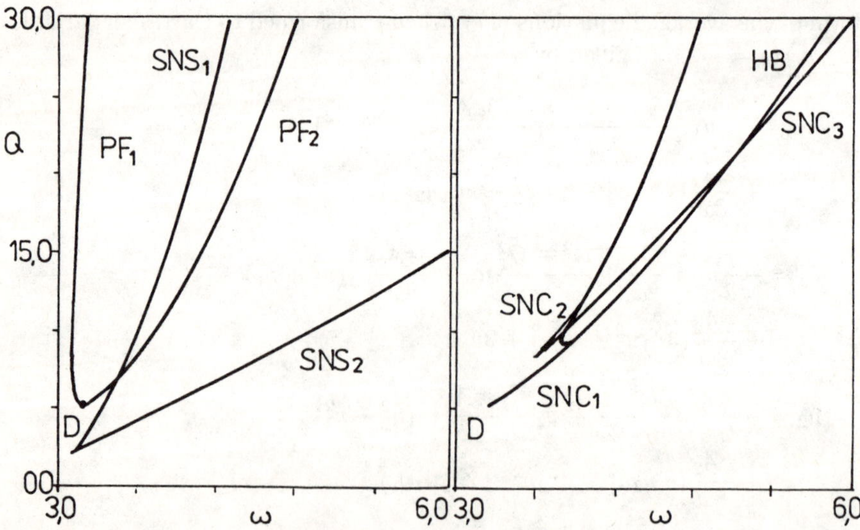

Fig. 3. Saddle-node and pitchfork bifurcation sets for single-mode solutions (a) for ($Q_2 = 0.0$, $c = 0.195$) and saddle-node and Hopf bifurcation sets for coupled-mode solutions (b) for ($Q_2 = 0.0$, $c = 0.195$)

These saddle-node and pitchfork bifurcation sets, for the single-mode solutions, can be obtained in the parameter space by combining equations (21) with the expressions obtained from equations (27). A representative set of these graphs for (1,2) and (3,1) modes are shown in Fig. 3a for $c = 0.195$. Note that as the force amplitude Q_1 is increased for a fixed damping, the single-mode solution first develops multiplicity and only then pitchfork bifurcations arise. This can also be shown to be the case by careful examination of the constant terms in equations (27a) and (27b).

The geometry of solutions in the phase space (u_1, v_1, u_2, v_2) is quite interesting. First note that $Q_2 = 0$ implies that the (u_1, v_1) surface, that is, $(u_2, v_2) = (0.0, 0.0)$ is an invariant of the vector field. If initial conditions are chosen in the (u_1, v_1) plane, the motion governed by solutions of equations (26) remains confined to it, that is, the dynamics of the plate is a single-mode motion. For single-mode constant solutions, the instability boundary defined by equation (27a) corresponds to disturbances restricted to the (u_1, v_1) plane. The instability condition from equation (27b) only arises when disturbances outside the (u_1, v_1) plane are allowed for. Thus, pitchfork bifurcation from single-mode to coupled-mode constant solutions only arises due to coupled-mode disturbances.

A similar stability analysis can be carried out for the coupled-mode steady-state constant solutions. Now both zero and purely imaginary pairs of eigenvalues are possible criteria for the loss of stability. A zero eigenvalue can lead to a saddle-node bifurcation and the associated multiple coupled-mode responses, whereas, a purely imaginary eigenvalue leads to Hopf bifurcation and the possibility of limit cycle solutions [12], [14] for the amplitude equations. Pitchfork bifurcation points are found to arise only at the points where the coupled-mode solutions meet the single-mode solutions and this set is already identified above. The saddle-node and the Hopf bifurcation sets for the coupled-mode responses were obtained using AUTO [45] (see Fig. 3b). Among the points at which the various bifurcation sets intersect (see both Fig. 3a and 3b) only the point D has special significance since it corresponds to a double-zero eigenvalue and is therefore a codimension-2 point [12, 14]. More complicated bifurcation phenomena are expected for values of parameters near the codimension-2 point and results of analytical investigations will be reported in a future piece of work.

The system response depends on four parameters Q_1, ω, c and $(\Omega_1^2 - \Omega_2^2)$. The bifurcation sets shown in Fig. 3 correspond to zero internal mistuning $(\Omega_1^2 = \Omega_2^2)$ and a fixed value of damping. The parameters Q_1 and c play opposite roles and, in fact, Q_1 can be eliminated by an additional scaling. It is therefore expected and indeed can be seen that the bifurcation sets at other damping values are qualitatively similar to the ones shown here. Though physically more realistic, we have not yet studied the case of nonzero internal mistuning in sufficient detail. One can, however, clearly see from equations (27a) and (27b) that the internal mistuning only effects the pitchfork bifurcation points where coupled-mode solutions arise from single-mode solutions. The locations of these points controls the overall coupled-mode dynamics.

In Figs. 3a and 3b it is shown that, beginning with very small values of Q_1, as the amplitude of excitation is slowly raised, the plate response undergoes interesting and significant qualitative changes. Fig. 4 is a series of bifurcation diagrams depicting these changes with u_1 as a function of the excitation frequency ω. For small forcing amplitudes, the response is harmonic and single-valued, that is, for each forcing frequency the plate undergoes a unique harmonic motion in the (1,2) mode (Fig. 4a). At force levels above the cusp point on the SNS curve, the single-mode response undergoes saddle-node bifurcations and now three single-mode responses exist between the frequency boundaries SNS_1 and SNS_2 (Fig. 4b). The upper and the lower solution branches are stable whereas the middle branch is unstable. These saddle-node bifurcations result in the familiar "jump" phenomenon in single-mode response. The next qualitative change occurs when the pitchfork bifurcation set appears. For a very small interval of values of Q_1, the pitchfork bifurcations, which occur in the upper single-mode branch, are supercritical and all the coupled-mode motions are stable. Above the codimension-2 point (point D), the pitchfork bifurcation from the right boundary, PF_2, becomes subcritical with two possible coupled-mode motions now existing between the curves PF_2 and SNC_1 (Fig. 4c). The subcritical branch is unstable and is of the saddle-type with one real positive eigenvalue. A further increase the forcing amplitude results in two additional turning points in the coupled-mode branch, SNC_2 and SNC_3, so that two stable coupled-mode motions are possible. One of the coupled-mode solutions then develops Hopf bifurcation points that asymptotically approach the saddle-node bifurcation points SNC_2 and SNC_1 as Q_1 becomes large. Examples of such response curves are shown in Fig. 4d. Over the frequency interval bounded by the two branches of the Hopf bifurcation set, it is expected, from the Hopf bifurcation theorem [12, 14], that the amplitude equations will possess limit cycle solutions. These solutions will be explored in some detail in the next section.

A similar stability analysis can be performed for the case of $Q_1 = 0$ and $Q_2 \neq 0$. Analytical expressions for the results are not given here, but the corresponding bifurcation sets are shown in Figures 5a and 5b. The set in Figure 5a is for the single-mode branch, now in the plane defined by (u_2, v_2). In Fig. 5b the bifurcation sets for the coupled-motions is given. There are many qualitatively distinct response diagrams determined by the forcing amplitude Q_2. The most significant difference from the case where the (1,2) mode is excited occurs is caused by the presence of a codimension-2 point, identified as E in Fig. 5a and b. At this point, the saddle-node, the pitchfork, and the Hopf bifurcation sets meet. In fact the saddle-node and the pitchfork bifurcation sets are tangent without crossing each other. As the forcing amplitude Q_2 is increased this allows for one of the pitchfork points

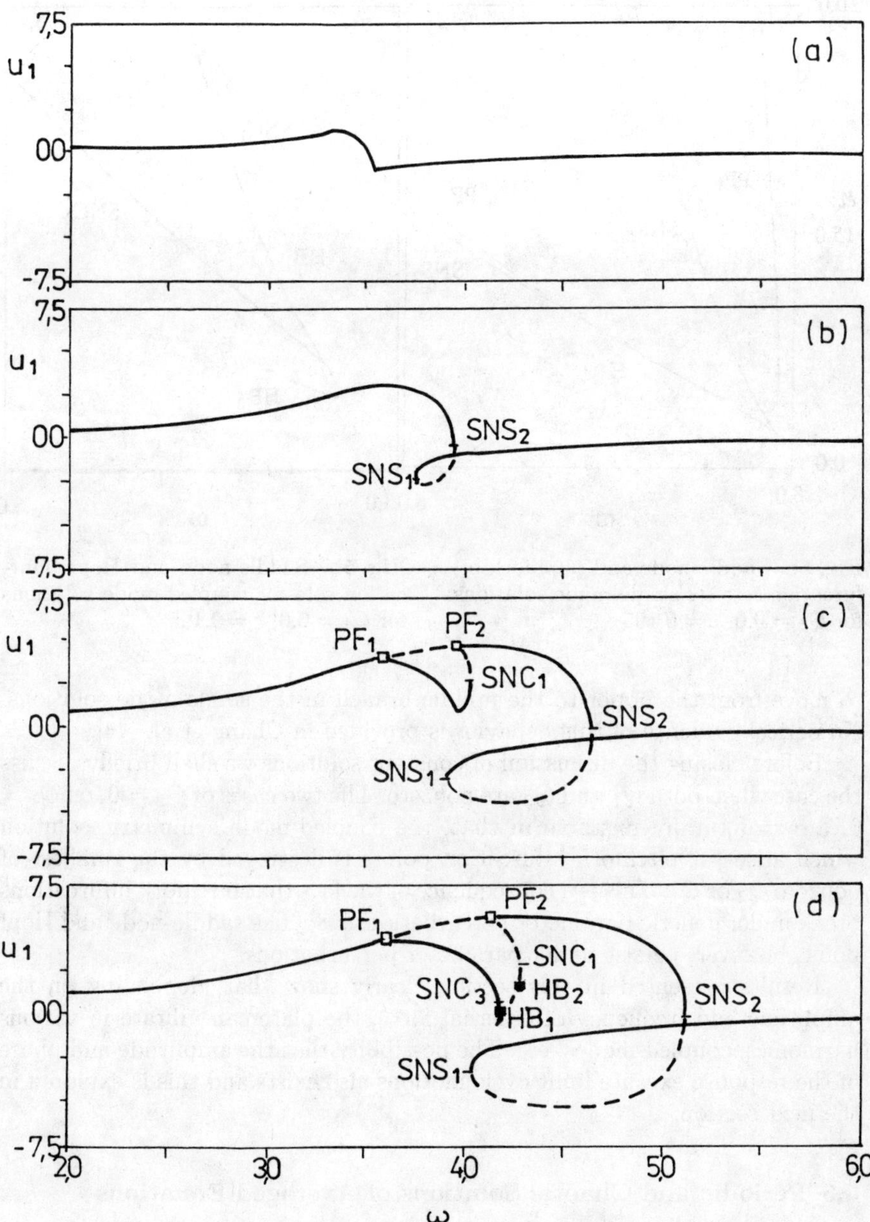

Fig. 4. Bifurcation response diagrams at various force levels; $c = 0.195$, $Q_2 = 0.0$. a) $Q_1 = 1.5$, b) $Q_1 = 4.0$, c) $Q_1 = 7.5$, d) $Q_1 = 10.0$

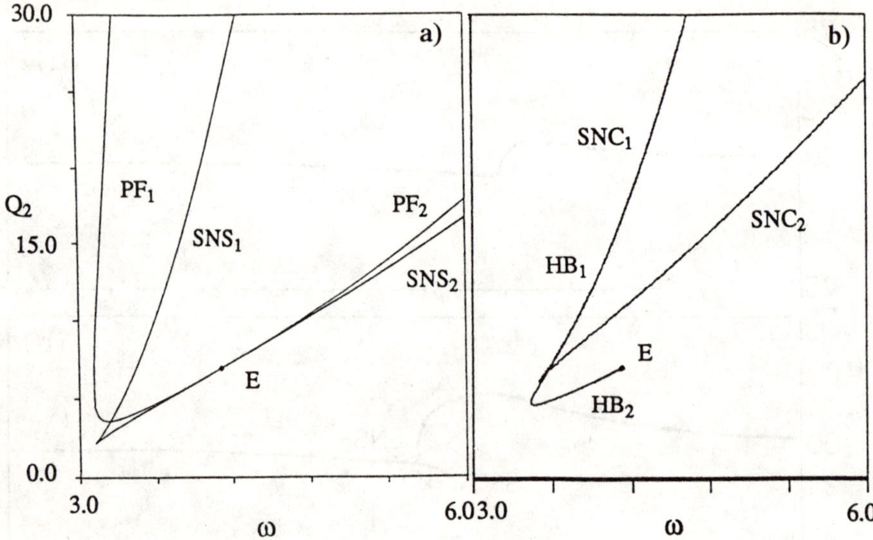

Fig. 5a. Saddle-node and pitchfork bifurcation sets for single-mode solutions for $Q_1 = 0.0$, $c = 0.195$

Fig. 5b. Saddle-node and Hopf bifurcation sets for coupled-mode solutions for $Q_1 = 0.0$, $c = 0.195$

to move from the upper to the middle branch in the single-mode solutions. Numerical evidence of this behavior is provided in Chang et al. [44].

Before closing the discussion of constant solutions we shall briefly discuss the case when both Q_1 and Q_2 are nonzero. The two cases of $Q_2 = 0$, on $Q_1 = 0$, are structurally unstable in that, the coupled-mode symmetric solution which arises at pitchfork bifurcation points is destroyed by the smallest of nonzero Q_2 or Q_1. This is a consequence of the fact that pitchfork bifurcations break under generic parameter perturbations [48]. The saddle-node and Hopf point, however, persist under parameter perturbations.

Results presented in this section clearly show that, depending on the amplitude and frequency of external force, the plate can vibrate in various harmonics, coupled-modes, etc. The possibility that the amplitude and phase of the response execute limit cycle motions also exists and this is explored in the next section.

3.5 Periodic and Chaotic Solutions of Averaged Equations

A numerical study of periodic solutions of the averaged equations has been performed using direct time integration as well as AUTO [45]. As in the previous section, we present the results for the cases when (i) $Q_1 \neq 0$, $Q_2 = 0$, (ii) $Q_1 = 0$, $Q_2 \neq 0$, and (iii) $Q_1 \neq 0$, $Q_2 \neq 0$, separately. Each of the three cases exhibits qualitatively different behavior as described below.

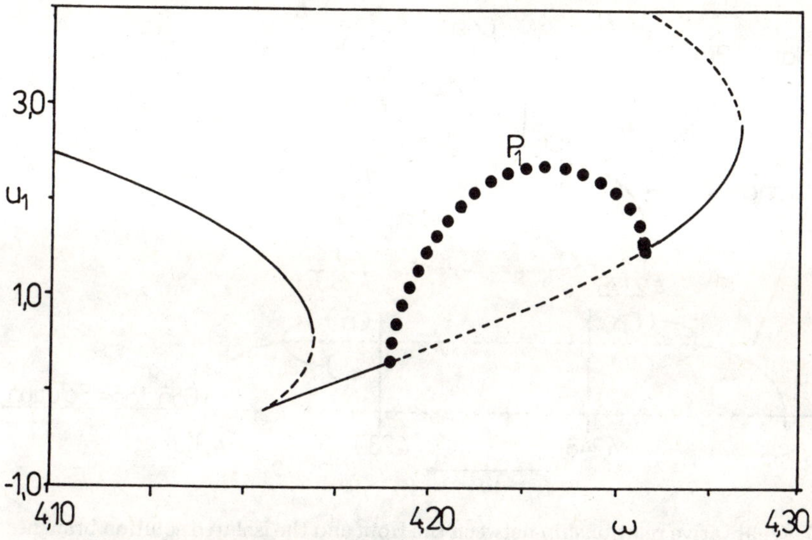

Fig. 6. Response amplitude for the limit cycle solution for u_1 as a function of excitation frequency; $Q_1 = 10.0$, $Q_2 = 0.0$, $c = 0.20$

(i) $Q_1 \neq 0$, $Q_2 = 0$: For sufficiently low Q_1, the response of the averaged equations is limited to equilibrium points in the directly excited (1,2) mode. For higher levels of Q_1, however, the (3,1) mode also contributes to the response. As the excitation Q_1 increases further, some of the coupled-mode steady-state constant solutions lose stability due to Hopf bifurcation and the averaged system develops periodic solutions from the Hopf bifurcation points. These periodic solutions, denoted as P_1 solutions, correspond to amplitude- and phase-modulated motions of the rectangular plate and result in a slow oscillation of the nodal pattern. In Fig. 6, the solutions of the averaged equations (26) for $c = 0.20$ are shown. The P_1 solutions are stable (denoted by solid circles) over the whole frequency interval connecting the two Hopf points. With a further increase in Q_1, these P_1 solutions become unstable via period-doubling bifurcations and develop P_2 solutions. At a certain value of Q_1, a cascade of period-doubling arises leading to chaotic solutions.

While numerically investigating the Hopf solution branch, a new periodic solution branch was discovered. This branch of periodic solutions arises due to a saddle-node bifurcation with periodic solutions as the primary solution. That is, a stable and an unstable limit cycle arise due to a saddle-node bifurcation for a sufficiently low level of damping and the branch exists over a small frequency interval. As the damping c is reduced, the stable periodic solution branch undergoes a sequence of period-doubling bifurcations which ultimately lead to chaotic attractors. For $c = 0.19$, $Q_1 = 10.0$, the isolated branch arises at $\omega \simeq 4.238$, goes through bifurcations and ultimately terminates at $\omega \simeq 4.291$. In Fig. 7 the qualitative relationship between the isolated

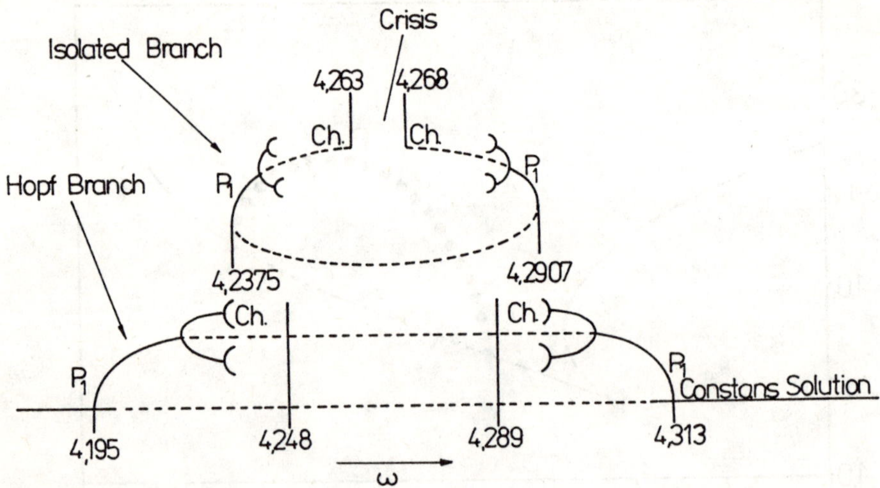

Fig. 7. Qualitative relationship between the Hopf and the isolated solution branches; $Q_1 = 10.0$, $Q_2 = 0.0$, $c = 0.19$

branch and the branch originating at Hopf points, $\omega \simeq 4.195$ and $\omega \simeq 4.313$ is shown. Over the frequency intervals (4.2375, 4.248) and (4.289, 4.2907), stable steady-state solutions are found to exist in both the branches, and phase plots of some representative solutions are shown in Fig. 8. The chaotic solutions in the isolated branch are found to undergo 'boundary crisis' [22], [49], at $\omega \simeq 4.263$ and 4.268 (see Fig. 9), where the chaotic attractor touches the stable manifold of the saddle-type coupled-mode equilibrium point (denoted by CM) and ceases to exist. Near the above listed frequencies, the averaged equations exhibit transient chaos where the solution, when initiated in the neighborhood of the chaotic solution, traces the ghost of the previous attractor for some time and is then quickly attracted by the single-mode constant solution (SM).

The fact that an isolated solutions branch exists can also be verified using tools of numerical bifurcation analysis. AUTO [45] is one of the powerful packages available for bifurcation analysis and continuation of solutions for ordinary differential equations. It can also compute periodic solution branches, given approximate string points, and can help construct 'saddle-node' bifurcations sets in two parameter space. Numerical results for the continuation of periodic solutions starting at the two Hopf points are shown in Fig. 10 (for $c = 0.18755$). Four turning points are found in each of the curves starting from the left and the right Hopf points. These points are identified by numbers 1-4 and 5-8, respectively. As the frequency ω is varied the turning points 2, 4, 6 and 8 correspond to the locations where the isolated branches are created, whereas, the points 1, 3, 5 and 7 correspond to frequencies where they merge with other periodic solution branches. Thus, as damping is increased

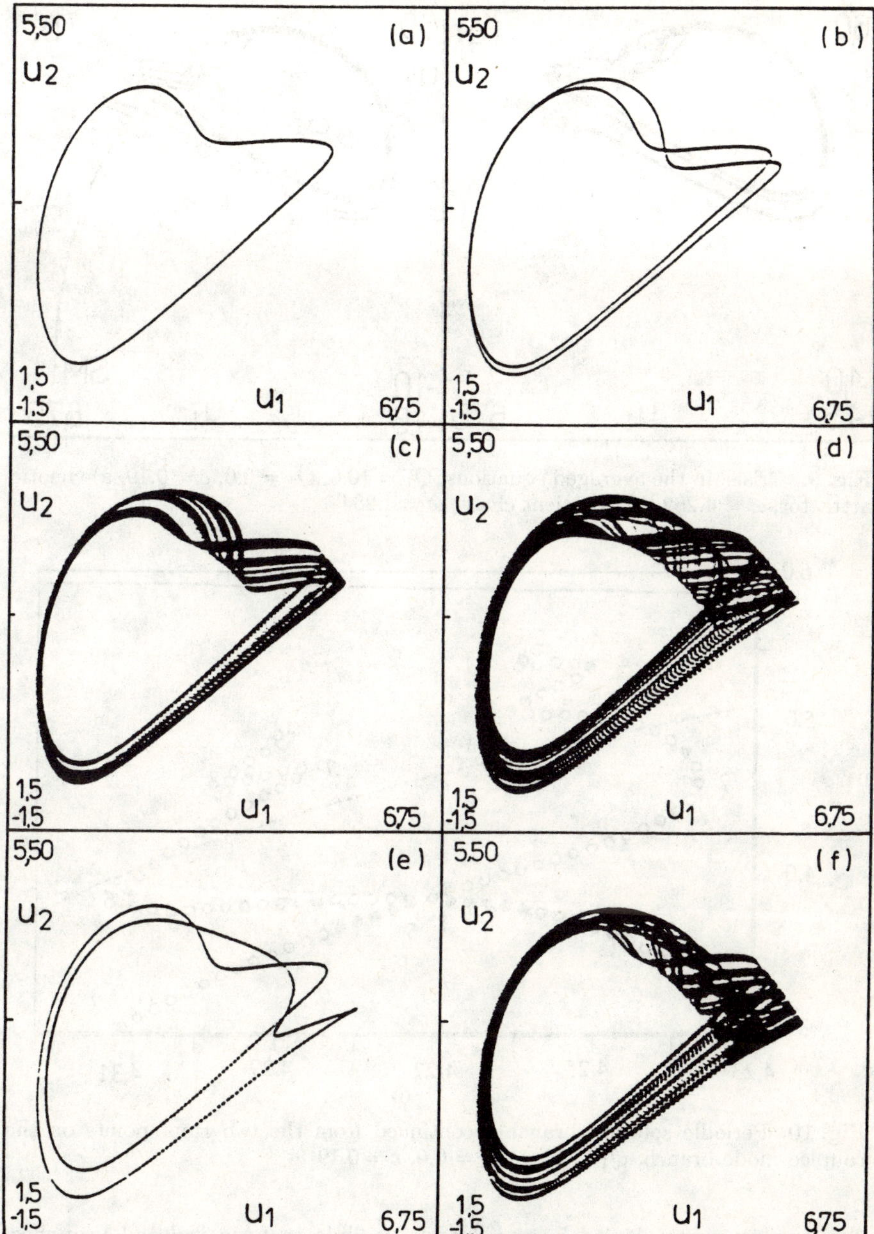

Fig. 8. Phase plot for the steady-state solutions in the isolated branch; $Q_1 = 10.0$, $Q_2 = 0.0$, $c = 0.19$. a) $\omega = 4.238(\mathrm{P}_1)$, b) $\omega = 4.241(\mathrm{P}_2)$, c) $\omega = 4.243$(Ch.), d) $\omega = 4.2485$(Ch.), e) $\omega = 4.249(\mathrm{P}_3)$, f) $\omega = 4.255$ (Ch.)

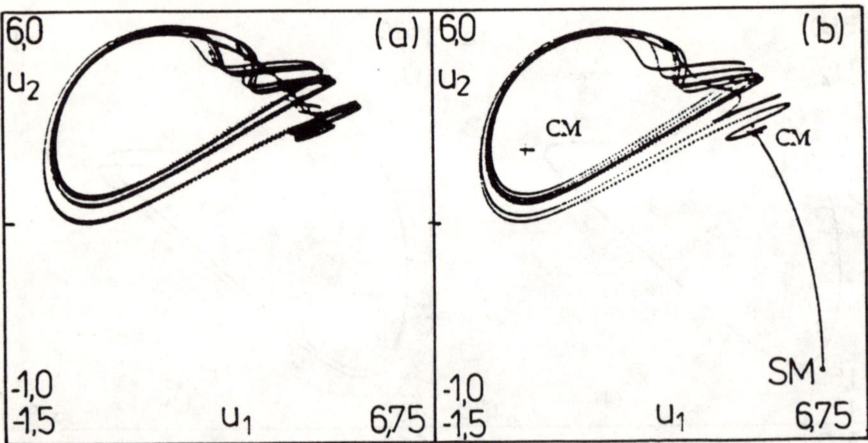

Fig. 9. 'Crisis' in the averaged equations; $Q_1 = 10.0$, $Q_2 = 0.0$, $c = 0.19$. a) chaotic attractor, $\omega = 4.262$, b) transient chaos, $\omega = 4.264$

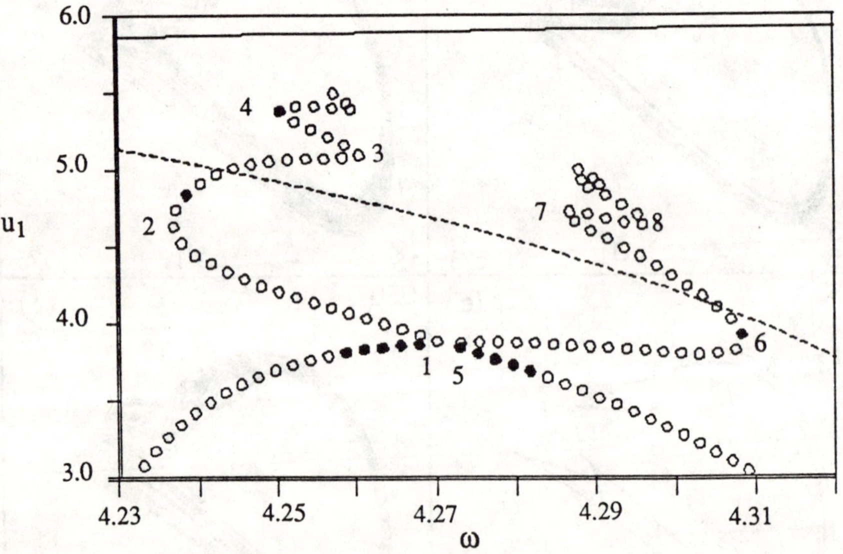

Fig. 10. Periodic solution branches continued from the two Hopf points on the coupled-mode branch; $Q_1 = 10.0$, $Q_2 = 0.0$, $c = 0.19$

the turning points 1 and 5, and 3 and 7 collide to form isolated branches. This bubble structure is typical of the transition to chaotic behavior observed in various dynamical systems [22, 50].

In Figure 11 the saddle-node bifurcation sets for the isolated periodic solution branches corresponding to the points 1-8 in Figure 10 are shown. For damping $c > 0.193$, no isolated branch exists and numerical simulations show

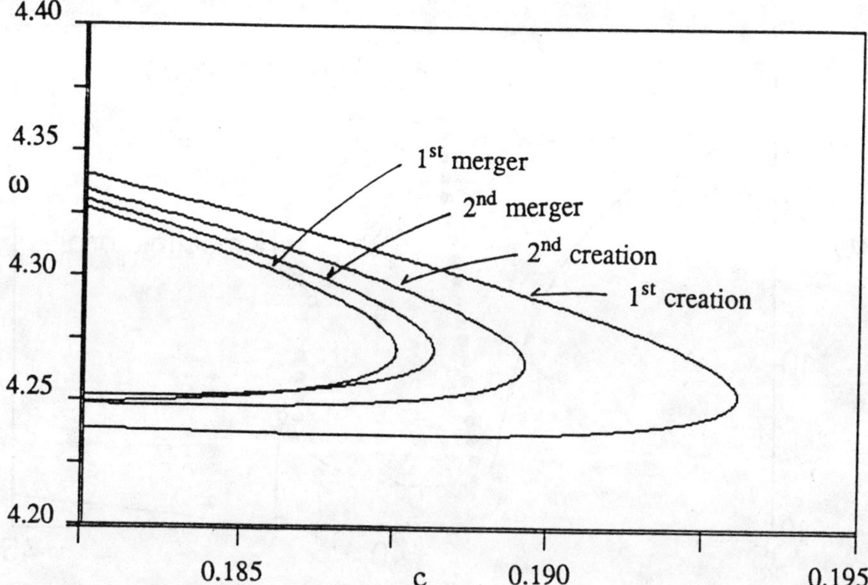

Fig. 11. Saddle-node bifurcation sets for the first and second isolated periodic solution branches in $(c-\omega)$ plane; $Q_1 = 10.0$, $Q_2 = 0.0$

that there are chaotic solutions in the Hopf branch. The set now confirms that at $c = 0.19$ (corresponding to the qualitative diagram in Fig. 7), the isolated branch has not yet merged with the Hopf branch. In fact, the bifurcation sets indicate that on lowering the damping further, another isolated branch is created which merges with the first isolated branch before the merging with the Hopf branch takes place. Thus, the cascade of isolated branch creations and mergers is quite complex.

Before closing this discussion let us point out that, because of the symmetry inherent in the system when $Q_2 = 0$, there is another image branch of coupled-mode solutions in which the solutions undergo an identical evolution as the system parameters are varied. As is shown in the next section, the response exhibited by the averaged equation in the case of $Q_1 = 0$ is quite different from the one presented here.

ii) $Q_1 = 0$, $Q_2 \neq 0$: The bifurcation sets for the single-mode and coupled-mode solutions are, for this case, shown in Fig. 5. For a fixed damping ($c = 0.195$), the Hopf unstable region in the coupled-mode branch only arises when $Q_2 \geq 4.5$. For values of Q_2 slightly above $Q_2 = 4.5$, there are two Hopf points in the solution branch, the bifurcating limit cycles (P_1 solutions) are found to be supercritical and the P_1 solutions join the two Hopf points. This behavior is very similar to the one observed in case (i) above. Note now that the averaged equations (26) with $Q_1 = 0$ enjoy symmetry under the transformation $(u_1, v_1, u_2, v_2) \to (-u_1, -v_1, u_2, v_2)$ and thus the coupled-

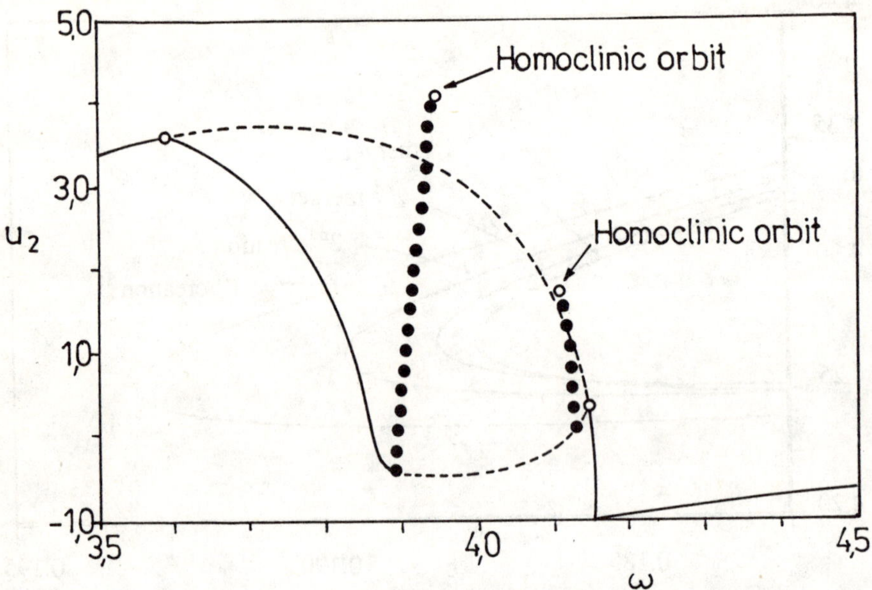

Fig. 12. Response amplitude u_2 as a function of q excitation frequency; $Q_1 = 0.0$, $Q_2 = 5.5$, $c = 0.195$

mode solutions exist in pairs or are themselves symmetric about the invariant (u_2, v_2) plane. There are two identical Hopf branches.

For higher force (Q_2), the P_1 solution branch, instead of undergoing a period-doubling bifurcation, as is the behavior in case (i), develops homoclinic orbits and AUTO is unable to continue periodic solutions beyond those points. In Fig. 12 the response curves for $Q_2 = 5.5$ are shown. In the frequency interval (3.946 – 4.106), no results for periodic solutions are found.

iii) $Q_1 \neq 0$, $Q_2 \neq 0$: The response amplitudes of periodic solutions bifurcating from the Hopf points are shown in Fig. 13. In the isolated solutions branch, periodic solutions could only be continued from the left Hopf point and the solutions approach a homoclinic orbit via a series of turning points. In the primary solutions branch, however, the Hopf points are joined by the limit cycle solutions. The limit cycles are unstable over a frequency interval due to period-doubling, and ultimately lead to the usual chaotic solutions. Thus, quite interestingly, the isolated branch exhibits a response similar to the case (ii) above, when the (3,1) mode is directly excited. The primary branch exhibits behavior similar to the case (i) above where the (1,2) mode is directly excited.

Before closing the discussion of limit cycles and chaotic solutions exhibited by the averaged equations for a two-mode approximation of the von Karman plate equations, it should be pointed out that the limit cycle solutions of the averaged equations (17) or (26) imply motion on a two-torus

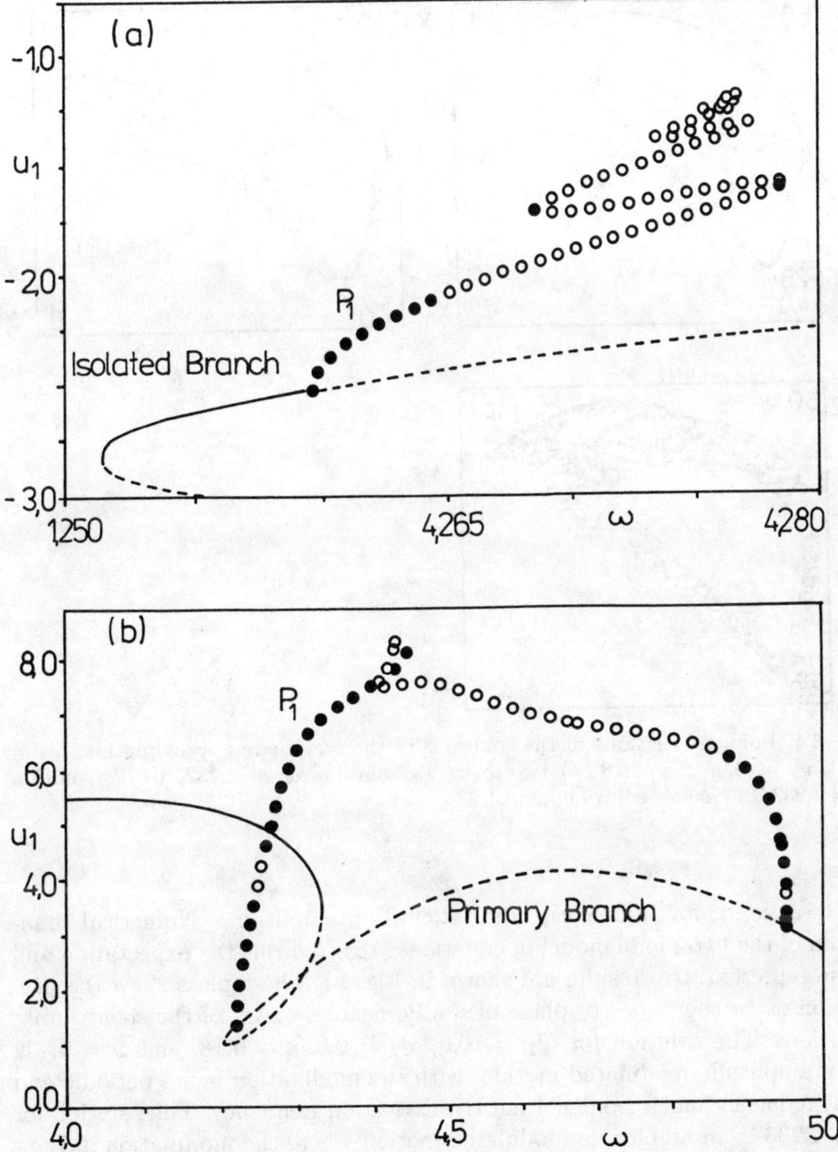

Fig. 13. The periodic solution branches; $Q_1 = 10.0$, $Q_2 = 5.0$, $c = 0.19$. a) isolated branch, b) primary branch

for the coupled oscillators (equations (15)) via the integral manifold theorem (see Bajaj and Johnson [22]). For parameter values close to those for which the averaged equations exhibit chaotic motions, it is expected [22] that, the coupled oscillators also exhibit chaotic behavior. This conclusion should be

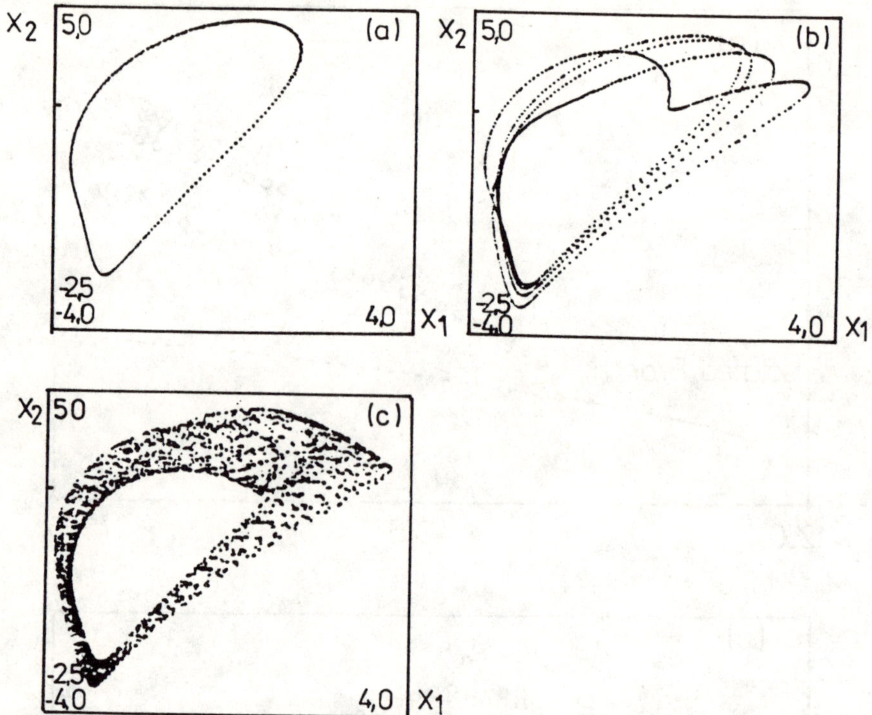

Fig. 14. Poincaré sections of the solutions of the two-mode approximation; $Q_1 = 10.0$, $q_2 = 0.0$, $c = 0.18$. a) two-torus T_1 solution, $\omega = 4.232$, b) T_4 solution $\omega = 4.233$, c) chaotic solution, $\omega = 4.234$

valid, at least, for sufficiently small excitation amplitudes. Numerical simulation of the two-mode model in equations (15) confirms the expectation and some representative results are shown in Fig. 14. These plots show the projection on to the $X_1 - X_2$ plane of the Poincaré sections of the steady-state solutions. The solution for $Q_1 = 10.0$, $Q_2 = 0.0$, $c = 0.18$, and $\omega = 4.232$ is an amplitude-modulated motion, with the modulation being periodic and of a frequency much smaller than the excitation frequency. The solution for $\omega = 4.233$ is an amplitude-modulated motion where the modulation has undergone a period-doubling twice, resulting in a so-called T_4 solution. The section in Fig. 14 represents the solution for $\omega = 4.234$ where the regular torus has finally cascaded to the chaotic attractor. Again, the chaotic features arise in the modulation of the basic harmonic motion.

4. Summary and Conclusions

In this work we have reviewed recent literature on the resonant response of weakly nonlinear multi-degree-of-freedom models of structural systems. The focus was placed on conditions under which the structures possess internal resonances, so that more than one mode of vibration participates in the response. The structural members considered include beams, strings, and plates and shells. The latter part of the work is devoted to a study of the N-mode approximation of a resonantly excited thin rectangular plate with uniform in-plane tension. The plate is modelled by the von Karman equations, and it is shown that for the N-mode model, only the resonantly excited mode and the mode in 1:1 internal resonance have a non-zero amplitude in the lowest order approximation.

A careful bifurcation analysis of the averaged equations is carried out as a function of the excitation amplitudes and frequency, and as a function of the damping present in the plate. Various saddle-node, pitchfork, and Hopf bifurcation sets are constructed and it is shown that the response of the plate depends very significantly on the mode which is directly excited. In the parameter regions, where the coupled-mode constant solutions are unstable due to a Hopf bifurcation, the steady-state solutions are explored by continuation of periodic solutions and by using direct time integration. The limit cycle, as well as chaotic solutions of the averaged equations are shown to predict qualitatively similar amplitude-modulated motions for the original two-mode model for neighboring values of parameter.

Acknowledgements. The authors wish to acknowledge the financial support provided by the U.S. Army Research Office through the grant DAAL 03–90–G–0220. Dr. Gary L. Anderson is the technical monitor.

References

1. A.H. Nayfeh, D.T. Mook: Nonlinear oscillations. Wiley-Interscience, New York 1979
2. A.H. Nayfeh, B. Balachandran: Modal interactions in dynamical and structural systems. App. Mech. Rev., **42**, 5175–5201 (1989)
3. P.R. Sethna: Coupling in certain classes of weakly nonlinear vibrating systems. In Nonlinear Differential Equations and Nonlinear Mechanics, edited by J.P. Lasalle and S. Lefschetz, Academic Press, New York, 58–70 (1963)
4. J.M. Johnson A.K. Bajaj: Amplitude modulated and chaotic dynamics in resonant motion of strings. J.Sound and Vib., **128**, 87–107 (1989)
5. N.B. Tufillaro: Nonlinear and chaotic string vibrarions. Am. J. Phys., **57**, 408–414 (1989)
6. O. O'Reilly, P.J. Holmes: Non-linear, non-planar and non-periodic vibrations of a string. J. Sound and Vib., **153**, 413–435 (1992)
7. X.L. Yang P.R. Sethna: Nonlinear phenomena in forced vibrations of a nearly square plate – antisymmetric case. J. Sound and Vib., **155**, 413–441 (1992)

8. S. Sridhar, D.T. Mook, A.H. Nayfeh: Nonlinear resonances in the forced responses of plates. Part I: Symmetric responses of circular plates. J. Sound and Vib., **41**, 359–373 (1975)
9. S. Sridhar, D.T. Mook, A.H. Nayfeh: Nonlinear resonances in the forced responses of plates. Part II: Asymmetric responses of circular plates. J. Sound and Vib., **59**, 159–170 (1978)
10. P.F. Pai, A.H. Nayfeh: Nonlinear non-planar oscillations of a cantilever beam under lateral base excitations. Int. J. Non-Lin. Mech., **25**, 455–474 (1990)
11. P.J. Holmes, J. Marsden: A partial differential equation with infinitely many periodic orbits: chaotic oscillations of a forced beam. Arch. Rat. Mech. Anal., **76**, 135–165 (1981)
12. J. Guckenheimer, P.J. Holmes: Nonlinear oscillations, dynamical systems, and bifurcations of vector fields. Springer-Verlag, New York 1983
13. S. Wiggins: Global bifurcations and chaos – analytical methods. Springer-Verlag, New York 1988
14. S. Wiggins: Introduction to applied nonlinear dynamical systems and chaos. Springer-Verlag, New York 1991
15. K. Yagasaki: Chaotic dynamics of a quasi-periodically forced beam. ASME J. Appl. Mech., **59**, 161–167 (1992)
16. X.L. Yang, P.R. Sethna: Local and global bifurcations in parametrically excited vibrations of nearly square plates. Int. J. Non-Lin. Mech., **26**, 199–220 (1991)
17. J.W. Miles: Stability of forced oscillations of a vibrating string. J. Acous. Soc. Am., **38**, 855–861 (1965)
18. J.C. Chen, C.D. Babcock: Nonlinear vibration of cylindrical shells. AIAA J., **13**, 868–876 (1975)
19. D.A. Evensen: Nonlinear flexural vibrations of thin circular rings. ASME J. Appl. Mech., **33**, 553–560 (1966)
20. J.W. Miles: Resonant, nonplanar motion of a stretched string. J. Acous. Soc. Am. **75**, 1505–1510 (1984)
21. A.K. Bajaj J.M. Johnson: Asymptotic techniques and complex dynamics in weakly non-linear forced mechanical systems. Int. J. Non-Lin. Mech., **25**, 211–226 (1990)
22. A.K. Bajaj, J.M. Johnson: On the amplitude dynamics and 'crisis' in resonant motion of stretched strings. Phil. Trans. Soc. Lond., **A338**, 1–41 (1992)
23. T.C.A. Molteno, N.B. Tufillaro: Torus doubling and chaotic string vibrations: experimental results. J. Sound and Vib., **137**, 327–330 (1990)
24. O. O'Reilly: Global bifurcations in the forced vibration of a damped string. Preprint (1991)
25. M.R.M. Crespo da Silva, C.C. Glynn: Nonlinear flexural-flexural-torsional dynamics of inextensional beams. I. Equation of motion. J. Struct. Mech., **6**, 437–448 (1978)
26. M.R.M. Crespo da Silva, C.C. Glyn: Nonlinear flexural-flexural-torsional dynamics of inextensional beams. II. Forced motions. J. Struct. Mech., **6**, 449–461 (1978)
27. M.R.M. Crespo da Silva: On the whirling of a base-excited cantilever beam. J. Acous. Soc. Am., **67**, 704–707 (1980)
28. A. Meawal: Chaos in harmonically excited elastic beam. ASME J. Appl. Mech., **53**, 625–632 (1986)
29. A.H. Nayfeh, P.F. Pai: Non-linear non-planar parametric responses of an inextensional beam. Int. J. Non-Lin. Mech., **24**, 139–158 (1989)
30. J.M. Restuccio, C.M. Krousgrill, A.K. Bajaj: Nonlinear nonplanar dynamcs of a parametrically excited inextensional elastic beam. Nonlinear Dynamics, **2**, 263–289 (1991)

31. A. Maewal: Nonlinear harmonic oscillations of gyroscopic structural systems and the case of a rotating ring. ASME J. App. Mech., **48**, 627–633 (1981)
32. A. Maewal: Miles' evolution equations for axisymmetric shells: simple strange attractor's in structural dynamics. Int. J. Non-Lin. Mech., **21**, 433–438 (1987)
33. J.W. Miles: Resonantly forced surface waves in a circular cylinder. J. Fluid Mech., **149**, 15–31 (1984)
34. J.W. Miles: Resonant motions of a spherical pendulum. Physica D, **11**, 309–323 (1984)
35. S.P. Maganty, W.B. Bickford: Large amplitude oscillations of thin circular rings. ASME J. Appl. Mech., **54**, 315–322 (1987)
36. S.P. Maganty, W.B. Bickford: Influence of internal resonance on the non-linear oscillations of a circular ring under primary resonance conditions. J. Sound and Vib., **122**, 507–521 (1988)
37. A.H. Nayfeh, R.A. Raouf: Nonlinear forced response of infinity long circular cylindrical shells. ASME J. Appl. Mech., **54**, 571–577 (1987)
38. R.A. Raouf, A.H. Nayfeh: One-to-one autoparametric resonances in infinitely long cylindrical shells. Computers and Structures, **35**, 163–173 (1990)
39. A.H. Nayfeh, R.A. Raouf, J.F. Nayfeh: Nonlinear response of infinitely long circular cylindrical shells to subharmonic radial loads. ASME J. Appl. Mech., **58**, 1033–1041 (1991)
40. M. Sathyamoorthy: Nonlinear vibration analysis of plates: a review and survey of current developments. Appl. Mech. Rev., **40**, 1553–1561 (1987)
41. J. Hadian, A.H. Nayfeh: Modal interaction in circular plates. J. Sound and Vib., **142**, 279–292 (1990)
42. K. Yasuda, T. Torii: Multi-mode response of a square membrane. JSME Inter. J., **30**, 963–969 (1987)
43. K. Yasuda, T. Asano: Nonlinear forced oscillations of a rectangular membrane with degenerate modes. Bull. JSME, **29**, 3090–3095 (1986)
44. S.I. Chang, A.K. Bajaj, C.M. Krousgrill: Non-linear vibrations and chaos in harmonically excited rectangular plates with one-to-one internal resonance. Nonlinear Dynamics (to appear).
45. E. Doedel: AUTO: Software for continuation and bifurcation problems in ordinary differential equations. Report, Dept. of Appl. Math., Cal Tech., Pasadena (1986)
46. A.H. Nayfeh: Introduction of perturbation techniques. Wiley-Interscience, New York 1981
47. P.J. Holmes: Chaotic motions in a weakly nonlinear model for surface waves. J. Fluid Mech., **162**, 365–388 (1986)
48. B.J. Matkowsky, E.L. Reiss: Singular perturbations of bifurcations. SIAM J. Appl. Math., **33**, 230–255 (1977)
49. C. Grebogi, E. Ott, J.E. Yorke: Crisis, sudden changes in chaotic attractors, and transient chaos. Physica D, **7**, 181–200 (1983)
50. E. Knobloch, N.O. Weiss: Bifurcations in a model of magneto convection. Physica D, **9**, 379–407 (1983)

Appendix

The expressions for coefficients a_{mnrs}, b_{mnrs}, c_{mnrs} and d_{mnrs} in equation (9) are given as follows:

$$a_{mnrs} = \begin{cases} \frac{mnrs - m^2s^2}{4\left((m-r)^2+\kappa^2(n-s)^2\right)^2} & \text{if } m \neq r \text{ or } n \neq s, \\ 0 & \text{if } m = r \text{ and } n = s, \end{cases} \tag{A1}$$

$$b_{mnrs} = \frac{mnrs + m^2s^2}{4\left((m-r)^2 + \kappa^2(n+s)^2\right)^2}, \tag{A2}$$

$$c_{mnrs} = \frac{mnrs + m^2s^2}{4\left((m+r)^2 + \kappa^2(n-s)^2\right)^2}, \tag{A3}$$

$$d_{mnrs} = \frac{mnrs - m^2s^2}{4\left((m+r)^2 + \kappa^2(n+s)^2\right)^2}. \tag{A4}$$

The expressions for S_{x0mn}, S_{y0mn} and $S_{xy0mnrs}$ in equation (11) are given as follows:

$$S_{x0mn} = \frac{\pi^2}{8(1-\nu^2)\kappa^4}(m^2 + \nu\kappa^2n^2), \tag{A5}$$

$$S_{y0mn} = \frac{\pi^2}{8(1-\nu^2)\kappa^4}(\nu m^2 + \kappa^2n^2), \tag{A6}$$

$$S_{xy0mnrs} = \begin{cases} \frac{2}{(1+\nu^2)\kappa^4} \frac{mnrs}{(m+n)(m-r)(n+s)(n-s)} & \text{if } (m \pm r) \text{ and} \\ \quad -4(a_{mnrs} + b_{mnrs} & (n \pm s) \text{ are odd,} \\ \quad +c_{mnrs} + d_{mnrs}), & \\ 0 & \text{if } (m \pm r) \text{ or} \\ & (n \pm s) \text{ are even.} \end{cases} \tag{A7}$$

The expressions for Ω_{kl} and $L_{mnrsijkl}$ in equation (12) are given as follows:

$$\Omega_{kl} = \left((k^2 + \kappa^2l^2) + D\pi^4(k^2 + \kappa^2l^2)^2\right)^{1/2}, \tag{A8}$$

and

$$\begin{aligned} L_{mnrsijkl} = 4\pi^2\varepsilon\Big(&\frac{1}{4}\delta_r^m\delta_s^n\delta_k^i\delta_l^j(i^2S_{x0mn} + j^2S_{y0mn}) + \\ &+ 2ijS_{xy0mnrs}T_{xy0ijkl} + \\ &+ (i^2T_{xmnrsijkl} + 2ijT_{xymnrsijkl} + j^2T_{ymnrsijkl})\Big), \end{aligned} \tag{A9}$$

where S_{x0mn}, S_{y0mn} and $S_{xy0mnrs}$ are given above, and

$$T_{xy0ijkl} = \begin{cases} \frac{4}{\pi^2} \frac{kl}{(i+k)(i-k)(j+l)(j-l)} & \text{if } (i \pm k) \text{ and } (j \pm l) \text{ are odd,} \\ 0 & \text{if } (i \pm k) \text{ or } (j \pm l) \text{ is even,} \end{cases} \tag{A10}$$

$$
\begin{aligned}
T_{xmnrsijkl} = -\pi^2\big((n-s)^2 a_{mnrs} U_1 U_3 + (n+s)^2 b_{mnrs} U_1 U_4 \\
+ (n-s)^2 c_{mnrs} U_2 U_3 + (n+s)^2 d_{mnrs} U_2 U_4\big) ,
\end{aligned} \tag{A11}
$$

$$
\begin{aligned}
T_{ymnrsijkl} = -\pi^2\big((m-r)^2 a_{mnrs} U_1 U_3 + (m-r)^2 b_{mnrs} U_1 U_4 \\
+ (m+r)^2 c_{mnrs} U_2 U_3 + (m+r)^2 d_{mnrs} U_2 U_4\big) ,
\end{aligned} \tag{A12}
$$

$$
\begin{aligned}
T_{xymnrsijkl} = \pi^2\big((m-r)(n-s) a_{mnrs} V_1 V_3 \\
+ (m-r)(n+s) b_{mnrs} V_1 V_4 \\
+ (m+r)(n-s) c_{mnrs} V_2 V_3 \\
+ (m+r)(n+s) d_{mnrs} V_2 V_4\big) ,
\end{aligned} \tag{A13}
$$

with

$$
\begin{aligned}
U_1 &= \frac{1}{4}\left(\delta_{k+r}^{i+m} - \delta_{r}^{i+k+m} + \delta_{k+m}^{i+r} - \delta_{m}^{i+k+r}\right) , \\
U_2 &= \frac{1}{4}\left(\delta_{k+m+r}^{i} - \delta_{m+r}^{i+k} + \delta_{k}^{i+m+r}\right) , \\
U_3 &= \frac{1}{4}\left(\delta_{l+s}^{j+n} - \delta_{s}^{j+l+n} + \delta_{l+n}^{j+s} - \delta_{n}^{j+l+s}\right) , \\
U_4 &= \frac{1}{4}\left(\delta_{l+n+s}^{j} - \delta_{n+s}^{j+l} + \delta_{l}^{j+n+s}\right) ,
\end{aligned} \tag{A14}
$$

$$
\begin{aligned}
V_1 &= \frac{1}{4}\left(\delta_{k+r}^{i+m} - \delta_{r}^{i+k+m} - \delta_{k+m}^{i+r} + \delta_{m}^{i+k+r}\right) , \\
V_2 &= \frac{1}{4}\left(-\delta_{k+m+r}^{i} + \delta_{m+r}^{i+k} + \delta_{k}^{i+m+r}\right) \\
V_3 &= \frac{1}{4}\left(\delta_{l+s}^{j+n} - \delta_{s}^{j+l+n} - \delta_{l+n}^{j+s} + \delta_{n}^{j+l+s}\right) , \\
V_4 &= \frac{1}{4}\left(-\delta_{l+n+s}^{j} + \delta_{n+s}^{j+l} + \delta_{l}^{j+n+s}\right) ,
\end{aligned} \tag{A15}
$$

where the Kronecker's delta is defined as $\delta_b^a = \begin{cases} 1 & \text{if } a = b, \\ 0 & \text{if } a \neq b. \end{cases}$

Non-Linear Behavior of a Rectangular Plate Exposed to Airflow

J. Awrejcewicz [1], J. Mrozowski [1] and M. Potier-Ferry [2]

[1] Technical University of Lódź,
Division of Dynamics and Control,
90–924 Lódź, Poland
[2] Université de Metz,
Laboratoire de Physique et Mecanique des Materiaux,
Ile du Sauley,
57045 Metz, France

Abstract

Oscillations of a plate subjected to a aerodynamic force are considered. The first part of the investigation is devoted to the establishment of the mathematical model and derivation of the governing partial differential equations. Also the condition for the Hopf bifurcation threshold is defined, corresponding to the flutter oscillations. In the second part the analytical predictions are verified numerically. Further numerical investigations show the existence of chaotic motion. Two different scenarios leading to chaos are examined.

1. Introduction

A rectangular plate can serve as a basic element of construction in various technical applications. One of them is aeroelasticity, which is a combination of two classical fields: fluid and solid mechanics. Dowell [1] was one of the first investigators who pointed out that chaotic, self-excited oscillations may occur for plates or shells exposed to the aerodynamic force action (see also [2–4]).

In this paper a thin simply supported rectangular plate exposed to the aerodynamic force is considered. This can serve as a mechanical model for many real systems which can be found in civil engineering (bridges, roadways) and the air industry (wings of planes). The plate has a concentrated mass in the point (x_0, y_0) and is submitted to the constant velocity airflow acting perpendicularly to its freely supported edges. One would expect that for certain parameters a previously stable configuration of the plate becomes unstable and periodic oscillations could appear. This phenomenon, known as flutter, is well described by the aeroelasticity theory [5–8]. However, because of the occurrence of non-linear terms in the governing equations, we expect much more complicated dynamical behavior of the investigated system. As it will be shown later a period doubling scenario and intermittency scenario leading to chaotic motion of the plate have been detected.

This paper contains the following parts. First a mathematical model, i.e. three partial non-linear autonomous differential equations with the appropriate boundary conditions, is introduced. Then the Galerkin procedure is applied in order to obtain the two second order ordinary differential equations. These equations allowed us to determine the Hopf bifurcation threshold. Periodic orbits born after Hopf bifurcation have been analyzed numerically and two further routes leading to chaotic orbits have been investigated in some detail showing some interesting phenomena.

The direction of the airflow and the external compressive force are parallel to the freely supported ends of the plate. This one dimensional model has allowed us to obtain a solution in the form of the sinusoidal series with respect to the coordinate x. Dowell's investigation has been reduced to the consideration of the influence of two essential control parameters, i.e. the velocity of the airflow and the magnitude of the external compressive force, on the transition from steady stationary state (equilibrium) to a periodic motion (flutter). Dowell has numerically proved the occurrence of chaotic motion exhibited by the analyzed plate. Contrary to his approach, the current model has a concentrated mass m_1 located at the point (x_0, y_0) of the plate. Finally, the direction of the aerodynamic force is perpendicular to the freely supported ends of the plate.

This theoretical approach develops the ideas presented in [9,10]. The developement of the averaged governing ordinary differential equations is followed by analytical formulation of the Hopf condition resulting in periodic oscillations of the analyzed system. Then by taking arbitrarily two control parameters, two different routes to chaos, transitional and steady state chaotic behavior, as well as the development of chaos with the change of control parameters are analyzed.

2. Mathematical Model

A rectangular long, thin, flat plate of length a, width b ($b \ll a$) and thickness h small in comparison with its other dimensions, is considered. It is assumed that the plate consists of a perfectly homogeneous isotropic material. The concentrated mass m_1 is located at the point (x_0, y_0). The airflow with velocity U acts on the plate in the direction parallel to the y axis. The rectangular Cartesian coordinates x, y, z, where x and y lie in the middle plane of the plate and z is measured from the midsurface of the plate are used (Fig. 1).

It is assumed that the plate obeys the classical Hooke's low, but the geometrical nonlinearities which appear as a result of the rotations of structure elements, play an essential role in our further consideration.

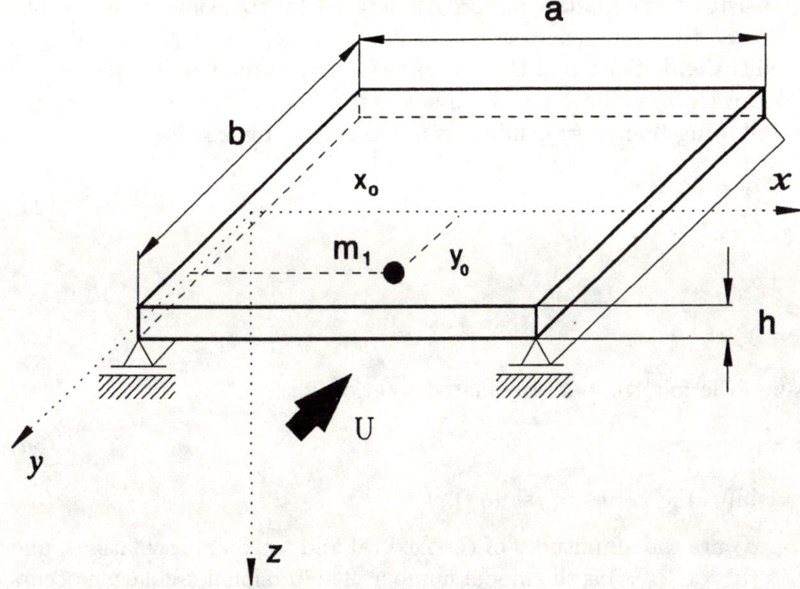

Fig. 1. Model of the analyzed system

Based on the theory of virtual work the following equilibrium equations are derived

$$\frac{\partial N_x}{\partial x} + \frac{\partial N_{xy}}{\partial y} = 0\,, \tag{1a}$$

$$\frac{\partial N_y}{\partial y} + \frac{\partial N_{xy}}{\partial x} = 0\,, \tag{1b}$$

$$\begin{aligned} &D\frac{\partial^4 w}{\partial x^4} + 2D\frac{\partial^4 w}{\partial x^2 \partial y^2} + D\frac{\partial^4 w}{\partial y^4} - N_x\frac{\partial^2 w}{\partial x^2} - 2N_{xy}\frac{\partial^2 w}{\partial x \partial y} - N_y\frac{\partial^2 w}{\partial y^2} \\ &+m\frac{\partial^2 w}{\partial t^2} + m_1\delta(x - x_0, y - y_0)\frac{\partial^2 w}{\partial t^2} + \frac{\rho U^2}{M}\left(\frac{\partial w}{\partial y} + \frac{1}{U}\frac{\partial w}{\partial t}\right) = 0\,, \end{aligned} \tag{1c}$$

where: N_x, N_y, N_{xy} are the membrane forces per unit length; u, v, w are displacement components in the midsurface in the x, y, z directions, respectively; D is the flexural stiffness of the plate; m is the plate mass per unit area; m_1 – the concentrated mass; $\delta(x_0, y_0)$ is the Dirac delta; ρ – the air density; U – air velocity; M – the Mach number and t – the time.

The first two equations describe the static equilibrium conditions of a $dxdy$ element along the x and y axes. The third equation represents the dynamic equilibrium along the z axis. The first three terms of the equation (1c) can be obtained using the classical linear plate theory. The further three terms are a result of the geometrical non-linearity. The terms containing masses m and m_1 are the inertial terms. The last term describes the aerodynamic force which consists of two parts: the force generated by the convection velocity $U(\partial w/\partial y)$ and the force generated by the imposed velocity $\partial w/\partial t$. The details dealing with the derivation of the formulas for the aerodynamic force can be found in Dowell [8] and are not discussed here.

The following boundary conditions are assumed: for $x = 0$,

$$\begin{aligned} &w = M_x = 0\,, \\ &N_x = N_{xy} = 0\,; \end{aligned} \tag{2a}$$

for $y = \pm b/2$:

$$N_y = N_{xy} = 0\,. \tag{2b}$$

We assume the following approximated solution form

$$w = a_1 w_1 + a_2 w_2\,; \tag{3a}$$

$$w_1 = \sin(qx)\,, \qquad w_2 = y\sin(qx)\,, \tag{3b}$$

where a_1, a_2 are the amplitudes of the flexural and torsional oscillations, and $q = n\pi/a$ $(n = 1, 2, \ldots)$ is the modal number. The torsional oscillation occurs because the plate is long and the direction of the airflow is perpendicular to the freely supported edges of the plate.

In order to define the unknown membrane forces N_x, N_y, N_{xy} in equation (1c), equations (1a), (1b) and the boundary conditions are used. Because the left-hand sides of equations (1a) and (1b) correspond to the projection of all forces acting on a $dxdy$ element of the plate along the x and y axes, respectively, using the virtual work principle one obtains:

$$\int_\Omega \left[(N_{x,x} + N_{xy,y})\delta u + (N_{xy,x} + N_{y,y})\delta v\right] d\Omega = 0\,. \tag{4}$$

Based on the non-linear theory of plates we have

$$\begin{aligned} N_x &= C\left(u,_x + \frac{1}{2}w,_x^2 + \nu v,_y + \frac{\nu}{2}w,_y^2\right), \\ N_y &= C\left(v,_y + \frac{1}{2}w,_y^2 + \nu u,_x + \frac{\nu}{2}w,_x^2\right), \\ N_{xy} &= C\frac{1-\nu}{2}\left(u,_y + v,_x + w,_x w,_y\right). \end{aligned} \tag{5}$$

where $(\ ),_x$ denotes $\frac{\partial(\ )}{\partial\partial x}$ and so on.

Introducing eq. (5) into equation (4) the following integral equation is obtained

$$
\begin{aligned}
& C\int_{\Omega}\left\{\left[u,_{xx}+\nu v,_{xy}+\frac{1-\nu}{2}(u,_{yy}+v,_{xy})\right]\delta u\right. \\
& \left.+\left[\frac{1-\nu}{2}(u,_{yx}+v,_{xx})+v,_{yy}+\nu u,_{xy}\right]\delta v\right\}dxdy \\
& +C\int_{\Omega}\left\{\left[w,_{x}w,_{xx}+\nu w,_{y}w,_{yx}+\frac{1-\nu}{2}(w,_{xy}w,_{y}+w,_{x}w,_{yy})\right]\delta u\right. \\
& +\left[\frac{1-\nu}{2}(w,_{xx}w,_{y}+w,_{x}w,_{yx})+w,_{y}w,_{yy}\right. \\
& \left.\left.+\nu w,_{x}w,_{xy}\right]\delta v\right\}dxdy=0\,,
\end{aligned} \tag{6}
$$

where Ω is the plate area.

The last equation is of the following structure

$$\langle \boldsymbol{A}\boldsymbol{u},\delta\boldsymbol{u}\rangle+\langle \boldsymbol{B}(w,w),\delta\boldsymbol{u}\rangle=0\,, \tag{7}$$

where $\langle\cdot,\cdot\rangle$ denotes the scalar product and:

$$\boldsymbol{u}=\begin{bmatrix}u\\v\end{bmatrix},\qquad \delta\boldsymbol{u}=\begin{bmatrix}\delta u\\ \delta v\end{bmatrix},$$

$$\boldsymbol{A}=\begin{bmatrix}\frac{\partial^2}{\partial x^2}+\frac{1-\nu}{2}\frac{\partial^2}{\partial y^2} & \nu\frac{\partial^2}{\partial y\partial x}+\frac{1-\nu}{2}\frac{\partial^2}{\partial x\partial y}\\ \frac{1-\nu}{2}\frac{\partial^2}{\partial y\partial x}+\nu\frac{\partial^2}{\partial x\partial y} & \frac{1-\nu}{2}\frac{\partial^2}{\partial x^2}+\frac{\partial^2}{\partial y^2}\end{bmatrix},$$

$$\boldsymbol{B}(w,w)=\begin{bmatrix}w,_{x}w,_{xx}+\nu w,_{y}w,_{yx}+\frac{1-\nu}{2}(w,_{xy}w,_{y}+w,_{x}w,_{yy}\\ \frac{1-\nu}{2}(w,_{xx}w,_{y}+w,_{x}w,_{yx})+w,_{y}w,_{yy}+\nu w,_{x}w,_{xy}\end{bmatrix}.$$

It should be noted that $\boldsymbol{Au}$ is a linear expression because of the u vector, however the expression $\boldsymbol{B}(w,w)$ is of a bilinear form. Because the virtual displacements can be taken arbitrarily, equation (7) is equivalent to the equation

$$\boldsymbol{Au}+\boldsymbol{B}(w,w)=0\,. \tag{8}$$

Taking into account the assumed form of the solution (3) and the properties of the bilinear form we have

$$\boldsymbol{B}(w,w)=a_1^2\boldsymbol{B}(w_1,w_1)+2a_1a_2\boldsymbol{B}(w_1,w_2)+a_2^2\boldsymbol{B}(w_2,w_2)\,. \tag{9}$$

We split the assumed solution for $\boldsymbol{u}$ into a form similar to that given by $\boldsymbol{B}(w,w)$, i.e.

$$\boldsymbol{u}=a_1^2\boldsymbol{u}^{(1)}+2a_1a_2\boldsymbol{u}^{(12)}+a_2^2\boldsymbol{u}^{(2)}\,. \tag{10}$$

On substitution from (9) and (10) into (8)

$$
\begin{aligned}
a_1^2\left[\boldsymbol{Au}^{(1)}+\boldsymbol{B}(w_1,w_1)\right]+2a_1a_2\left[\boldsymbol{Au}^{(12)}+\boldsymbol{B}(w_1,w_2)\right]& \\
+a_2^2\left[\boldsymbol{Au}^{(2)}+\boldsymbol{B}(w_2,w_2)\right]=0\,.&
\end{aligned} \tag{11}
$$

The amplitudes a_1 and a_2 are time dependent and the above equation is fulfilled only when the expression given in the square brackets is equal to zero. Thus equation (8) can be replaced by a set of the following three equations

$$\boldsymbol{A}\boldsymbol{u}^{(1)} + \boldsymbol{B}(w_1, w_1) = 0 , \tag{12}$$

$$\boldsymbol{A}\boldsymbol{u}^{(12)} + \boldsymbol{B}(w_1, w_2) = 0 , \tag{13}$$

$$\boldsymbol{A}\boldsymbol{u}^{(2)} + \boldsymbol{B}(w_2, w_2) = 0 . \tag{14}$$

The algebraic form of the equation describing a stress-displacement dependence is as follows:

$$\boldsymbol{N} = \boldsymbol{L}[\boldsymbol{\varepsilon}] = \boldsymbol{L}\left[\boldsymbol{e}(\boldsymbol{u}) + \frac{1}{2}\Delta(w, w)\right] , \tag{15}$$

where

$$\boldsymbol{N} = \begin{bmatrix} N_x \\ N_y \\ N_{xy} \end{bmatrix} \quad \text{– vector of membrane forces} , \tag{15a}$$

$$\boldsymbol{L} = \begin{bmatrix} C & C\nu & 0 \\ C\nu & C & 0 \\ & 0 & C\frac{1-\nu}{2} \end{bmatrix} \quad \text{– linear operator} , \tag{15b}$$

$$\boldsymbol{\varepsilon} = \begin{bmatrix} \varepsilon_x \\ \varepsilon_y \\ \varepsilon_{xy} \end{bmatrix} \quad \text{– displacement vector} , \tag{15c}$$

$$\boldsymbol{e} = \begin{bmatrix} e_x \\ e_y \\ e_{xy} \end{bmatrix} = \begin{bmatrix} u,_x \\ v,_y \\ u,_y + v,_x \end{bmatrix} \quad \text{– linear part of displacement vector} , \tag{15d}$$

$$\boldsymbol{\Delta} = \begin{bmatrix} \Delta_x \\ \Delta_y \\ \Delta_{xy} \end{bmatrix} = \begin{bmatrix} w,_x^2 \\ w,_y^2 \\ 2w,_x w,_y \end{bmatrix} \tag{15e}$$

– nonlinear part of displacement vector .

Again splitting equation (15) into a form similar to that of (8) it can be replaced by

$$\boldsymbol{N} = a_1^2\boldsymbol{N}^{(1)} + 2a_1a_2\boldsymbol{N}^{(12)} + a_2^2\boldsymbol{N}^{(2)} , \tag{16}$$

where

$$\boldsymbol{N}^{(1)} = \boldsymbol{L}[\boldsymbol{\varepsilon}^{(1)}] = \boldsymbol{L}\left[\boldsymbol{e}(\boldsymbol{u}^{(1)} + \frac{1}{2}\boldsymbol{\Delta}(w_1, w_1)\right] , \tag{17}$$

$$\boldsymbol{N}^{(12)} = \boldsymbol{L}[\boldsymbol{\varepsilon}^{(12)}] = \boldsymbol{L}\left[\boldsymbol{e}(\boldsymbol{u}^{(12)} + \frac{1}{2}\boldsymbol{\Delta}(w_1, w_2)\right] , \tag{18}$$

$$\boldsymbol{N}^{(2)} = \boldsymbol{L}[\boldsymbol{\varepsilon}^{(2)}] = \boldsymbol{L}\left[\boldsymbol{e}(\boldsymbol{u}^{(2)} + \frac{1}{2}\boldsymbol{\Delta}(w_2, w_2)\right] . \tag{19}$$

A separate solution of each of the equations (12), (13) and (14) is required. Starting with equation (12) the following solution is assumed

$$N_x^{(1)} = N_{xy}^{(1)} = N_y^{(1)} = 0 . \tag{20}$$

This solution satisfies equations (1a) and (1b) and the boundary conditions (2a) and (2b). Equations (20) are equivalent to the equations

$$u,_x^{(1)} + \frac{1}{2} w_{1,x}^2 = 0 , \tag{21a}$$

$$u,_y^{(1)} + v,_x^{(1)} + w_{1,x} w_{1,y} = 0 , \tag{21b}$$

$$v,_y^{(1)} + \frac{1}{2} w_{1,y}^2 = 0 . \tag{21c}$$

Since $w_{1,y} = 0$, it follows that

$$v^{(1)} = v^{(1)}(x) . \tag{22}$$

In particular, the following solution of the set of equations (21) can be assumed to be

$$\begin{cases} u^{(1)} = u^{(1)}(x) , \\ v^{(1)} = 0 . \end{cases} \tag{23}$$

This fulfils equations (21b,c). From equation (21a) the following relationship is obtained

$$u^{(1)}(x) = C - \frac{1}{2} \int_0^x \left(\frac{\partial w_1}{\partial x}(x') \right)^2 dx' \tag{24}$$

where C is an integral constant.

The above considerations imply that it is possible to find solutions for u, v (e.g. (23)) which fulfil equations (21). This proves that the assumed solution (20) is not a contradictory one, so the solution $\boldsymbol{N}^{(1)} = 0$ is the exact solution of (12).

Similarly to the above $\boldsymbol{N}^{(12)} = 0$ is the exact solution of equation (13).

Applying the same procedure to equation (14), it is noted that the solution $\boldsymbol{N}^{(2)} = 0$ does not fulfil this equation. Thus it is necessary to assume another type of solution, i.e.

$$\varepsilon_x^{(2)} = \varepsilon_y^{(2)} = 0 , \qquad \varepsilon_{xy}^{(2)} \neq 0 . \tag{25}$$

Equations (25) are equivalent to the equations

$$\begin{cases} u,_x^{(2)} + \frac{1}{4}q^2y^2(1+\cos(2qx)) = 0\,, \\ v,_y^{(2)} + \frac{1}{2}\sin^2(qx) = 0\,. \end{cases} \tag{26}$$

After integration of equations (26) we get

$$\begin{cases} u^{(2)}(x,y) = -\frac{1}{4}q^2y^2\left(x + \frac{1}{2q}\sin(2qx)\right) + f(y)\,, \\ v^{(2)}(x,y) = -\frac{1}{2}y\sin^2(qx) + g(x)\,. \end{cases} \tag{27}$$

Now it is necessary to determine the unknown functions $f(y)$ and $g(x)$ which appear in eq. (27). From the previous analysis $\boldsymbol{N}^{(1)} = \boldsymbol{N}^{(12)} = 0$ and because of the assumed solution (25) ($N_x^{(2)} = N_y^{(2)} = 0$) the equation (4) reduces to the form

$$\int_\Omega \left(N_{xy,y}^{(2)}\delta u + N_{xy,x}^{(2)}\delta v\right) d\Omega = 0\,. \tag{28}$$

It follows from (19), (15 a,b,d,e) and from (27), (3b) that

$$N_{xy}^{(2)} = C\frac{1-\nu}{2}\left[-\frac{1}{2}q^2y\left(x + \frac{1}{2q}\sin(2qx)\right) + f,_y + g,_x\right]\,. \tag{29}$$

After integrating (28) by parts and using equation (29)

$$\int_\Omega \left[-\frac{1}{2}q^2y\left(x + \frac{1}{2q}\sin(2qx)\right) + f,_y + g,_x\right](\delta u,_y + \delta v,_x)dxdy = 0\,. \tag{30}$$

The virtual displacements can be chosen freely e.g. $\delta w = \delta v = 0$, $\delta u = \delta f(y)$. Integrating eq. (30) by parts:

$$\begin{aligned} &\frac{q^2a^2}{4}\int_{-b/2}^{+b/2}\delta f(y)dy + \frac{q^2a^2}{4}\left[-y\delta f(y)\right]_{-b/2}^{+b/2} - a\int_{-b/2}^{+b/2} f,_{yy}\,\delta f(y)dy + \\ &\quad + a\left[f,_y\,\delta f(y)\right]_{-b/2}^{+b/2} + (g(a) - g(0))\left[\delta f(y)\right]_{-b/2}^{+b/2} = 0\,. \end{aligned} \tag{31}$$

Equating the terms which appear after the integrals in equation (31) (for the plate area Ω) and the expressions for the edges $y = \pm b/2$ separately to zero we get the following set of equations

$$\frac{q^2a^2}{4} - af,_{yy} = 0\,, \tag{32a}$$

$$-b\frac{q^2a^2}{8} + af,_y(b/2) + g(a) - g(0) = 0\,, \tag{32b}$$

$$b\frac{q^2a^2}{8} + af,_y(-b/2) + g(a) - g(0) = 0\,. \tag{32c}$$

After integrating (32a) we have

$$f(y) = \frac{q^2 a}{8} y^2 + Cy + D \,. \tag{33}$$

Introducing the following virtual displacements: $\delta w = \delta u = 0$, $\delta v = \delta g(x)$ and repeating the procedure described earlier a form of the second function, that we are looking for, is obtained

$$g(x) = Ax + B \,. \tag{34}$$

Four unknown constants appear in functions $f(y)$ and $g(x)$. In order to define them, the equations (32b,c) and two equations similar to them, obtained for virtual displacements $\delta w = \delta u = 0$, $\delta v = \delta g(x)$, are used. Thus

$$C + A = 0 \,. \tag{35}$$

There are four unknowns A,B,C,D which must satisfy only one equation (35). Thus three of them can be arbitrarily chosen, e.g.

$$B = C = D = 0 \,. \tag{36}$$

From (36) and (35) $A = 0$, which leads to

$$N_{xy}^{(2)} = \frac{Eh}{2(1+\nu)} \frac{q^2 y}{2} \left(\frac{a}{2} - x - \frac{1}{2q} \sin(2qx) \right) , \tag{37}$$

and from (16)

$$N_{xy} = a_2^2 N_{xy}^{(2)} \,. \tag{38}$$

Finally, the approximate expression for the unknown membrane forces is:

$$\begin{cases} N_x = 0 \,, \\ N_y = 0 \,, \\ N_{xy} = a_2^2 \frac{Ehq^2}{4(1+\nu)} y \left(\frac{a}{2} - x - \frac{1}{2q} \sin(2qx) \right) \,. \end{cases} \tag{39}$$

Now the Galerkin method is used to reduce the problem to the nondimensional ordinary differential equations, and the following amplitude equations are obtained:

$$\begin{cases} \ddot{A}_1 + \beta_1 \ddot{A}_2 + \varepsilon \dot{A}_1 + A_1 + \gamma A_2 = 0 \,, \\ \alpha \ddot{A}_2 + \beta_2 \ddot{A}_1 + \varepsilon \dot{A}_2 + A_2 + \eta A_2^3 = 0 \,, \end{cases} \tag{40}$$

where

$$\begin{aligned} \alpha &= \frac{2+3\beta}{2(1+2\beta)} \,; \quad \beta_1 = \frac{\beta}{2(h/b)(1+2\beta)} \,; \quad \beta_2 = \frac{6\beta(h/b)}{(1+2\beta)} \,; \\ \varepsilon &= \frac{\rho U_s}{q^2 \sqrt{mD(1+2\beta)}} \,; \quad \gamma = \frac{\rho U^2}{MhDq^4} \quad \eta = \frac{3(1-\nu)}{2h^2 q^2} \,. \end{aligned} \tag{40a}$$

The set (39) consists of the non-linear and ordinary autonomous differential equations. The only non-linear term appears in the third equation of (39) and is generated by the membrane force N_{xy}.

3. Treshold Determination of Periodic Oscillations

Periodic oscillations occur when due to a change of the control parameter a pair of complex conjugate eigenvalues crosses the imaginary axis with nonzero velocity. This phenomenon is often called the Hopf bifurcation [12–16]. The aim of this approach is to find the Hopf condition analytically. First the system of equations (39) is transformed to the following one:

$$\begin{cases} \dot{x}_1 = x_3 \,, \\ \dot{x}_2 = x_4 \,, \\ \dot{x}_3 = c_{31}x_1 + c_{32}x_2 + c_{33}x_3 + c_{34}x_4 + r_3 x_2^3 \,, \\ \dot{x}_4 = c_{41}x_1 + c_{42}x_2 + c_{43}x_3 + c_{44}x_4 + r_4 x_2^3 \,, \end{cases} \tag{41}$$

where:

$$\begin{aligned} c_{31} &= -\frac{\alpha}{\alpha - \beta_1\beta_2} = -\frac{(2+3\beta)(1+2\beta)}{2+7\beta} \,, \\ c_{32} &= \frac{\beta_1 - \gamma\alpha}{\alpha - \beta_1\beta_2} = \left[\frac{\beta}{h/b} - \gamma(2+3\beta)\right]\frac{1+2\beta}{2+7\beta} \,, \\ c_{33} &= -\frac{\varepsilon\alpha}{\alpha - \beta_1\beta_2} = -\frac{\rho U_s(2+3\beta)\sqrt{1+2\beta}}{q^2(2+7\beta)\sqrt{mD}} \,, \\ c_{34} &= \frac{\beta_1\varepsilon}{\alpha - \beta_1\beta_2} = \frac{qU_s\beta\sqrt{1+2\beta}}{q^2(2+7\beta)(h/b)\sqrt{mD}} \,, \\ c_{41} &= \frac{\beta_2}{\alpha - \beta_1\beta_2} = \frac{12\beta(h/b)(1+2\beta)}{2+7\beta} \,, \\ c_{42} &= \frac{\beta_2\gamma - 1}{\alpha - \beta_1\beta_2} = \frac{[12(h/b)\beta\gamma - 4\beta - 2](1+2\beta)}{2+7\beta} \,, \\ c_{43} &= \frac{\beta_2\varepsilon}{\alpha - \beta_1\beta_2} = \frac{12\beta(h/b)\rho U_s\sqrt{1+2\beta}}{q^2(2+7\beta)\sqrt{mD}} \,, \\ c_{44} &= -\frac{\varepsilon}{\alpha - \beta_1\beta_2} = -\frac{2\rho U_s\sqrt{1+2\beta}}{q^2(2+7\beta)\sqrt{mD}} \,, \\ r_3 &= \frac{\beta_1\eta}{\alpha - \beta_1\beta_2} = \frac{3}{2}\frac{\beta(1-\nu)}{(h/b)h^2q^2}\frac{(1+2\beta)}{(2+7\beta)} \,, \\ r_4 &= -\frac{\eta}{\alpha - \beta_1\beta_2} = -\frac{3(1-\nu)(1+2\beta)^2}{h^2q^2(2+7\beta)} \,. \end{aligned} \tag{42}$$

The solution at the equilibrium point $\boldsymbol{x}_0 = \boldsymbol{0}$ is locally perturbed and the linearized set of differential equations, whose Jacobian is given below is obtained:

$$\boldsymbol{J}(\boldsymbol{0}) = \begin{bmatrix} 0 & 0 & 1 & 0 \\ 0 & 0 & 0 & 1 \\ c_{31} & c_{32} & c_{33} & c_{34} \\ c_{41} & c_{42} & c_{43} & c_{44} \end{bmatrix} . \tag{43}$$

The eigenvalues of the matrix $\boldsymbol{J}$ can be found from the characteristic equation

$$\lambda^4 + \alpha_3\lambda^3 + \alpha_2\lambda^2 + \alpha_1\lambda + \alpha_0 = 0 \,, \tag{44}$$

where:

$$\begin{aligned} \alpha_3 &= -c_{33} - c_{44} \,, \\ \alpha_2 &= c_{33}c_{44} + c_{33}c_{42} - c_{32}c_{43} - c_{31} \,, \\ \alpha_1 &= c_{33}c_{42} - c_{32}c_{43} - c_{34}c_{43} + c_{31}c_{44} - c_{41}c_{34} \,, \\ \alpha_0 &= c_{31}c_{42} - c_{32}c_{41} \,. \end{aligned} \tag{44a}$$

According to the Hopf bifurcation theorem [14], there is one pair of conjugate purely imaginary eigenvalues $\lambda = \pm i\omega$ at the critical point. Substituting $\lambda = +i\omega$ into the characteristic equation (44) leads to

$$\begin{cases} \omega^4 - \alpha_2\omega^2 + \alpha_0 = 0 \\ -\alpha_3\omega^3 + \alpha_1\omega = 0 \end{cases} \,. \tag{45}$$

(Substitution of $\lambda = -i\omega$ leads to the same result.)

Eliminating the parameter ω from (45):

$$\alpha_1^2 - \alpha_1\alpha_2\alpha_3 + \alpha_0\alpha_3^2 = 0 \,, \tag{46}$$

with the additional inequality to be fulfilled

$$\alpha_1/\alpha_3 > 0 \,. \tag{47}$$

Taking into account the equations (44a) and (46) the Hopf condition in the parameter plane $\gamma - \beta$ is obtained.

As an example the following geometrical and physical parameters of the system are taken as fixed: length a = 10 m, width b = 1 m, thickness h = 0.005 m, material density $\rho_m = 7.85 \cdot 10^3$ kg/m^3, modulus of elasticity $E = 2.1 \cdot 10^{11}$ N/m^2, Poisson's ratio $\nu = 0.3$. The graphic representation of the equation (46), is shown in Fig. 2.

In the above figure the dashed area corresponds to the values of parameters, for which the plate is in the stationary steady state (static equilibrium position). In the dotted region the steady state becomes unstable, and periodic motion occurs.

In order to check the agreement between the numerical and analytical solution a fixed value of β (= 0.3) and a varaiable γ (starting at 400) are used. For $\gamma = 400$ the system approaches a steady state very quickly. The increase of γ causes oscillations and lengthens the time needed to reach the steady state with the trend to infinity at the bifurcation point. The value of γ at the critical Hopf point determined analytically was equal to $\gamma = 1268$, which is in very good agreement with the numerical solution.

Shortly past the Hopf bifurcation curve the periodic orbit appears. The amplitude of oscillations and the corresponding frequencies grow with an

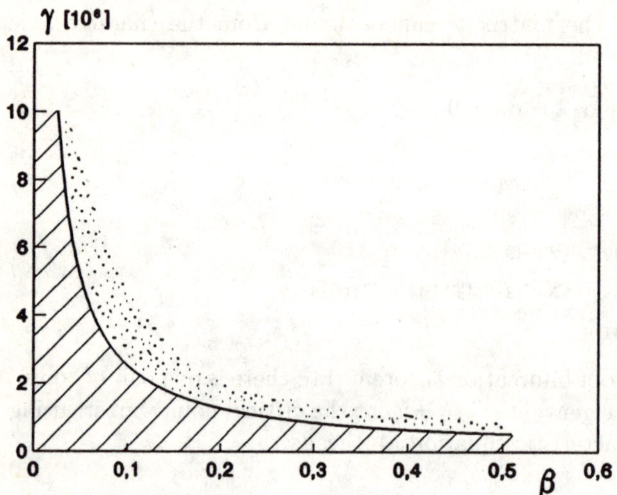

Fig. 2. The Hopf bifurcation diagram

increase of the parameter γ. Similarly increasing γ causes the transient state to become shorter. This is illustrated in Figures 4b, c.

4. Dynamics Past the Hopf Bifurcation Point

The analytical methods applied in the investigation of non-linear oscillatory systems usually lead to complicated analytical formulas and generally do not give accurate enough results. The reason is that the system of non-linear equations can only rarely be analytically integrated and the present system is a complex one (two degrees of freedom). To avoid the occurrence of such potential problems only numerical techniques have been used.

Again two control parameters β and γ were used to observe numerically the behavior of the periodic orbits found earlier.

The analysis was carried out with a few arbitrarily chosen rectangular plates with different geometrical and physical properties. All the considered cases were investigated in a similar way. For a fixed value of the parameter β, the parameter γ was increased beginning with the values close to the Hopf bifurcation curve. For each pair of β and γ values time histories, phase portraits, Poincaré maps or power spectra were observed. All the above mentioned numerical tools enabled the tracing of dynamical behavior of the corresponding analyzed system. The result was that for all investigated cases the scenario of qualitative changes in dynamics was similar.

The numerical analysis has shown, that the periodic motion appears for a broad region of the control parameters above the Hopf bifurcation curve. Starting with the value of the γ close to the Hopf bifurcation point and increasing it causes an increase of the oscillation amplitude as well as an increase of the frequencies observed in the power spectra (to this aim Fast Fourier Transformation was used). Values of the parameter γ exist for which first period doubling occurs and a new periodic orbit with a period twice longer than the previous one appears. The basic periodic orbit becomes unstable. This is the first step of the so called Feigenbaum scenario, i.e. a cascade of period doubling bifurcation. It has been observed numerically using Poincaré maps and power spectra (see Figures 3 a-f).

The sequence of power spectra clearly proves the doubling of period. Each succeeding picture contains new frequencies half the value of the previous frequencies. For a certain value of γ the density of the frequency peaks becomes so high, that it becomes impossible to distinguish them. This indicates that the system is close to the chaotic motion threshold. The corresponding Poincaré maps also prove the period doubling scenario. Based on the observation of Poincaré maps, it is evident that each point of the previous Poincaré map splits into two points on the next map. Thus when two successive Poincaré maps are considered the number of points after bifurcation doubles.

Increasing the parameter γ further above the chaotic motion threshold the reverse phenomenon was observed, the so called period halving scenario. The successive frequency peaks vanish into the broad band power spectrum background. From the corresponding Poincaré maps the development of the strange chaotic attractor can be traced. In Fig. 4a eight separate parts of a strange chaotic attractor which correspond to the smallest frequency found in the broad band spectrum, i.e. $f/8$ can be distinguished. This property is observed in each succeeding figure. Two adjacent elements join together, then finally a fully developed "one element" compact strange attractor is obtained. Fourier spectra show one basic frequency. The numerical simulation shows that there are some rare intervals of the parameter γ for which the system behaves regularly, so called periodic windows. Outside these intervals the chaotic dynamics of the system are exhibited. This phenomenon is illustrated in Figure 4e.

Increasing the parameter γ further it is seen that the scenario beginning as period doubling and proceeding to the full development of chaos repeats a few times.

For a big enough value of γ we observed another route to chaos. The periodic motion is unpredictably interrupted by sudden jumps of chaos. This phenomenon is called intermittency and was first introduced by Manneville-Pomeau during their analysis of one parameter maps. Contrary to their investigation, the current study observes such a scenario in a real mechanical system. The described intermittency scenario is illustrated in Figs. 5 and 6. As can be seen in Fig. 5a, with the background of regular motion irregular bursts are clearly distinguishable. One of the marked intervals in Fig. 5a is magnified and shown in Fig. 5b. Again the intermittency structure is

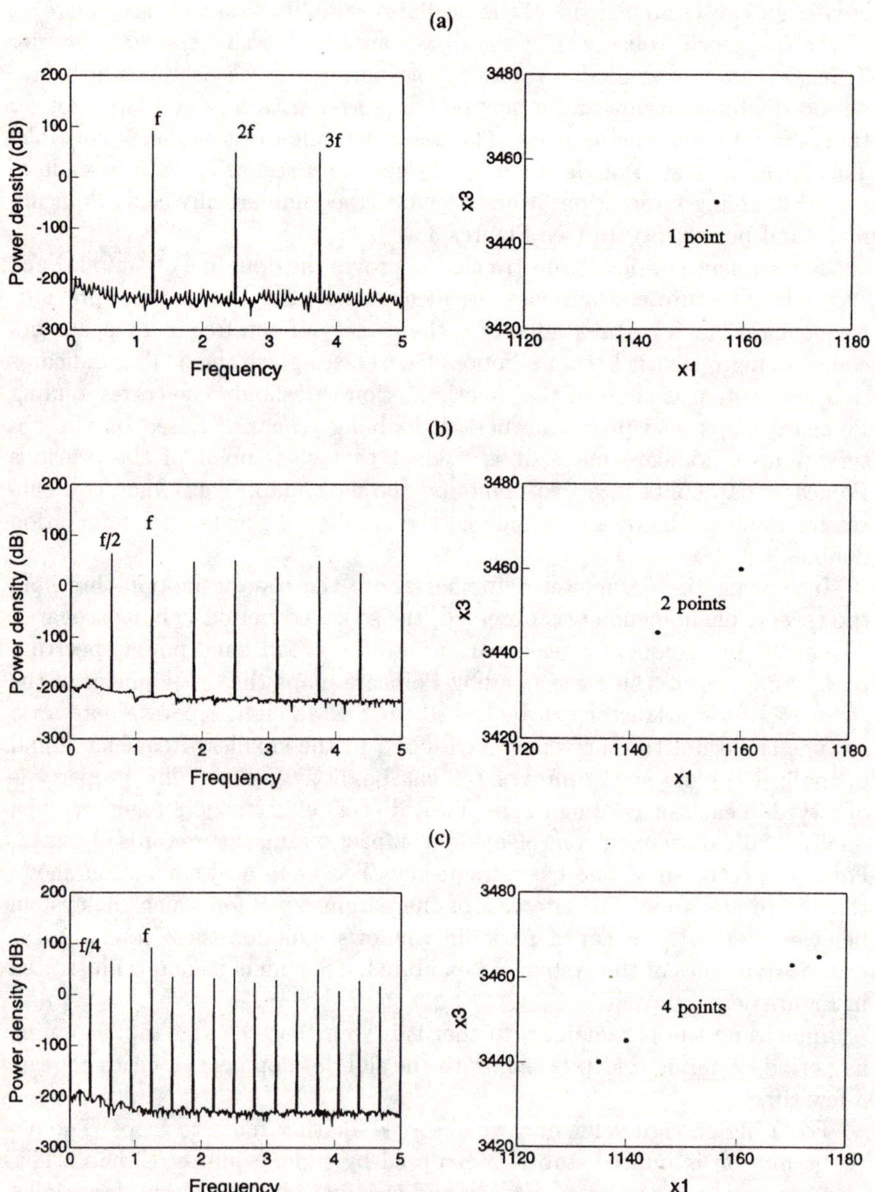

Fig. 3. Power spectra and Poincaré maps illustrating period doubling scenario leading to chaos: (a) $\gamma = 13900$; (b) $\gamma = 14000$; (c) $\gamma = 14020$

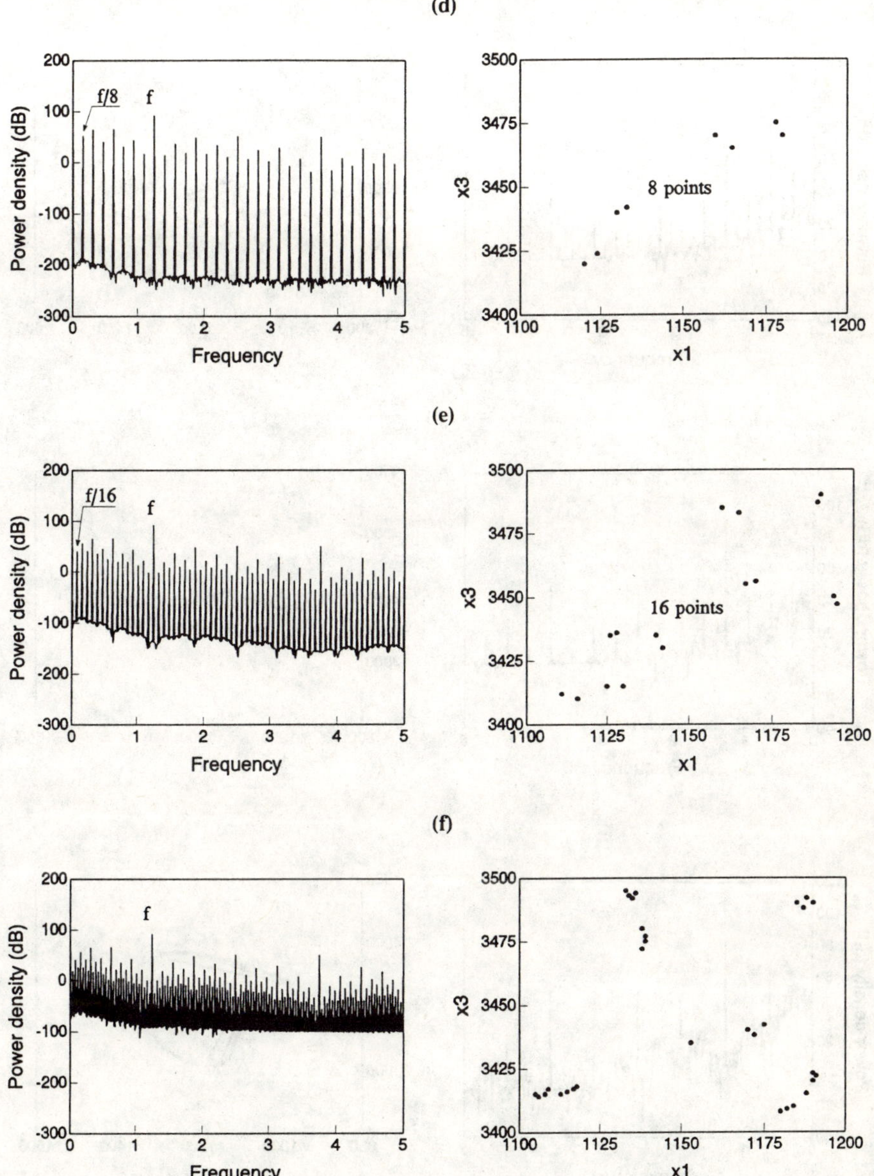

Fig. 3. Continued: (d) $\gamma = 14025$; (e) $\gamma = 14028$; (f) $\gamma = 14028.3$

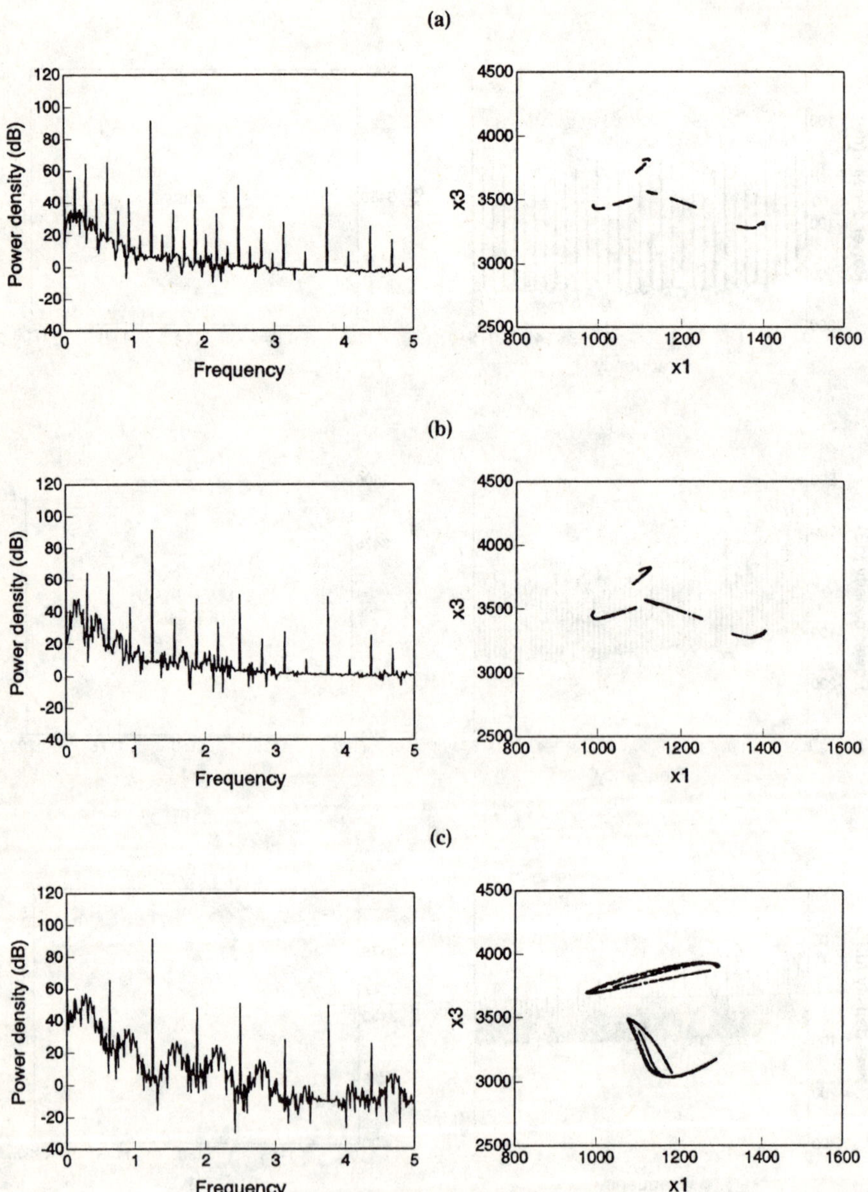

Fig. 4. Power spectra and Poincaré maps illustrating the development of chaos with the increase of the control parameter γ: (a) $\gamma = 14029$; (b) $\gamma = 14030$; (c) $\gamma = 14035$

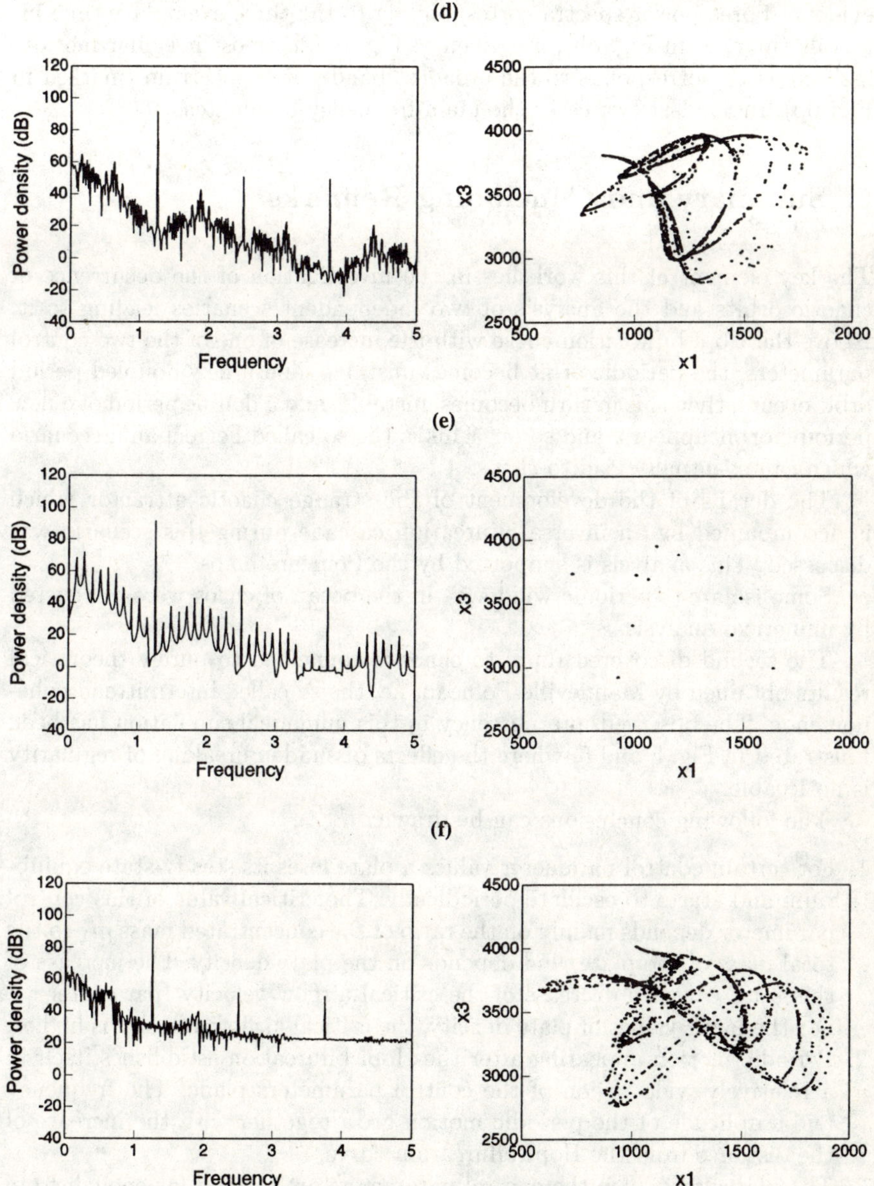

Fig. 4. Continued: (d) $\gamma = 14045$; (e) $\gamma = 14050$; (f) $\gamma = 14055$

evident. Three power spectra corresponding to the three exemplary time intervals (marked in Fig. 5b), are given in Fig. 6. The most irregular motion, marked as II, corresponds to the broadest band power spectrum (marked in Fig. 6b). In the last two cases the main frequency dominates.

5. Summary and Concluding Remarks

The key element of this work lies in the investigation of the occurrence of chaotic orbits and the analysis of two independent scenarios leading to it. Above the Hopf bifurcation curve with the increase of one of the two control parameters, the periodic orbit becomes unstable, and a new doubled period orbit occurs, then this in turn becomes unstable, and a double period of a new periodic orbit appears, and so on. This is the so called Feigenbaum scenario which can eventually lead to chaos.

The details of the development of the strange chaotic attractor which is accompanied by an inverse bifurcation cascade during this scenario was discussed. This analysis is supported by the Poincaré maps.

Some isolated "periodic windows" in the ocean of chaos were discovered by numerical analysis.

The second discovered route to chaos is connected to purely theoretical results obtained by Manneville-Pomeau, i.e. the so called intermittency phenomenon. The observed intermittency in this numerical simulation has been illustrated in Fig. 5 and 6, where the effects of sudden breaking of regularity is noticeable.

The following conclusions can be drawn:

1. For certain control parameter values a plate loses its steady-state equilibrium and starts to oscillate periodically. The critical value of this control parameter depends mainly on the ratio of the concentrated mass m_1 to the total mass of the plate, and depends on the plate density. The increase of this ratio causes a decrease of the critical airflow velocity (parameter γ). For the lower values of plate density, the critical airflow velocity is higher.
2. Periodic motion appearing after the Hopf bifurcation establishes itself in a relatively wide region of the control parameters plane. The frequency and amplitude of the periodic motion grow together with the increase of the distance from the Hopf bifurcation curve.
3. The plates for which the control parameters are located far enough from the Hopf bifurcation curve, exhibit an irregular motion. The transition from periodicity to chaos is accomplished by the period doubling cascade. A sequence: period doubling cascade — chaotic motion repeats itself a few times with a continuous change of the control parameter.

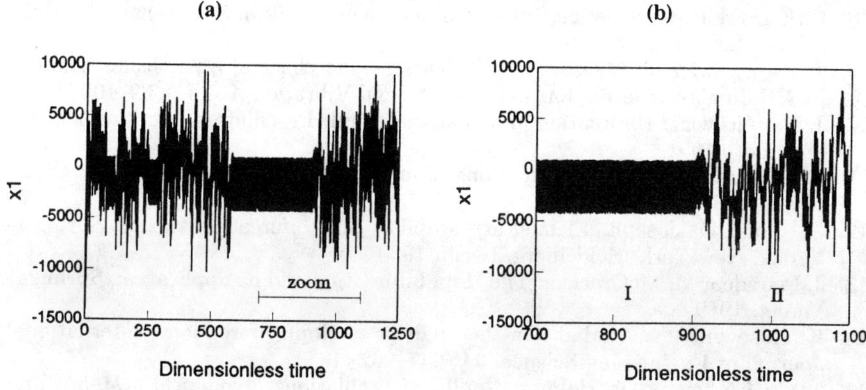

Fig. 5. The example of intermittency chaos (a) with zoom of marked interval (b)

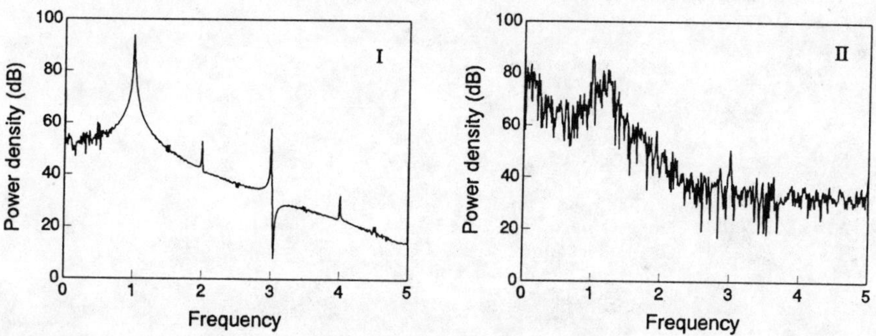

Fig. 6. Power spectra corresponding to the marked intervals in Fig. 5b

References

1. Don O. Brush, Bo O. Almroth: Buckling of bars, plates and shells. McGraw-Hill Book Company, 1975
2. V.V. Bolotin: Nonconservative problems of the theory of elastic stability. Pergamon Press, Oxford, London, New York, Paris, 1963
3. Chia Chuen-Yuan: Nonlinear analysis of plates. McGraw-Hill International Book Company, 1980.
4. P.J. Holmes: Bifurcations to divergence and flutter in flow-induced oscillations: a finite dimensional analysis. Journal of Sound and Vibration, **53**(4), 471–503 (1977)
5. E.H. Dowell: Nonlinear oscillations of a fluttering plate. AIAA Journal, **4**(7), 1267–1275 (1966)
6. E.H. Dowell: Nonlinear oscillations of a fluttering plate. II. AIAA Journal, **5**(10), 1856–1862 (1967)
7. E.H. Dowell: Flutter of buckled plate as an example of chaotic motion of a deterministic autonomous system. Journal of Sound and Vibration, **85**(3), 333–344 (1982)

8. E.H. Dowell: Aeroelasticity of plates and shells. Noordhoff International Publishing, Leyden 1975
9. J. Awrejcewicz, J. Mrozowski: Bifurcations and chaos of a particular Van der Pol-Duffing's oscillator. Journal of Sound and Vibration, **132** (1), 89–100 (1989)
10. J. Awrejcewicz: Bifurcation and chaos in coupled oscillators. World Scientific, Singapore 1991
11. E.L. Reiss: Cascading bifurcations. Journal of Applied Mathematics, **43**(1), 57–65 (1983)
12. G. Iooss, D. Joseph: Elementary stability and bifurcation theory. Springer-Verlag, New York, Heidelberg, Berlin 1980
13. J. Marsden, M. McCracken: The Hopf bifurcation and its application. Springer-Verlag, 1989
14. K. Huseyin, A.S. Atadan: On the analysis of Hopf bifurcations. International Journal of Engineering Science, **21**, 247–262 (1983)
15. J. Guckenheimer, P. Holmes: Nonlinear oscillations, dynamical systems, and bifurcations of vector fields. Springer-Verlag, New York, Berlin, Heidelberg, Tokyo 1986
16. R. Seydel: From equilibrium to chaos. Elsevier Science Publishers, New York, Amsterdam, London 1988

ND | Springer Series in **Nonlinear Dynamics**

Springer-Verlag and the Environment

We at Springer-Verlag firmly believe that an international science publisher has a special obligation to the environment, and our corporate policies consistently reflect this conviction.

We also expect our business partners – paper mills, printers, packaging manufacturers, etc. – to commit themselves to using environmentally friendly materials and production processes.

The paper in this book is made from low- or no-chlorine pulp and is acid free, in conformance with international standards for paper permanency.

Printing: Saladruck, Berlin
Binding: Buchbinderei Lüderitz & Bauer, Berlin